中 国 古 生 物 志

总号第 200 册　新丙种第 31 号

中国科学院　南京地质古生物研究所　编辑
　　　　　　古脊椎动物与古人类研究所

甘肃临夏盆地晚中新世副竹鼠类

王伴月　邱占祥　著

（中国科学院古脊椎动物与古人类研究所）

科学出版社

北　京

内 容 简 介

本书对在甘肃临夏回族自治州内发现的副竹鼠类进行了详细研究，共记述了两个属 [副竹鼠属（*Pararhizomys*）和假竹鼠新属（*Pseudorhizomys* gen. nov.）] 的 7 个种。其中包括 1 个已知种 [三趾马层副竹鼠（*Pararhizomys hipparionum*）]，2 个新种（*Pa. huaxiaensis*、*Pa. longensis*）和 4 个新属、新种（*Ps. indigenus*、*Ps. gansuensis*、*Ps. planus*、*Ps. pristinus*）。书中对副竹鼠类各属、种的头骨和颅后骨骼进行了详细描述、对比；运用分支系统学分析，将副竹鼠类归入鼠超科中盲鼹鼠科（Spalacidae）的拟速掘鼠亚科（Tachyoryctoidinae），并为副竹鼠类创建了一个新族——副竹鼠族（Pararhizomyini）。书中对假竹鼠和副竹鼠的头部和前肢部分肌肉作了复原和肌能分析，讨论了它们的生活习性，认为副竹鼠类为穴居的啮齿类。副竹鼠族的起源、演化发展和绝灭与晚中新世—早上新世时东亚中部的地理和气候环境的变化有关，特别与青藏高原隆升有密切关系。

本书是新近纪地层古生物研究人员的重要参考书，可供国内外地质科研人员、大专院校地质系师生和自然博物馆科研人员参考使用。

图书在版编目（CIP）数据

中国古生物志. 新丙种第31号（总号第200册）：甘肃临夏盆地晚中新世副竹鼠类/王伴月，邱占祥著. —北京：科学出版社，2018. 6

ISBN 978-7-03-057512-8

I. ①中⋯ II. ①王⋯ ②邱⋯ III. ①古生物–中国 ②晚中新世–副竹鼠类–甘肃

IV. Q911.72

中国版本图书馆CIP数据核字（2018）第110142号

责任编辑：胡晓春 孟美岑／责任校对：樊雅琼
责任印制：肖 兴／封面设计：黄华斌

科 学 出 版 社 出版

北京东黄城根北街16号
邮政编码：100717
http://www.sciencep.com

中国科学院印刷厂 印刷

科学出版社发行 各地新华书店经销

*

2018年6月第 一 版   开本：A4（880×1230）
2018年6月第一次印刷   印张：17 3/4
字数：575 000

定价：228.00元

《中国古生物志》新丙种出版品目录

前　言

　　2000 年 5 月中国科学院古脊椎动物与古人类研究所与甘肃省临夏回族自治州和政县合作，开始进行该地区新生代地层和哺乳动物化石的研究和展陈筹备工作。和政县，自上世纪末即以盛产"龙骨"闻名。在挖掘"龙骨"的农民中，以赵永昌和赵荣父子最为知名。和政县政府征集的哺乳动物化石标本主要出自赵氏父子之手。这些化石主要是新近纪大型哺乳动物化石，如大唇犀、三趾马、铲齿象、长颈鹿、肉食类等等。但我们在赵氏父子家中存留的"龙骨"中也发现了少量的小哺乳动物化石，如松鼠、豪猪、鼢鼠、副竹鼠、兔形类等。由于小哺乳动物化石在地层年代鉴定中具有特别重要的意义，我们一直很注意对这类化石的收集，但收效一直很小。加之，根据工作安排，开始一段时间内，我们把主要的精力放在了对龙担第四纪初期动物群的研究与和政古动物化石博物馆的筹展工作上，后来又插入了对延宕已久的中国巨犀化石专著的最后完成。这种无暇他顾的情况一直持续到 2007 年底。2008 年 5 月我们检视了和政博物馆此前收集到的所有小哺乳动物化石，发现副竹鼠类化石的材料保存得最多、最好，不但有保存较好的头骨、下颌骨，甚至还有颅后骨骼，其中大多有产地和层位的记录。这些材料对于本来所知甚少的副竹鼠类的研究具有很高的价值。这样，从 2009 年开始，副竹鼠类化石成了我们研究的重点。最初这一工作主要由本书前一作者独自承担，计划先对副竹鼠属（*Pararhizomys*）进行研究，并已于 2011 年初完成了关于副竹鼠属的属型种（*Pararhizomys hipparionum*）和另 2 新种的两篇文稿，准备在《古脊椎动物学报》上发表。在审稿过程中，学报编辑部的史立群女士提出，是否将临夏盆地所有副竹鼠类化石作为一个整体进行研究，以专著形式发表。这一建议也得到了啮齿类专家李传夔先生的支持。在认真考虑和斟酌后，我们决定接受史立群女士的意见，转为对临夏所产全部副竹鼠类化石进行综合研究，并以专著形式出版。

　　在对副竹鼠类进行全面综合研究时，随着材料的增加和涉及问题的范围扩大，出现了一些新的难题。首先是能收集到的有关小哺乳动物头骨系统解剖学的资料有限（仅大鼠和豚鼠的资料较完整）。在参考大鼠、豚鼠等解剖的基础上，有时不得不与解剖学研究程度较高的其他一些哺乳动物，如家畜（马、牛、羊、猪和狗）、灵长类（人和长臂猿）和肉食类（大熊猫、猫、狗）等的解剖学资料和标本进行对比。其次，关于副竹鼠类的高阶元分类，一直是一个悬而未决的问题。这批较好的副竹鼠类的标本显示了该类较全面的形态特征，为进一步探讨它们的分类位置提供了较好的基础。但在当今关于动物分类学的原理、原则和方法的认识分歧如此之大，而且似乎在可预见的未来很难得到解决的情况下，这一工作如何进行？这些无疑都大大增加了工作量。经过认真的考虑和斟酌，从 2014 年 8 月开始，本书后一作者（邱）也参加了这一研究工作。

　　在本书撰写过程中，得到了中国科学院古脊椎动物与古人类研究所（以下简称古脊椎所）所内同仁和国内外同行多方面的帮助。在此期间与古脊椎所的李传夔、邱铸鼎、郑绍华、张兆群、吴文裕、邓涛、倪喜军、李强、毕顺东等研究员，以及美国哈佛大学皮博迪博物馆的弗林（Flynn）博士等所进行的广泛的讨论，对我们研究工作的深入和提高起到了重要的作用。甘肃省和政古动物化石博物馆的何文馆长和陈善勤副馆长慷慨地将馆藏标本及有关信息供作者自由使用；古脊椎所的邱铸鼎、郑绍华、张兆群、吴文裕和李强等将他们最近采集的尚未研究的有关标本供作者观察和比较，特别是邱铸鼎和内蒙古队的其他成员，他们不但将有关标本供作者观察、比较，并慷慨地允许作者在他们有关文章发表之前，引用并讨论他们的观点；古脊椎所的李传夔、倪喜军、张兆群和吴秀杰将他们多年收集的一些现生哺乳动物和人的骨骼标本提供给本书作者使用；美国亚利桑那大学（University of Arizona）

的 Lindsay 博士赠送有关标本的模型；这些都使我们大受裨益。古脊椎所的邱铸鼎、张兆群和倪喜军先生在百忙之中抽出时间帮助审阅本书的中文文稿，提出了许多宝贵意见；美国洛杉矶自然历史博物馆（Natural History Museum of Los Angeles County）的王晓鸣博士，在百忙之中抽出时间帮助修改英文摘要。在此一并表示衷心感谢！

张丽芬女士和王平先生精心修理化石，高伟、张文定先生和张兆霞女士为本书摄制照片，已故的沈文龙先生为本书清绘线条图，司红伟女士帮助修整插图；在此表示诚挚的谢意！

本书是在古生物《志书》编研及门类系统总结（编号：2013FY113000）、中国科学院前沿科学重点研究项目（编号：QYZDY-SSW-DQC022）、中国科学院 B 类先导培育项目（编号：XDPB05）和国家自然科学基金重点项目（编号：41430102）资助下完成和出版的。

目　　录

一、导 言

（一）副竹鼠类的研究历史及临夏新材料发现的意义

副竹鼠属（*Pararhizomys*）是一类生活在晚中新世—早上新世东亚中纬度地区的现已绝灭的土著啮齿动物。该属是德日进和杨钟健于 1931 年创建的，属型种是 *Pararhizomys hipparionum*（中译名是三趾马层副竹鼠）。建属的材料仅是产自陕西府谷的一段左下颌（Teilhard de Chardin et Young, 1931）。主要依据下颊齿的形态，德日进和杨钟健认为该属与竹鼠（*Rhizomys*）有较近的关系。虽然如此，他们只是将其归入到鼠超科（Muroidea）中。此后一直没有关于该类化石的报道。直到 37 年后的 1968 年，波兰学者 Kowalski 才又报道了产自蒙古西部的一件归入该种的含有下颌骨的不完整的头骨。Kowalski 将它直接归入了竹鼠科（Rhizomyidae）。又过了 37 年，张兆群等（Zhang et al., 2005）才报道了产自陕西和甘肃的较多的材料。张等将这些材料分为副竹鼠属的两个种：除三趾马层副竹鼠外还建了一个新种，秦副竹鼠（*Pararhizomys qinensis*）。前者的材料由部分上颌和一段左下颌组成，后者包括一头骨的前部和两枚上臼齿。随后，邱铸鼎和李强（Qiu et Li, 2008）报道了产自青海柴达木盆地深沟的一枚右 m2，鉴定为 *Pararhizomys* sp.。近年，李强（Li, 2010）又报道了产自我国内蒙古宝格达乌拉上中新统的三趾马层副竹鼠化石。这批材料较多，但也仅包括一段左下颌，一些单个牙齿和几枚零散的颅后骨骼。

关于 *Pararhizomys* 的科一级的分类位置，目前主要有两种不同的看法。大部分古生物学家将其归入到竹鼠科中（Simpson, 1945；Young et Liu, 1950；Kowalski, 1968），或鼠科中的竹鼠亚科（Rhizomyinae，McKenna et Bell, 1977）。只有 Flynn（1982, 1990）明确地将其从竹鼠科中排除，而认为副竹鼠是盲鼹鼠科（Spalacidae）的成员。张兆群等（Zhang et al., 2005）和李强（Li, 2010）都未对副竹鼠的科级分类地位提出明确的看法，而只将其归入鼠超科内，科未定。

总之，副竹鼠属自建属以来已过了 80 多年，但发现的材料很少，至今仍然没有发现完整的头骨、下颌和较完整的颅后骨骼；也没有与其相近的其他属的材料发现，至今仍为一属二种。这使我们对这一类啮齿动物的形态特征、演化历史、分类地位和生活习性等方面了解至今仍然停留在很低的水平。

近年来，我们在甘肃临夏盆地采集到一批副竹鼠类化石。材料中包括若干相当完整的头骨，其中部分还含有下颌，有的还有颅后骨骼。绝大多数标本都有清楚的产出地点和层位的记录。根据我们的研究，这批材料至少包括两个不同的属和七个不同的种。这是副竹鼠类目前已知最好的一批材料，对于我们了解副竹鼠类的形态特征，探讨其系统演化和分类位置及生活习性等都提供了较全面的证据。

（二）材 料

1. 本书所研究的化石标本

本书所研究的副竹鼠类化石标本包括头骨、下颌骨和颅后骨骼等，分属 22 个个体。这些个体的标本在系统描述部分都有详细的记载（详见本书表 2 和表 10），在此不另述。化石产自甘肃省临夏回族自治州的和政县、广河县和东乡族自治县的 15 个地点，详见图 1 和表 1。

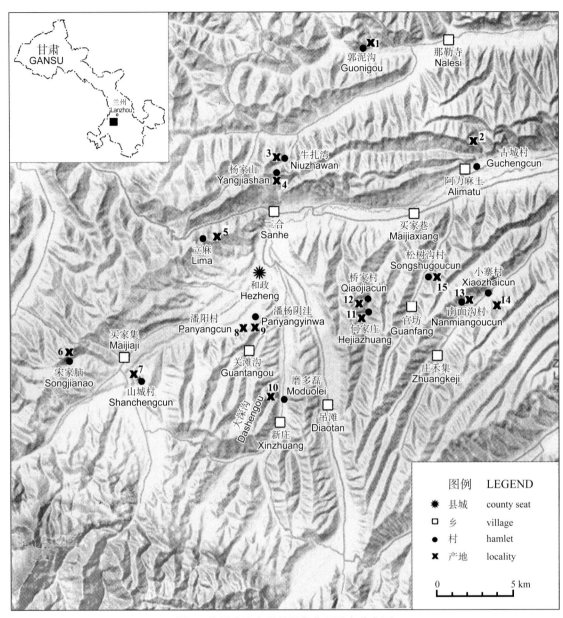

图 1 临夏盆地产副竹鼠类化石地点分布图

Fig. 1 Map showing the localities yielding pararhizomyine fossils in the Linxia Basin

左上方的方形插图中黑方块（■）显示研究区域在甘肃省内的位置（Study area is shown by a black square in the inserted quadrangle），图中 1–15 化石地点的详细信息见表 1（For further information of localities 1–15 see Table 1）

2. 对比标本

本书作者除尽量广泛地参阅系统关系较接近的啮齿类文献外，还直接观察了下列化石和现生门类的标本：

1）古脊椎所所藏标本。计有：竹鼠科 [竹鼠（*Rhizomys sinensis* 和 *R. pruinosus*）]、鼠科 [褐家鼠（*Rattus norvegicus*）、黑线姬鼠（*Apodemus agrarius*）、鼢鼠（*Myospalax myospalax*）]、鼠平科 [麝鼠（*Ondatra zibethicus*）、田鼠（*Microtus fortis*，*M. clackei* 和 *M. mongolicus*）]、仓鼠科 [大仓鼠（*Tscheskia triton*）]、松鼠科 [黑尾场拨鼠（*Cynomys ludovicianus*）、旱獭（*Marmota monax* 和 *M. himalayana*）、花鼠（*Tamias sibricus*）]、北美豪猪科 [北美豪猪（*Erethizon dorsatum*）、勃氏堆土鼠（*Thomomys bottae*）、赤色衣囊鼠（*Geomys bursarius*）、海狸鼠科 [海狸鼠（*Myocastor coypus*）]、沙鼠科 [子午

表1 临夏盆地产副竹鼠类化石地点信息

表1 临夏盆地产副竹鼠类化石地点信息

Table 1 Information of the localities yielding pararhizomyine fossils in the Linxia Basin

编号 No.	古脊椎所野外地点 IVPP Loc.	赵荣编号 Zhao's Loc.	产副竹鼠类化石地点 Localities yielding pararhizomyines	卫星定位 GPS
1	LX 200042	Z 32	东乡族自治县那勒寺乡郭泥沟东北（60°）约300 m 300 m NE (60°) of Guonigou, Nalesi Village, Dongxiang Autonomous County	35°33'01.1"N, 103°26'15.6"E 2234 m asl*
2	LX 200007	Z 11	广河县阿力麻土乡古城村（寺沟）北约1.5 km 1.5 km N of Guchengcun (Sigou), Alimatu Village, Guanghe County	35°29'41.8"N, 103°30'06.1"E 2140 m asl
3	LX 200503	Z 65/ Z 86	和政县三合乡牛扎湾 Niuzhawan, Sanhe Village, Hezheng County	35°29'08.7"N, 103°21'20.3"E 2343 m asl
4	LX 200004	Z 6	和政县三合乡杨家山南坡 South slope of Yangjiashan, Sanhe Village, Hezheng County	35°28'37.9"N, 103°21'25.9"E 2340 m asl
5	LX 200207	Z 52	和政县三合乡立麻（麻沟）东约1 km 1 km E of Lima (Magou), Sanhe Village, Hezheng County	35°26'43.5"N, 103°19'07.8"E 2365 m asl
6	LX 200502	Z 69（下）	和政县买家集乡宋家脑西北约100 m 100 m NW of Songjianao, Maijiaji Village, Hezheng County	35°22'28.6"N, 103°12'51.4"E 2496 m asl
7	LX 200041	Z 3/ Z64	和政县买家集乡山城村西北（290°）约300 m 300 m NW (290°) of Shanchengcun, Maijiaji Village, Hezheng County	35°22'00.8"N, 103°15'58"E 2352 m asl
8	LX 200037	Z 4	和政县关滩沟乡潘杨村 [潘杨阴洼西南（220°）约800 m] Panyangcun [800 m SW (220°) of Panyangyinwa], Guantangou Village, Hezheng County	35°23'35.6"N, 103°20'38.6"E 2270 m asl
9	LX 201001		和政县关滩沟乡潘杨阴洼南约500 m 500 m S of Panyangyinwa, Guantangou Village, Hezheng County	35°23'43.3"N, 103°20'47.5"E 2313 m asl
10	LX 200011	Z 1	和政县新庄乡大深沟北坡（磨多磊村西约600 m） North slope of Dashengou (600 m W of Moduoleicun), Xinzhuang Village, Hezheng County	35°21'23.9"N, 103°21'22.4"E 2276 m asl
11	LX 200046	Z 17	广河县买家巷乡何家庄西南（200°）约150 m（现属和政县吊滩乡桦林三队） 150 m SW (200°) of Hejiazhuang, Maijiaxiang Village, Guanghe County (now belonging to Hualin Third Brigade, Diaotan Village, Hezheng County)	35°23'57.5"N, 103°25'25.5"E 2207 m asl
12	LX 200047	Z 15	广河县买家巷乡桥家村西南（240°）约200 m 200 m SW (240°) of Qiaojiacun, Maijiaxiang Village, Guanghe County	35°24'12.5"N, 103°25'28.4"E 2218 m asl
13	LX 200020	Z 25	广河县庄禾集乡 ** 南面沟村（南面沟北坡） Nanmiangoucun (north slope of Nanmiangou), Zhuangkeji Village, Guanghe County	35°24'23.1"N, 103°29'33.8"E 2205 m asl
14	LX 200019	Z 26	广河县庄禾集乡小寨村东南（150°）约1 km 1 km SE (150°) of Xiaozhaicun, Zhangkeji Village, Guanghe County	35°23'56.6"N, 103°30'20.8"E 2206 m asl
15	LX 200030	Z 10	广河县官坊乡松树沟村东坡 East slope of Songshugoucun, Guanfang Village, Guanghe County	35°25'05.5"N, 103°28'05.1"E 2327 m asl

* asl. 海拔（above sea level）。

** 在地图中有两种写法："庄禾集"和"庄窠集"，当地人的发音为"Zhuangkeji"。在英译名中我们只使用了"Zhuangkeji"，而没有使用"Zhuangheji"。

沙鼠（*Meriones meridianus*）]和跳鼠科 [五趾跳鼠（*Allactaga sibirica*）]，以及一些大哺乳动物（包括马、家犬、猫、大熊猫和人）的头骨和颅后骨骼标本。

2）古脊椎所邱铸鼎等提供的 *Tachyoryctoides* 属的头骨化石标本。

3）古脊椎所倪喜军提供的现生山河狸（*Aplodontia rufa*）和中华始鼢鼠（*Eospalax fontanierii*）的完好的骨架标本。

4）古脊椎所李传夔提供的现生河狸（*Castor*）头骨标本。

5）古脊椎所张兆群提供的现生盲鼹鼠（*Spalax*）头骨标本。

6）古脊椎所吴秀杰提供的现代人的骨骼标本。

7）美国亚利桑那大学的 Lindsay 博士赠送的产自蒙古国的 *Pararhizomys hipparionum* 标本的模型。

（三）研 究 方 法

1. 骨骼的划分及解剖学术语

本书采用了 Hildebrand 和 Goslow（2001）所建议的方案，即将骨骼（skeleton）分为头部骨骼（head skeleton = cranial part）和体部骨骼（body skeleton = postcranial part）两部分。从演化和发育生物学的角度看，头部骨骼由软颅（chondrocranium）和由脏颅（splanchnocranium）及真皮骨（dermal bones）产生出的脑颅（cranium）和包括上、下颌（maxillae 和 mandible）在内的面部（face）、容纳鼻和耳的部分、以及舌骨（hyoid bones）组成。在哺乳动物化石中，由于下颌 [仅以齿骨（dentary）为代表] 和舌骨经常单独保存，而所有其他部分（包括脑颅、上颌和容纳鼻和耳的部分）则经常愈合为一体，而被习惯地称作头骨（skull）。这是狭义的头骨的含义，虽很适用，却不准确（Hildebrand et Goslow, 2001: 115）。这样，当下颌与颅部及面部等其他骨骼一起保存时，我们将其称作头骨含下颌。下颌本身包括左、右两半下颌（left and right hemimandibles）。在中文中我们简单地将它们称作左、右下颌。按照多数古生物学家的习惯用法，每半下颌又分为前部含牙部分的水平支（horizontal ramus）和其后的上升支（ascending ramus）。颅后骨骼包括除头部之外的所有部分，包括脊柱、胸骨、肋骨和肢骨（limb bones）等。

在解剖学术语上，本书基本采用了 Greene（1935）、杨安峰和王平等（1985）基于大鼠（= 褐家鼠，*Rattus norvegicus domestica*）所推荐的术语体系，并参照了 Cooper 和 Schiller（1975）对豚鼠，Hebel 和 Stromberg（1976）对实验鼠，王增涛等（2009）对 Wistar 大鼠的解剖，Bugge（1970，1971，1974，1985）对啮齿类和 Wahlert（1974）对始啮鼠类的头骨上的部分术语及由 Constantinescu 和 Schaller 主编的第三版《图解家畜解剖名词》（Illustrated Veterinary Anatomical Nomenclature, 2012），对少数几个术语作了修改和补充。

2. 头骨中各孔的位置及其功能意义

（1）啮齿类动物头部动脉及其供血型式

Greene（1935）对褐家鼠头部血管记述时使用的术语主要是基于 1927 年版的 B.A.N.（Emmel's Basle Anatomical Nomenclature）。其中有些术语，特别是有关颈内动脉者，与后来的用法不尽相同。图 2 是 Greene（1935）对颈内动脉的分支和有关孔的名称的用法。

到 20 世纪 70 年代，人们对啮齿类头部动脉系统和供血型式有了很多新的认识。这主要体现在丹麦解剖学家 Bugge 1970–1974 年间连续发表的 6 篇文章中。1974 和 1985 年他又两次对其观点进行了综述。Bugge 共解剖了 78 种现生啮齿动物的 478 件头骨标本。在大量解剖工作的基础上，Bugge 提出，啮齿类的头部动脉的最原始的基本模式是，除椎动脉直接供应脑外，颈内动脉、由颈内动脉分出的镫骨动脉和颈外动脉是供应头部（包括下颌）血液的三条主动脉，而不是颈内和颈外两条动脉。这三条主动脉，特别是颈内和镫骨动脉，在演化过程中发生了非常大的变化，有些部分，甚至可以完全消失，其供血则通过在不同位置上新生的吻合支（anastomotic branches）由另外的动脉取代。例如，在鼠亚科（包括褐家鼠）中，这三条动脉都发育，而在大部分松鼠和河狸类中，其颈内动脉就大部退化消失，在盲鼹鼠及竹鼠中则是镫骨动脉退化。Bugge 使用了一些独特的术语和模式图来表示这些变化。在副竹鼠类所有的化石标本中我们发现都只有颈内动脉孔，而没有镫骨动脉孔，这种情况和盲鼹鼠及竹鼠最为接近。图 3 是 Bugge 所提出的头部动脉的原始模式图和在盲鼹鼠及竹鼠中的模式图。

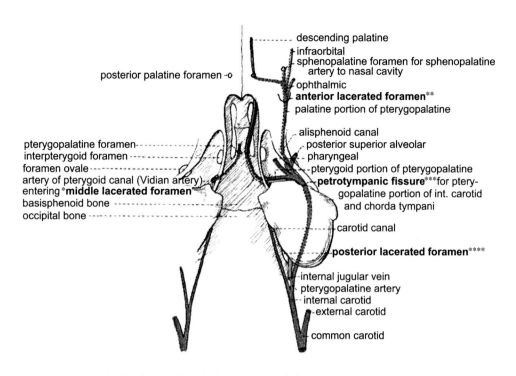

图 2　褐家鼠与颅骨腹面有关的颈内动脉分支（据 Greene, 1935, Fig. 209）

Fig. 2　Branches of the internal carotid artery in relation to the ventral aspect of the cranium of *Rattus norvegicus* (after Greene, 1935, Fig. 209)

本书中使用的孔的名称（terms applied in present monograph）：∗: posterior foramen of pterygoid canal（翼管后孔），∗∗: anterior alar fissure（前翼裂），∗∗∗: middle lacerate foramen（中破裂孔），∗∗∗∗: jugular foramen（颈静脉孔）；有关动脉型式及名称的不同意见见下节（different opinions as to cephalic arterial pattern and nomenclature, *vide infra*）

（2）头骨中常见各孔的名称及位置

　　Wahlert（1974: 363）在前人（Hill, 1935；Guthrie, 1963, 1969 和 Bugge, 1970, 1971 等）工作的基础上，首次明确提出"孔（这里专指头骨通过血管和神经的诸孔——本书作者注）的存在或缺失、其相对位置、数量和相对大小，在确定系统关系中都是有用的特征。"在该文中，Wahlert 以始啮型啮齿动物（protrogomorphous rodents）为代表列举了 44 个孔（管），并对其位置及功能做了介绍。后来 Voss（1988）及 Musser 和 Heaney（1992）对鼠形超科中渔鼠族（Ichthyomyini）和某些鼠科动物头骨各孔的位置及形态也有比较详细的记述。

　　根据副竹鼠类头骨的特点和保存情况，我们采用了 Wahlert 所记述的 44 个孔（管）中的 26 个，对其中的部分孔（管）的功能，也根据 Bugge 的观点作了少许修订和补充，并增加了 4 个新的名称；我们还采用了 Wahlert 提到但未正式命名的小翼孔；另外，Wahlert 并未记述岩鼓裂，而在副竹鼠及若干相近的鼠类头骨中岩鼓裂确实存在。我们记述的头骨孔合计为 32 个。下列带 ∗ 者为新建名称；括号内的数字为 Wahlert 原用序数；常用的术语缩写为：a. arteria，b. branch，c. canal，for. foramen，n. nerve，r. ramus，v. vena。

　　1 (2). inf 一门齿孔（incisive for.），亦称腭前孔（anteiror palatine for.）或腭裂（palatine fissure）：位于门齿和颊齿之间的齿隙的中矢线两侧的一对细长孔；为腭降动脉（descending palatine a.）[属于 Bugge 称之为眶下动脉支（r. infraorbitalis），而 Greene 称之为翼腭动脉（pterygopalatine a.）之一支] 自腭部进入鼻腔的通道。Greene 在图中显示三叉神经鼻腭支（nasopalatine b. of n. V）进入门齿孔（Greene, 1935: 16, fig. 11），但在文中又称该神经支进入蝶腭孔（Greene, 1935: 117）。从上下文看，Bugge 显然是把蝶腭神经支（sphenopalatine b.）错写成鼻腭神经支（nasopalatine b.），

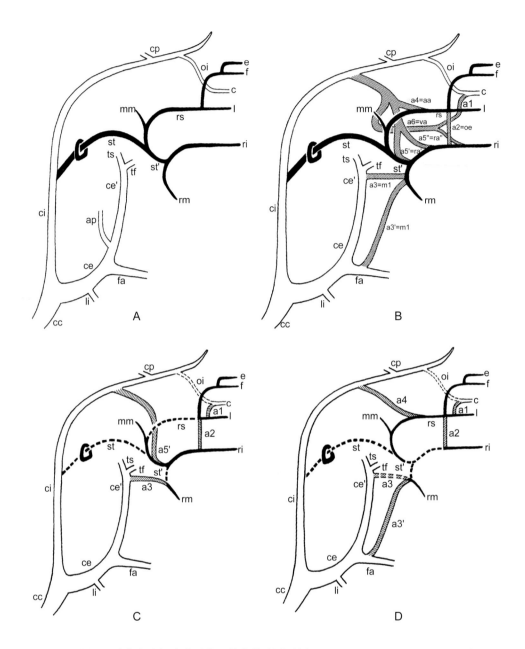

图 3 啮齿类头部动脉系统和供血模式图 （据 Bugge, 1974, Figs. 1, 2A, 9B et C）

Fig. 3 Cephalic arterial systems of rodents (after Bugge, 1974, Figs. 1, 2A, 9B et C)

A. 原始基本模式（primitive basic pattern）, B. 所有可能的吻合支的位置（positions of all possible anastomoses, a1–a6）,

C. *Spalax leucodon*, D. *Rhizomys pruionosus*

缩写（Abbreviations）: a1. 睫状动脉近段（proximal part of the ciliary artery）, a2 = oe. 外眼动脉（*a. ophathalmica externa*）, a3 = m1 和a3' = m1. 上颌动脉第一段（first part of the maxillary artery）, a4 = aa. 吻合动脉（*a. anastomoticus*）, a5' = ra' 和a5" = ra". 吻合支（*r. anastomoticus*）, a6 = va. 维杜斯动脉（Vidian artery）; ap. 耳后动脉（*a. auricularis posterior*）, c. 睫状动脉（*a. ciliaris*）, cc. 颈总动脉（*a. carotis cammunis*）, ce. 颈外动脉近段（proximal part of *a. carotis externa*）, ce'. 颈外动脉远段（distal part of *a. carotis externa*）, ci. 颈内动脉（*a. carotis interna*）, cp. 后交通动脉（*a. communicans posterior*）, e. 筛动脉（*a. ethmoidalis*）, f. 额动脉（*a. frontalis*）, fa. 面动脉（*a. facialis*）, l. 泪动脉（*a. lacrimalis*）, li. 舌动脉（*a. lingualis*）, mm. 内脑膜动脉（*a. meningea media*）, oi. 内眼动脉（*a. ophthalmica interna*）, ri. 眶下支（*r. infraorbitalis*）, rm. 下颌支（*r. mandibularis*）, rs. 眶上支（*r. supraorbitalis*）, st. 蹬骨动脉（*a. stapedia*）, st'. 蹬骨动脉远段（distal part of *a. stapedia*）, tf. 面横动脉（*a. transversa faciei*）, ts. 颞浅动脉（*a. temporalis superficialis*）

又把其路径也写错了。Wahlert（1974）及杨安峰和王平等（1985: 10）在记述腭裂时都提到该裂是门齿管的通路。这大概是对的。虽然我们并不知道与门齿管相连的犁鼻器（vomeronasal organ，或 Jacobson's organ）是否也存在（目前只知在现生家鼠和少数田鼠中存在）。

2 (5). iof —眶下孔（infraorbital for.）：在鼠形超科中为位于吻部两侧、眼眶前方的一对很大的孔，为眶下管的前开口。由三部分组成：上方为附着内层咬肌（*m. masseter medialis*）[Voss（1988）称为颧下颌肌（*m. zygomaticomandibularis*）]的区域，中部有容纳鼻泪管（nasolacrimal duct）的部位，而外下方为裂隙状，系三叉神经第二支，即上颌神经（maxillary n.）的眶下支（infraorbital n.）和同名动脉穿过的通道。

3 (8). nlf —鼻泪孔（nasolacrimal for.）：在鼠超科中呈裂隙状，位于吻部两侧、眶下管的前内侧，是泪囊（*saccus lacrimalis*）和鼻泪管的后部向上和向内的开口。Voss 根据他对渔鼠族的解剖，证实三叉神经的第二支（maxillary n.）之前上齿槽神经（anterior superior alveolar n.）和鼻泪管一起进入吻的内部（Voss, 1988: 292）。Greene（1935: 117）认为褐家鼠的前上齿槽神经通过眶下裂隙后，再穿过上颌骨进入吻部。Wahlert（1974: 371）则认为，在始啮型啮齿类中，前上齿槽神经和同名动脉是从眶下管底部的另一个孔进入吻部的。Wahlert 将其称为前齿槽孔（anterior alveolar for., aa, no. 6）。我们在副竹鼠类头骨眶下管的前下方没有发现该孔。这可能表明，在副竹鼠类中，前上齿槽神经是通过鼻泪孔而非 Wahlert 所说的前齿槽孔进入吻部的。

*4. psaf —后上齿槽孔（posterior superior alveolar for.）：位于眶下管后端的底面，蝶腭孔（sphenopalatine for.，见下）之前下方。Greene（1935: 117）提到，后上齿槽神经支（posterior superior alveolar n.），在进入眶下管时与三叉神经之上颌神经分开，分成很多细丝，通过若干小孔进入上颌窦。Greene 并没有为这些小孔单独起名。我们在褐家鼠头骨上见到其中也有较大的孔，位于 M1 的上方。在副竹鼠类中在相应的位置上我们也发现有一个较大的孔。这里将它称作后上齿槽孔。

5 (11). spf —蝶腭孔（sphenopalatine for.）：位于眶窝内，大约在 M1 的上方；为镫骨或颈内动脉之眶下动脉支（infraorbital r.）或翼腭动脉（pterygopalatine a.）之蝶腭动脉（sphenopalatine a.）以及三叉神经上颌神经支之同名神经（即蝶腭神经）[Greene（1935: 117）错误地称之为鼻腭支（nasopalatine n.）]进入鼻腔的通道；也是眼下静脉之蝶腭静脉（sphenopalatine v.）之通道。

6 (16). dpf —背腭孔（dorsal palatine for.）：在腭骨的眶面，位于蝶腭孔后下方的小孔；为镫骨或颈内动脉之眶下动脉支（infraorbital r.）或蝶腭动脉（sphenopalatine a.）之腭降动脉（descending palatine a.）和三叉神经上颌神经支之同名神经小支进入腭管（palatine canal）的入口，出腭后孔进入腭面。

7 (12). etf —筛孔（ethmoidal for.）：通常是两个，位于额骨之眶面，在视神经孔的前上方。褐家鼠只有一个筛孔，被 Greene（1935: 116, 184）、杨安峰和王平等（1985: 8, 115）称为前筛孔（anterior ethmoidal for.）。他们认为前筛孔是三叉神经第一支，即眼神经（ophtalmic n.）之鼻睫支（nasociliary n.）进入鼻腔筛部之通道，同时也是颈内动脉之翼腭动脉的眼上动脉（superior ophthalmic a.）进入鼻腔筛部之通道。Wahlert（1974）认为在有两个筛孔的啮齿类中，筛动脉和静脉通过后边的筛孔，而三叉神经之眼神经支的鼻睫支（nasociliary n.）通过前边的筛孔。

8 (14). opf —视神经孔（optic for.）：位于蝶骨眶蝶部内的一个孔；为第二对脑神经（视神经, optic n.）自脑部伸出至眶窝的通道。

9. aafi —前翼裂（anterior alar fissure）：系 Wahlert（1983）所创，用以取代 Hill（1935: 124）和他本人（Wahlert, 1974: 372, no. 18）之蝶裂（sphenoidal fissure）。Wahlert 在创建此名时指出："Hill 声称这一开孔相当于灵长类中的眶上裂（superior orbital fissure）和圆孔（*foramen rotundum*）"，而没有观察到还包括"翼蝶管之前端"（Wahlert, 1983: 2）。实际上，Hill 在使用该名时，已经明确指出，"该孔也包括翼蝶管和蝶翼管的前开孔（anterior opening of the alisphenoid and sphenopterygoid canals）"。不过，我们还是赞成使用前翼裂一名，因为蝶裂在人类解剖中已为惯用名称（见下），和圆孔是分开的，而且由于在人类中没有翼蝶管，当然也不会包含其前开孔。在鼠形类中这三个孔（蝶裂、圆孔和翼蝶管前孔）经常是合在一起的。第三（动眼，oculomotor n.）、四（滑车，trochlear n.）、六（外展，abducent n.）和第五（三叉，trigeminal）脑神经的眼神经支（ophthalmic n.）和上颌神经支（maxillary n.），还有颈内动脉的翼腭动脉（pterygopalatine a.）的腭部（palatine portion），包括眼动脉（ophthalmic a.）、蝶腭动脉（sphenopalatine a.）等都从这里伸出。

9a (19). sf —蝶裂（sphenoidal fissure）：这里是狭义的，而非 Hill（1935）和 Wahlert（1974）之广义的蝶裂。蝶裂在人体解剖中有时被称前破裂孔（anterior lacerate foramen）[Greene（1935: 14, Fig. 9）也称其为前破裂孔]，早期也被称为眶上裂（superior orbital fissure），它们都是同义词。在有些副竹鼠的标本上，前翼裂可见分成上、下两个孔，其中上者应为蝶裂。

9b (20). fr —圆孔（*for. rotundum*, = round for.）：经常与翼蝶管的前开孔组成一个大孔，位于蝶裂之下，第五（三叉）脑神经之上颌神经支由此伸出。

10. fap —小翼孔（*foramen alare parvum*）：在前翼裂最下端的后方、紧靠圆孔 + 翼蝶管前孔后缘处的一小孔。Wahlert（1974: 372–373, no. 20）在讨论圆孔时曾指出，三叉神经第二支的"颧骨神经支（zygomatic b.）可有一分开的孔。"Evans 和 Christensen 在犬的解剖中提到，有时在眶裂和前翼孔之间会有一小孔，通过颧神经（Evans et Christensen, 1979: 127–129, Fig. 4-17），将其称为小翼孔（*foramen alare parvum*）；但在另一处（914, Fig. 15-7）又将其称为颧神经孔（zygomatic foramen）。我们在 V 16294 和 V 16297 头骨的两侧，V 16301 和 V 16302 头骨的左侧都见到了该小孔（在 *Rhizomys* 和 *Aplodontia* 中也见有此孔）。我们称其为小翼孔。

11 (3). ppf —腭后孔（posterior palatine for.）：为腭管在腭面的后开口，位于上颌骨 - 腭骨缝附近，在中矢线两侧形成一对或两对较大的长形孔。Wahlert（1974: 371）曾提到，腭降动脉和神经及静脉通过该孔，但没有说明是什么神经和静脉。Greene（1935: 16, Fig. 11; 185）、杨安峰和王平等（1985: 115）指出：腭降动脉和同名神经穿过上颌骨经后腭孔进入腭面（在其他哺乳动物中大多有一至数条腭神经）。

12 (4). pmf —后上颌孔（posterior maxillary for. or notch）：位于 M3 之后，在上颌骨 - 腭骨缝附近的小孔或切迹；根据 Wahlert（1974: 371）和 Greene（1935: 223）的记述，该孔为腭降静脉支（descending palatine v.）进入眶部与眼静脉（ophtalmic v.）会合的通道。

13 (23). mf —咬肌神经孔（masticatory foramen）：Hill（1935）在创建此名称时把三叉神经的第三分支，即下颌神经中的咬肌支（masseteric b.）称作嚼肌支（masticatory b.）。本书在将其译成中文时，仍按咬肌支译之。该孔位于翼蝶骨内；下颌神经的咬肌支由此伸出。在副竹鼠类中，可见下颌神经中的另一小支——颞深神经支（deep temporal b.）——也由此孔伸出。

14 (23p). mfp —咬肌神经管后孔（posterior foramen of masseteric n. c.）：位于咬肌神经孔之后，在翼窝的后部，卵圆孔和中破裂孔之前；为三叉神经下颌神经咬肌神经支进入翼蝶骨的入口。

15 (24). bf —颊肌神经孔（buccinator for.）：位于咬肌神经孔的前下方，有时与咬肌神经孔合成一个长的裂隙状孔；三叉神经下颌神经的颊肌支（buccinator b.）由此伸出。

16 (24p). bfp —颊肌神经管后孔（posterior foramen of buccinator n. c.）：位于颊肌神经孔之后，在翼窝的后部，卵圆孔和中破裂孔之前；为下颌神经颊肌神经支进入翼蝶骨的入口。颊肌神经管通常与咬肌神经管共有一后开孔。

*17. dtf —颞深神经孔（deep temporal for.）：位于咬肌神经孔的后上方；为三叉神经下颌神经中颞深神经小支（deep temporal b.）的出口。在啮齿类中颞深神经通过的管道（颞深神经管，deep temporal b. c.）通常与咬肌神经管（masseteric b. c.）愈合。但在某些副竹鼠类中，颞深神经管有时与咬肌神经管完全分开，笔者建议给予该管及其前、后孔单独命名。

*18. dtfp —颞深神经管后孔（posterior foramen of deep temporal n. c.）：颞深神经孔之后，在翼窝的后部，卵圆孔和中破裂孔之前；下颌神经中颞深神经支（deep temporal b.）由卵圆孔出来之后又进入翼蝶骨的入口。

19 (26). fo —卵圆孔（*for. ovale*）：位于翼窝的后外侧，为三叉神经之下颌神经自脑部伸出之处；也是颈内动脉翼腭动脉（pterygopalatine a.）之腭部（palatine portion）进入翼蝶管的入口。在副竹鼠类中，卵圆孔开口在翼窝后缘，与中破裂孔汇合。

20 (28). mlf —中破裂孔（middle lacerate for.）：位于听泡（tympanic bulla）和蝶骨之间，翼窝之后 [Greene（1935）、杨安峰和王平等（1985）称其为岩鼓裂（petrotympanic fissure），实际上，中破裂孔是位于岩鼓裂前内侧的孔，与岩鼓裂有时相通，有时分开]；为颈内动脉蹬骨支（stapidial a.）[=Greene（1935）、杨安峰和王平等（1985）的翼腭支（pterygopalatine a.）] 和脑膜静脉（meningeal v.）的出口。在副竹鼠类中，颈内动脉的蹬骨支的近端很可能已经高度退化或仅为残迹。另外，在副竹鼠类中卵圆孔与中破裂孔汇合，三叉神经之下颌神经支也由此孔通过。

21. petf —岩鼓裂（petrotympanic fissure）：位于听泡前外侧和颞骨鳞部之间，前内端与中破裂孔或相通或被颞骨分开，外后端有时与臼后孔相通。在翼腭动脉（pterygopalatine a.）[=Bugge（1974）之蹬骨动脉（stapedial a.）] 发育的啮齿类中，该动脉由此伸出。在大多数情况下，第七脑神经（面神经）之鼓索神经也由此伸出。

*22. pcpf —翼管后孔（posterior for. of pterygoid canal）：位于翼窝内后角，内翼突的后基部；为颈内动脉翼腭支（pterygopalatine a.）的翼部（pterygoid portion）进入翼管（pterygoid c.）的入口。Greene（1935: 183, Fig. 209）、杨安峰和王平等（1985: 115）认为该翼部动脉通过中破裂孔进入翼管。笔者认为该孔不在蝶骨和听泡（或岩骨）之间的裂缝中，而在蝶骨内，故改称其为翼管后孔。在副竹鼠类中该孔已很退化。

23 (21). asc —翼蝶管（alisphenoid canal）：位于蝶骨外翼突的背侧，其后口（posterior for. of alisphenoid c., ascp）位于卵圆孔的前内下方，咬肌神经孔和颊肌神经孔后口的内侧，或与后两后孔汇合；翼腭动脉（pterygopalatine a.）的腭部（palatine portion）由此通过。

24 (29). eucf —欧氏管孔（for. of Eustachian c.）：位于听泡之前内角，听泡棘突之外侧，为欧氏管的出口。

25 (30). icf —颈内动脉孔（internal carotid for.）：位于听泡内侧接近前内角处，在鼓骨和基枕骨前端之间；为颈内动脉进入脑部的入口。

26 (32). juf —颈静脉孔（jugular for.）：位于听泡内后角，介于听泡与基枕骨之间，呈裂隙状，又称后破裂孔（posterior lacerate for.）；为第九（舌咽，glossopharygeal n.）、第十（迷走，vagus n.）、第十一（副，accessory n.）脑神经的出口，也是颈内静脉（internal jugular v.）的出口。在具 Bugge 的基干型头骨动脉的啮齿类（如褐家鼠等）中，蹬骨动脉（=Greene 之翼腭动脉）的入口处位于颈静脉孔的外下方。在听泡保存较好的副竹鼠类标本中，未见有蹬骨动脉进入听泡的孔。

27 (40). chuf —于吉埃氏管孔（for. of canal of Huguier）：位于听泡前侧面的一小孔。Wahlert（1974）在描述旱獭（*Marmota monax*）时发现其第七（面）神经（facial n.）的鼓索神经（chorda tympani n.）自听泡前外壁的一个小孔中伸出，然后和第五（三叉）神经的下颌神经的舌支（lingual b.）相连。在副竹鼠类中，这一小孔也在听泡的前外壁上。

28 (33). hyf —舌下神经孔（hypoglossal for.）：位于枕髁之前；为第十二（舌下，hypoglossal n.）脑神经的出口。

29 (34). pgf —臼后孔（postglenoid for.）：位于关节窝（glenoid fossa）[或称下颌窝（mandibular fossa）] 后端内方，是脑后部主要静脉的出口。在副竹鼠类中该孔为裂隙状，位于鼓骨和颞骨之间。在有些副竹鼠的标本中该裂隙可以和岩鼓裂（petrotympanic fissure）相连，位于其后端。

30 (37). styf —茎乳孔（stylomastoid for.）：位于外耳道的后下方，乳突之前；为第七（面）神经（facial n.）的出口。

31 (38). msf —乳突孔（mastoid for.）：位于项面中部，在枕骨与颞骨岩乳部之间的横向延伸的骨缝上；为脑后部静脉的通道。

（3）牙齿有关术语

颊齿冠面结构和术语如图 4 所示。

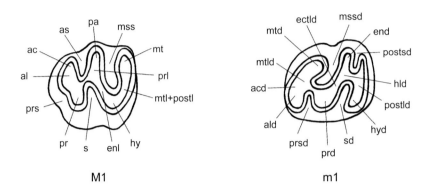

图 4　副竹鼠族的左 M1 和左 m1 冠面尖、脊和褶沟的术语

Fig. 4　Terminology of cusp(id)s, loph(id)s and reentrants of left M1 and left m1 in pararhizomyines

M1: ac. 前边尖（anterocone）, al. 前边脊（anteroloph）, as. 前边凹（anterosinus）[= abr. 颊侧前褶沟（anterior buccal reentrant）], enl. 内脊（entoloph）, hy. 次尖（hypocone）, mss. 中凹（mesosinus）[= pbr. 颊侧后褶沟（posterior buccal reentrant）], mt. 后尖（metacone）, mtl. 后脊（metaloph）, pa. 前尖（paracone）; postl. 后边脊（posteroloph）, pr. 原尖（protocone）, prl. 原脊（protoloph）, prs. 原凹（protosinus）, s. 内凹（sinus）[= lr. 舌侧褶沟（lingual reentrant）]

m1: acd. 下前边尖（anteroconid）, ald. 下前边脊（anterolophid）, ectld. 下外脊（ectolophid）, end. 下内尖（entoconid）, hld. 下次脊（hypolophid）, hyd. 下次尖（hypoconid）, mssd. 下中凹（mesosinusid）[= alr. 下舌侧前褶沟（anterior lingual reentrant）], mtd. 下后尖（metaconid）, mtld. 下后脊（metalophid）, postld. 下后边脊（posterolophid）, postsd. 下后边凹（posterosinusid）[= plr. 下舌侧后褶沟（posterior lingual reentrant）], prd. 下原尖（protoconid）, prsd. 下原凹（protosinusid）[= abr. 下颊侧前褶沟（anterior buccal reentrant）], sd. 下外凹（sinusid）[= pbr. 下颊侧后褶沟（posterior buccal reentrant）]

副竹鼠类臼齿的高冠类型很特别：上臼齿前侧齿冠高于后侧者，被称为前侧高冠（mesial hypsodonty），而下臼齿者则相反，被称为后侧高冠（distal hypsodonty）。本书统称这种高冠类型为前 - 后侧高冠型齿（mesiodistal hypsodonty）。此外，在副竹鼠类的一些齿冠较高的种类中，其上、下臼齿的舌侧的齿冠均高于颊侧者，本书称其为舌侧高冠型（lingual hypsodonty）。这种高冠型齿显然不同于 *Rhizomys* 的单面高冠型齿（unilateral hypsodonty），后者的上臼齿的舌侧齿冠高于颊侧、而下臼齿的则是颊侧齿冠高于舌侧 [这后一种高冠类型实际上应属舌 - 颊侧高冠（linguobuccal hypsodonty）]。

为了了解副竹鼠类门齿的微细结构的特点，我们对副竹鼠类的两个种（*Pararhizomys hipparionum* 和 *Pseudorhizomys gansuensis* gen. et sp. nov.）的下门齿作了切片观测。所观测的下门齿均采自已知种的下颌骨。方法是先将切割下的下门齿放入包埋盒中用 Tech 7200 光固化剂经 12 小时固化，将固化样品用 EXAKT 300CP 切割仪进行横向和纵向切割；用手将已切割的样品先后在 HERMES 的 P1000、P2500 和 P4000 等不同型号砂纸上研磨和抛光，然后将样品放在 0.1 mol/L 浓度的磷酸中浸泡约 60 秒；之后将样品用水漂洗、晾干。最后将镀金后的标本放在日立 S-3700N 扫描电子显微镜下观察。

3. 测量方法

头骨和颅后骨骼标本使用游标卡尺测量，测量精度接近 0.05 mm。牙齿在 Wild Heerbrung 显微镜下测量。测量精度接近 0.01 mm。

测量中一些通用的缩写：L，长（length）；W，宽（width）；H，高（height）；D，径（distance）；

APD，前后径（anteroposterior distance）；PE，近端（proximal end）；DE，远端（distal end）；
Max，最大（maximum）；Min，最小（minimum）；Ⓛ，左（left）；Ⓡ，右（right）；"+"，稍大（slightly larger）；"~"，大约（circa）。

（1）头骨(+ 下颌）

头骨测量时的定位和方法主要采用了潘清华等（2007：4–8，图3）的建议，笔者也作了部分修改和补充。其测量方位如下：

1. 枕鼻长（condylonasal length, CNL）：鼻骨前端—枕髁后缘
2. 颅基长（condylobasal length, CBL）：前颌骨前端—枕髁后缘
3. 头骨前部长（length of anterior part of skull, LAS）：前颌骨前端—M1 齿槽前缘
4. 头骨后部长（length of posterior part of skull, LPS）：M1 齿槽前缘—枕髁后缘
5. 腭长（palatal length, PL）：前颌骨前端—硬腭后缘
6. 上颌齿隙长（length of maxillary diastema, LMXD）：I2 齿槽后缘—M1 齿槽前缘
7. 腭桥长（length of palatal bridge, LPB）：门齿孔后缘—硬腭后缘
8. 腭桥前宽（anterior width of palatal bridge, AWPB）：左、右 M1 齿槽间最短距离
9. 腭桥后宽（posterior width of palatal bridge, PWPB）：左、右 M3 齿槽间最短距离
10. 门齿孔长（length of incisive foramen, LIF）：门齿孔的最大长度
11. 吻部长（rostrum length, RL）：前颌骨前端—颧弓前缘前颌骨 - 上颌骨缝处
12. 吻部前宽（anterior width of rostrum, AWR）：吻部前端最大宽度
13. 吻部后宽（posterior width of rostrum, PWR）：左、右颧弓前缘—前颌骨 - 上颌骨缝处间距
14. 吻部高（rostrum height, RH）：在紧接 I2 之后的鼻骨顶—前颌骨下缘间最短距离
15. 头骨中部高（height of middle part of skull, HMS）：紧靠 M1 前的头骨高度
16. 鼻骨长（nasal length, NL）：鼻骨最大长
17. 鼻骨前宽（anterior width of nasals, AWN）：左、右鼻骨前部最大宽
18. 鼻骨后宽（posterior width of nasals, PWN）：左、右鼻骨后部在鼻骨 - 额骨缝外端处宽
19. 眶间宽（interorbital width, IOW）：左、右泪结节外侧缘的间距
20. 眶后收缩处宽（width of postorbital constriction, WPOC）：眶后收缩处的最小宽
21. 颧弓处宽（zygomatic width, ZW）：左、右颧弓外侧缘的最大距离
22. 脑颅乳突处宽（width of cranial part at mastoid processes, WCM）：左、右乳突外侧缘处最大宽
23. 听泡长（length of auditory bulla, LAB）：听泡（除棘突外）的最大长
24. 听泡宽（width of auditory bulla, WAB）：听泡的最大横宽
25. 听泡间距（distance between two auditory bullae, DAB）：左、右听泡间的最短距离
26. 项面高（nuchal height, NH）：枕外隆凸点至枕髁下缘间的垂直距离
27. 枕骨大孔高（height of *foramen magnum*, HFM）：枕骨大孔最大垂直距离
28. 枕骨大孔宽（width of *foramen magnum*, WFM）：枕骨大孔最大横宽
29. 下颌骨长（mandibular length, MDL）：下颌骨最前端（不带门齿）—角突后端
30. 下颌骨高（mandibular height, MDH）：冠状突顶至下颌骨上升支下缘的垂直距离
31. 下颌髁突高（height of condyloid process of mandible, HCPM）：下颌髁顶面至下颌骨上升支下缘的垂直距离
32. 下颌切迹深（depth of mandibular notch, DMN）：冠状突顶至下颌切迹底的垂直距离
33. 下颌齿隙长（length of mandibular diastema, LMD）：i2 齿槽后缘— m1 齿槽前缘的距离
34. 下颌齿隙处高（height of mandiblar diastema, HMD）：下颌齿隙处的最小高
35. 下颌水平支高（height of horizontal ramus of mandible, HHR）：水平支在 m1 和 m2 间外侧的高

（2）颊齿

李强（Li, 2010: 50–51, Fig. 3）创立了一种冠高的替代指数（index H），即以牙齿侧面釉质曲线和褶沟底之间的最小距离（H）代表齿冠高度，"H值越大，表明齿冠越低，反之则齿冠越高"（Li, 2010: 48）。他将已知的 *Pararhizomys hipparionum* 化石产地的层位按时代排序后得出结论，认为内蒙古宝格达乌拉地点产 *Pa. hipparionum*（V 16306 等）的层位要比陕西喇嘛沟（Lamagou）产 V 14178 和府谷产 V 412（正模）的层位的时代要晚（详见 Li, 2010: 55–57, Fig. 8）。根据他的推论，前者的"H"替代指数应比后两者的小。然而，他的实际测量数据却与推论不符。譬如，在他的表 4 中，宝格达乌拉（V 16306）的 M2 和 m2 舌侧的冠高指数（LNH）分别为 1.7 mm 和 0.90–0.95 mm，而喇嘛沟（V 14178）的 M2 的 LNH 为 1.15 mm 和府谷正模（V 412）的 m2 的 LNH 为 0.8 mm。前者（1.7 mm 和 0.90–0.95 mm）明显分别大于后两者（1.15 mm 和 0.8 mm）。实际上，李强的"H"替代指数只代表褶沟底和釉质曲线间的最短距离，而不能确切地反映 *Pararhizomys* 所特有的前后侧高冠型齿的变化特点。此外，他的"H"替代指数测量的基准线是臼齿冠面两侧缘的线，而臼齿冠面和侧缘的方向会随着磨蚀程度的不同而变化，即使褶沟底和釉质曲线的顶点的位置不变，它们之间的距离也会随基准线方向的变化而变化，而使测量变得不准确。

笔者认为，郑绍华（1993: 102，图 49）对 *Rhizomys* 的颊齿的测量方法似乎更适合于测量 *Pararhizomys* 这样的前 - 后侧高冠型齿。本书对臼齿的测量主要采用郑绍华（1993）的测量方法，但作了部分修改和补充（见图 5）：

1. 臼齿长（molar length = mesiodistal length, L）：垂直于上臼齿齿冠后缘（或下臼齿齿冠前缘）的前 - 后最大距离
2. 臼齿宽（molar width, W）：垂直于臼齿长轴的臼齿齿冠的最大宽
3. 臼齿褶沟冠面长度（length of reentrant of molar on occlusal surface, LRM）：褶沟在颊齿冠面横向延伸的距离
4. 臼齿褶沟冠面宽度（width of reentrant of molar on occlusal surface, WRM）：褶沟在颊齿冠面纵向延伸的距离
5. 臼齿褶沟侧面深度（depth of reentrant of molar on lateral sides, DRM）：褶沟在颊齿侧面上向齿根方向延伸的距离

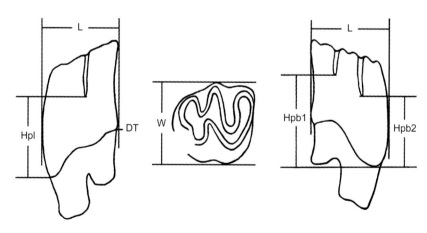

图 5　副竹鼠族臼齿的测量

Fig. 5　Measurements of molars in pararhizomyines

DT. 釉质曲线（dentine tract），Hp. 冠高替代指标（hypsodonty proxy），Hpb1. 颊侧前褶沟处冠高替代指标（hypsodonty proxy on anterior buccal reentrant），Hpb2. 颊侧后褶沟处冠高替代指标（hypsodonty proxy on posterior buccal reentrant），Hpl. 舌侧褶沟处冠高替代指标（hypsodonty proxy on lingual reentrant），L. 臼齿长（length），W. 臼齿宽（width）

6. 臼齿冠高替代指标（hypsodonty proxy, Hp）：与齿冠后面（上臼齿）或前面（下臼齿）平行，在臼齿侧面上从褶沟底至釉质曲线（DT）的最低点之间的直线距离

6a. 颊侧冠高替代指标（hypsodonty proxy on buccal side, Hpb）：包括，颊侧前褶沟 [= 上臼齿的前边凹（as）或下臼齿的下原凹（prsd）] 处的冠高替代指标（hypsodonty proxy on anterior buccal reentrant, Hpb1）；颊侧后褶沟 [= 上臼齿的中凹（mss）或下臼齿的下外凹（sd）] 处的冠高替代指标（hypsodonty proxy on posterior buccal reentrant，Hpb2）

6b. 舌侧褶沟的冠高替代指标（hypsodonty proxy on lingual reentrant, Hpl）是指上臼齿的内凹（s）或下臼齿的下中凹（mssd）处冠高替代指标

（3）颅后骨骼

颅后骨骼的解剖术语和测量方位主要依邱占祥和王伴月（2007）在犀类中采用者。为简便起见，前、后足的方位与骨架的其他部位的方位采用同一标准，即均以身体中矢面（sagittal plane）为基准，向中矢面的一侧为内侧（medial），远离中矢面的为外侧（lateral）。这样，足部中的第一掌骨（蹠骨）向中矢面的一侧为内侧，远离中矢面 [即面向第二掌骨（蹠骨）] 者为外侧。同样，McII (MtII) 的向 McI (MtI) 者为内侧，而向 McIII (MtIII) 者为外侧，等等。

化石编号缩写　HMV，甘肃省和政古动物化石博物馆古脊椎动物化石编号前缀；IVPP Loc. LX（行文中简化为 LX），中国科学院古脊椎动物与古人类研究所临夏盆地野外地点编号前缀；IVPP V（行文中简化为 V），中国科学院古脊椎动物与古人类研究所脊椎动物化石编号前缀；OV，中国科学院古脊椎动物与古人类研究所现生脊椎动物标本编号。

二、临夏盆地副竹鼠类化石系统描述

鼠超科 Muroidea Illiger, 1811
盲鼹鼠科 Spalacidae Gray, 1821
拟速掘鼠亚科 Tachyoryctoidinae Schaub, 1958
副竹鼠族（新族） Pararhizomyini tribe nov.

模式属 *Pararhizomys* Teilhard de Chardin et Young, 1931。

其他包括属 *Pseudorhizomys* gen. nov.。

分布与时代 东亚中纬度地区（包括中国内蒙古、陕西、甘肃、青海和西藏，蒙古国）；晚中新世—早上新世。

鉴别特征 中—大型鼠形类。头骨枕鼻长 50–70 mm。头骨具鼠型颧 - 咬肌结构，较低，具窄长的吻部和宽的颅部，面部与颅部长度相近。较宽的颧弓板斜向前外上方延伸，面向前外下方。外层咬肌附着处限于上颌骨内，有明显的弧形前缘嵴。颧弓后端向后延伸不达项嵴。上颌齿隙长，上颊齿列后移。鼻骨为后端趋尖的长楔形。鼻骨和前颌骨后端几乎在同一横线上。前颌骨具侧背嵴。前颌骨 - 上颌骨缝穿过门齿孔。门齿孔短小，约为上齿隙长的 1/4–1/3。眶下孔大，具腹侧裂隙。鼻泪窝位于眶下管内。眼眶很小。硬腭后部具明显的副翼窝。翼窝窄小而浅。咬肌和颊肌神经管很短。关节窝很长，向后延伸至项嵴，听泡参与关节窝的组成；外耳道很短，不成管状，位于关节窝的下后方。卵圆孔与中破裂孔相融合。未见蹬骨动脉孔。矢状嵴和项嵴很发达。无间顶骨。枕髁腹面较短宽，左、右枕髁彼此靠近。项面平，面向后上方，颞骨岩乳部在枕面出露较大。副乳突很小。

下颌为松鼠型。水平支短而粗壮。下颌骨齿隙长 ≥ 下颊齿列长。咬肌窝前伸至 m1 与 m2 交界之前下方；咬肌嵴向外伸张。颏孔位于 m1 前缘下方。颞肌窝明显，向下伸至冠状突内侧基部。髁突斜向后上方伸长。下门齿后端在上升支颊侧形成很明显的隆凸。下颌角尖突状，向后伸至髁突下方。翼内肌窝大而深。上升支后缘明显凹入。

齿式：1·0·0·3/1·0·0·3。臼齿为前 - 后侧高冠型齿，冠高中等，具齿根。臼齿冠面脊形，结构简单，无中脊：上臼齿无后边凹；M1 和 m1–2 具 2 条颊侧褶沟和 3 条横脊，M2–3 和 m3 具 1 或 2 条颊侧褶沟，上臼齿具 1 舌侧褶沟（内凹），m1–3 具 1 或 2 条舌侧褶沟（下后凹或有或无）；每条褶沟约伸达齿冠中部。M3 和 m3 趋于退化。门齿很强壮，釉质层在内、外侧的宽度都很窄。i2 唇面具 1 或 2 条纵嵴。

副竹鼠属 *Pararhizomys* Teilhard de Chardin et Young, 1931

模式种 *Pararhizomys hipparionum* Teilhard de Chardin et Young, 1931。

其他包括种 *Pararhizomys qinensis* Zhang, Flynn et Qiu, 2005（陕西蓝田）, *P. huaxiaensis* sp. nov. 和 *P. longensis* sp. nov.。此外尚有二未定种：*Pararhizomys* sp. I（青海德令哈深沟）和 *Pararhizomys* sp. II（内蒙古高特格）。

分布与时代　中国陕西、甘肃、青海和内蒙古，以及蒙古国西部；晚中新世—早上新世。

修订鉴别特征　前颌骨与鼻骨前端约在同一垂线上。鼻骨背面纵向和横向均圆凸。前颌骨 - 上颌骨缝起始于门齿孔后部，斜向前外方伸。前颌骨侧背嵴长而显著，与鼻骨 - 前颌骨缝前半部近于平行。前颌骨背部为长条形。前颌骨 - 上颌骨缝的背部直，纵向延伸，与上颌骨 - 额骨缝以钝角相交。眶下孔为横宽的椭圆形。颧弓前根和颧弓板的位置靠后，其背侧后缘，亦即眼眶前缘与鼻骨和前颌骨的后端约在同一横线上；颧弓板弧形后缘的最前点位于门齿孔之后。外层咬肌附着区的前缘为圆弧形嵴。颧弓圆弧形。颧骨短，其前端不伸达眼眶前缘。两顶骨组成中部微外凸的窄长方形，其前中央尖突插入两额骨之间。腭后孔每侧前后两个，非常小。副翼窝较大而深。硬腭后缘位于 M3 之后。中翼窝明显宽于翼窝。左、右内翼突向外方倾斜，向后彼此逐渐靠近。颞深神经孔与咬肌神经孔愈合，而颊肌神经孔则与它们分开。该三神经管（颞深神经管、咬肌神经管和颊肌神经管）的后孔愈合为一。颞骨岩乳部在项面上出露较大。下颌齿隙后部凹入深。下颌骨下缘自联合部后端至角突间较平直。颏孔位置低，低于咬肌窝的前端。下颌切迹较浅，其深小于下颌骨高的 1/3。I2 纵向强烈弯曲，始自上颌骨前缘近门齿孔后缘处；其前端稍向下后方弯曲，唇面具有一条很显著的纵棱。下门齿唇面具一或两条纵嵴。臼齿为前 - 后侧高冠型齿，仅具一条舌侧褶沟。M2–3 和 m3 仅具一颊侧褶沟（中凹），无前边凹。m3 约为长宽相近的圆方形，下中凹和下外凹通常横向延伸，彼此相对。

三趾马层副竹鼠 *Pararhizomys hipparionum* Teilhard de Chardin et Young, 1931

（图 6–15, 75；表 1–5, 15, 16）

正模　V 412（原地质调查所新生代研究室 No. c/90），左下颌部分水平支具 m1–3，产自原地质调查所新生代研究室第十地点，陕西省府谷县新民镇（原镇羌堡），上中新统"蓬蒂红色泥岩"，晚中新世保德期。

本书记述材料　1）V 16286.1，含下颌头骨，V 16286.2，一段左桡骨和 V 16286.3，一段右股骨，均产自 LX 200047（桥家村）；2）HMV 1923，不完整的头骨，产自 LX 200207（立麻）；

表 2　临夏盆地产副竹鼠属化石的地点和层位
Table 2　Localities and horizons yielding fossil *Pararhizomys* in the Linxia Basin

种 Species	标本 Specimens		编号		产出层位 Horizon	地方动物群 Local Fauna
			Catal. No.	IVPP Loc.		
三趾马层副竹鼠 *Pa. hipparionum*	归入标本 referred specimens	带下颌骨的头骨和部分肢骨 skull with mandible and partial limb bones	V 16286	LX 200047	柳树组中部下层 lower part of mid Liushu Fm.	大深沟 Dashengou
		头骨 skull	HMV 1923	LX 200207		
		部分头骨 partial skull	V 16287	无 no	层位不明 uncertain	不确定 uncertain
华夏副竹鼠（新种） *Pa. huaxiaensis* sp. nov.	正模 holotype	带下颌骨的头骨和部分颅后骨骼 skull with mandible & partial postcranial bones	HMV 1413	LX 200041	柳树组中部上层 upper part of mid Liushu Fm.	杨家山 Yangjiashan
	副模 paratype	头骨 skull	HMV 1924	LX 200503		
		部分头骨 partial skull	V 16291	LX 200030		
陇副竹鼠（新种） *Pa. longensis* sp. nov.	正模 holotype	不完整的含下颌骨的头骨 incomplete skull with mandible	V 16292.1	LX 200042	柳树组下部 lower part of Liushu Fm.	郭泥沟 Guonigou
	副模 paratype	头骨前部 anterior part of skull	V 16292.2			
		头骨前部 anterior part of skull	V 16292.3			

3）V 16287，不完整头骨，产出地为临夏盆地，详细地点和层位不明（详见表1，2）。

产地与层位（中国） 陕西：府谷新民镇（原镇羌堡），"蓬蒂红色泥岩"；府谷喇嘛沟，保德组；甘肃：秦安，"红色泥岩"；广河买家巷乡桥家村、和政县三合乡立麻（麻沟），柳树组中部下层；内蒙古：阿巴嘎旗宝格达乌拉，宝格达乌拉组；上中新统，中灞河—保德阶。

鉴别特征 前颌骨侧面上方的凹槽浅。颧弓板弧形后缘的最前点在M1之前。下颌齿隙长约等于下臼齿齿列长。臼齿齿冠为前-后侧高冠。M1–2和m1–3舌侧沟与颊侧沟的深度相近，而M3的中凹则深于内凹；在冠面上，M3的内凹通常斜向前颊侧延伸，与中凹部分重叠。m1下中凹强烈前弯，其颊部向前延伸超过下原凹。

区别特征 *Pa. hipparionum* 与 *Pa. qinensis* 的区别是具较长的上颌齿隙和具较长的M2等。

评注 到目前为止，我们对 *hipparionum* 这一种名的词性及词尾是否合乎动物命名法规的要求尚无定见。按照《International Code of Zoological Nomeclature》（4ᵗʰ Ed., 1999）第31条的规定，由人名组成的种名可以是主格或第二格。在本例中，如用主格则种名应为 *hipparion*，如用第二格，在我们查到的资料（张永辂，1983）中，以 n 结尾的名词，很可能是第三变格法的名词，这类名词的单数第二格词尾为 is，复数第二格词尾为 um 或 ium。*hipparion* 的复数第二格词尾是否是 *hipparionum* 目前尚无处查证。如果 *hipparionum* 系形容词，其词尾是中性，如和 *Pararhyzomys*（阳性）搭配，则应为 *hipparionus*。

1. 描述

（1）头骨（图6–9；表3）

头骨较大，具鼠型的颧-咬肌结构。头骨的外形与竹鼠（*Rhizomys*）及盲鼹鼠（*Spalax*）较类似：脑颅宽，具较平而前倾的项面。和竹鼠不同的是：其头骨低而窄，吻部窄长，上颌齿隙更长，颅部较短，具明显而长的矢状嵴等。和盲鼹鼠不同的是其项面不那么强烈前倾，顶骨长大于宽。

V 16287头骨虽然保存不完整，但其各骨的形态、骨缝及孔等保存更好。下面的描述主要依据这件标本（图6–7）。

背面观（图6 A，8 C） 吻部窄长，其两侧面彼此近于平行。鼻骨（nasal, N）为长楔形，后端狭窄。鼻骨-前颌骨缝仅前端稍向内弯，其后的绝大部分都较直，左、右鼻骨-前颌骨缝往后彼此靠近。鼻骨背面纵向和横向轻微圆凸，其前部横向圆凸较后部强。鼻骨后端向后伸达呈锯齿状、斜向前外方的前颌骨-额骨缝的最后点。前颌骨（premaxilla, Pm）在背面为长条形，其外侧有一很显著的侧背嵴（pmldc，图6 A，7 A，8 A），将吻部背面和侧面截然分开。该棱从眶下孔的前上角一直伸到上门齿槽的前外缘，其前半部大致与鼻骨-前颌骨缝平行。前颌骨-上颌骨缝在头骨背侧呈直线纵向延伸，其后端与斜向后外方延伸的上颌骨-额骨缝的内端呈钝角相连。泪骨（lacrimal, L）位于眼眶的内侧，在头骨背面出露得很小，呈结节状。泪骨结节（lacrimal tubercle, lt）明显。眶下孔（infraorbital foramen, iof）在V 16287的左侧保存完整。自背面可见部分眶下孔，由分成上、下两支的上颌骨颧突作为外壁围绕而成。和所有鼠形类一样，眶下孔很宽大，从前面看为横宽的椭圆形。在眶下孔的中部可见由眶下孔下部向上伸展出的纵向弧形骨质隔板，隔板内为袋状凹坑，其底部应为鼻泪孔（nasolacrimal foramen, nlf）。隔板外为腹侧裂隙（ventral slit）的上半部。眶下孔的腹侧裂隙为眶下神经和眶下动脉进入吻部侧面的通道。颧弓仅在HMV 1923标本的右侧保存完整（图8 B, C）。颧弓为圆弧形，使头骨的最大宽度位于颧弓中部，而不像在大部分鼠形类中那样位于颧弓的后部。颧弓前根上支后缘（即眼眶前缘）的最前点与鼻骨-额骨缝约在同一横线上。颧弓中部横向很薄，其横切面约为三角形：其上缘为薄锐的嵴；外侧面较宽平，面向侧上方；内面宽，而底面窄。颧弓后端向后延伸不达项嵴，两者间以明显的缺凹分开。颧骨（jugal, J）很短（图8A），其前端不达眼眶前缘，在其中部有很明显的眶后突（postorbital process，pop），作为眼眶的后界。头骨的眶后收缩明显，额骨上无明显的眶后突。额嵴（frontal crest，

图 6　三趾马层副竹鼠头骨（V 16287）

Fig. 6　Skull of *Pararhizomys hipparionum* (V 16287)

A. 背面（dorsal view），B. 前面（anterior view），C. 后面（posterior view）

缩写（Abbreviations）

头骨上的孔和其他结构（foramina and other structures）：eam. 外耳道（external auditory meatus），fc. 额嵴（frontal crest），fm. 枕骨大孔（*for. magnum*），glf. 关节窝（glenoid fossa），iof. 眶下孔（infraorbital for.），lt. 泪骨结节（lacrymal tubercle），msf. 乳突孔（mastoid for.），nc. 项嵴（nuchal crest, nc），nlf. 鼻泪孔（nasolacrimal for.），occ. 枕髁（occipital condyle），pmldc. 前颌骨侧背嵴（premaxillary laterodorsal crest），sc. 矢状嵴（sagital crest），tc. 颞嵴（temporal crest）

骨骼（bones）：F. 额骨（frontal），L. 泪骨（lacrimal），M. 上颌骨（maxilla），N. 鼻骨（nasal），Oc. 枕骨（occipital），P. 顶骨（parietal），Pm. 前颌骨（premaxilla），Pms. 岩乳部（petromastoid portion），Sq. 鳞骨（squamosal）

fc）很弱，从泪骨结节开始向后内方向延伸。左、右额嵴在眶后收缩部之前愈合形成明显的矢状嵴（sagital crest, sc）。顶骨（parietal, P）约为中部微向外凸的长方形，长大于宽，其前部稍变窄，具前中央突，该突伸入两额骨之间；顶骨表面平滑，只是在近矢状嵴处横向稍凹。颞窝在背面出露很宽大，颞嵴（temporal crest, tc）较短，其后部较平缓。颞区在外耳道和颧弓后根之间呈深弧形凹入。未见有间顶骨的痕迹。

　　侧面观（图 7 A，8 A）　头骨背缘轮廓稍呈波浪形：眶后收缩的上方稍凹，其前的部分呈圆弧形向前下方延伸，鼻骨背面为纵向圆凸的弧形；而后部矢状嵴处则稍向上凸起。鼻骨和前颌骨的前端几乎在同一垂直线上，或鼻骨稍更向前突出（图 8 A）。上颌齿隙远长于上颊齿列长。由于整个上颊齿列后移，头骨后部相对变短，使头骨的前、后部（以 M1 前缘为界）的长度相近（见表 3）。头骨眶后收缩之前的面部和之后的颅部的长度也相近。前颌骨的侧面部分较宽广。在其上部有一圆缓的纵棱显示出 I2 的隆凸的轮廓，在 V 16286 标本中，左侧 I2 在眶下孔内侧出露，表明 I2 是从眶下孔的内侧经过的。在前颌骨的侧背嵴和容纳门齿的骨隆凸之间形成一浅纵沟（图 8 A, g）。在头骨侧面，眶下孔之前的前颌骨 - 上颌骨缝约呈垂直向延伸。外层咬肌在颧弓板上的起始区限于上颌骨内，不伸达前颌骨。其前缘（anterior rim of area for *m. masseter lateralis*, araml）形成很明显的、向前圆凸的弧形嵴。在眶下管内壁的后方有一斜向后下方延伸的破裂区，大概与 Wahlert（1985a: figs. 2, 3）所描述的上颌骨与泪骨间的未骨化区相当（unossified area between maxilla and lacrimal, ua），是啮齿类中的近裔性状。与上颊齿列相比，眶部各孔的位置相对前移。V 16287 是眶部诸孔保存最好的，但由于眶内骨壁受到挤压和破损，仍有若干孔的形态保存不佳，其位置只能大体确定。蝶腭孔（sphenopalatine foramen, spf）位于眶部的最前下方，在 M1 齿槽缘之上方约 5 mm 处。背腭孔（dorsal palatine foramen, dpf）位于蝶腭孔的后下方、M2 前部上方，颈内动脉眶下动脉支中之腭降动脉小支自此进入腭管并伸达腭后孔（上述 2 孔在 HMV 1923 中看得最清楚，其背腭孔的位置较 V 16287 的稍高、稍后，可惜被颧弓遮住，见图 8 A）。筛孔（ethmoidal foramen, etf）在 V 16287 头骨的左侧保存得好，为一小孔，位于 M1 的上方和视神经孔的前上方。视神经孔（optic foramen, opf）很小，可能比筛孔稍大一点，位于 M2 和背腭孔的上方、蝶腭孔的后上方。前翼裂（anterior alar fissure, aafi）位于视神经孔的后下方、M2 和 M3 的上方。在 V 16287 标本中前翼裂明显分为两部分：前上者为蝶裂（sphenoidal fissure, sf）；后下者应为翼蝶管的前开孔和圆孔（*foramen rotundum*, fr）。在前翼裂后下角之后没有看到有小翼孔存在。包含上述各孔（筛孔、视神经孔、前翼裂、蝶腭孔和背腭孔）的眶部形成一个相当深的凹陷，其下界是相当高的颊齿齿槽囊隆起。在翼蝶骨上，在颧弓后根之下，有两个带浅沟的小孔：上者为咬肌神经孔（masticatory foramen, mf），其上有一浅沟绕过颧弓后根前缘伸向颞部，应为三叉神经下颌支之颞深神经小支（deep temporal n.）进入颞肌的通道，表明颞深神经小支是由咬肌神经孔出口，亦即颞深神经孔（deep temporal foramen, dtf）与咬肌神经孔已愈合；下者有一斜向前下方的浅沟，应为颊肌神经孔（buccinator foramen, bf）。翼管后孔（posterior foramen of pterygoid canal, pcpf）因 V 16287 的左、右两侧内翼突基部的外侧都保存不好，无法辨认，但在 HMV 1923 头骨两侧内翼突基部的外侧可见。该孔极小，可能已经很退化了（见图 8 B）。关节窝（glenoid fossa, glf；或下颌窝，mandibular fossa）纵向很长，向后伸至臼后孔（见下）之后，伸达项嵴，将外耳道挤到其下方。其前部较平或稍凸，由鳞骨组成，而后部较大，横向深凹，由鳞骨和听泡外侧

图 7　三趾马层副竹鼠头骨（V 16287）（立体照片）

Fig. 7　Skull of *Pararhizomys hipparionum* (V 16287) (stereopair)

A. 右侧面（right lateral view），B. 腹面（ventral view）

缩写（Abbreviations）

头骨上的孔和其他结构（foramina and other structures）：ab. 听泡（auditory bulla），araml. 外层咬肌附着区前缘（anterior rim of area for *m. masseter lateralis*），ascp. 翼蝶管后孔（posterior for. of alisphenoid canal），bf. 颊肌神经孔（buccinator for.），bfp. 颊肌神经管后孔（posterior for. of buccinator nerve canal），bt. 基结节（basilar tubercle），chuf. 吉埃氏管孔（for. for canal of Huguier），dpf. 背腭孔（dorsal palatine for.），dtf. 颞深神经孔（deep temporal for.），dtfp. 颞深神经管后孔（posterior for. of deep temporal nerve canal），eam. 外耳道（external auditory meatus），epp. 外翼突（external pterygoid process），eucf. 欧氏管孔（for. of Eustachian canal），fm. 枕骨大孔（*for. magnum*），fo. 卵圆孔（*for. ovale*），fr. 圆孔（*for. rotundum*），glf. 关节窝（glenoid fossa），hyf. 舌下神经孔（hypoglossal for.），icf. 颈内动脉孔（internal carotid for.），inf. 门齿孔（incisive for.），iof. 眶下孔（infraorbital for.），ipp. 内翼突（internal pterygoid process），juf. 颈静脉孔（jugular for.），mf. 咬肌神经孔（masticatory for.），mfp. 咬肌神经管后孔（posterior for. of the masseteric nerve canal），mlf. 中破裂孔（middle lacerate for.），mptf. 中翼窝（mesopterygoid fossa），msp. 乳突（mastoid process），mt. 咬肌结节（masseteric tubercle），occ. 枕髁（occipital condyle），opf. 视神经孔（optic for.），pgf. 臼后孔（postglenoid for.），pmf. 后上颌孔（posterior maxillary for.），pmldc. 前颌骨侧背嵴（premaxillary laterodorsal crest），pmp. 副乳突（paramastoid process），ppf. 腭后孔（posterior palatine for.），pptf. 副翼窝（parapterygoid fossa），ps. 腭沟（palatine sulcus），psaf. 后上齿槽孔（posterior superior alveolar for.），ptf. 翼窝（pterygoid fossa），spf. 蝶腭孔（sphenopalatine for.），sf. 蝶裂（sphenoidal fissure），styf. 茎乳孔（stylomastoid for.）

骨骼（bones）：Bo. 基枕骨（basioccipital），Bs. 基蝶骨（basisphenoid），F. 额骨（frontal），J. 颧骨（jugal），M. 上颌骨（maxilla），N. 鼻骨（nasal），P. 顶骨（parietal），Pl. 腭骨（palatine），Pm. 前颌骨（premaxilla），Pt. 翼骨（pterygoid），Sq. 鳞骨（squamosal）

· 18 ·

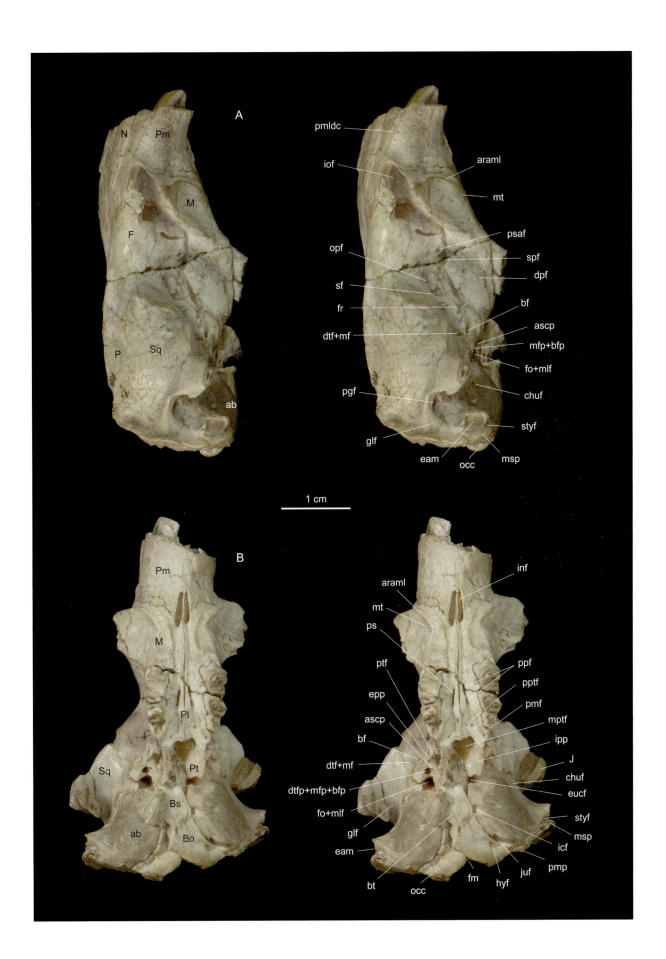

组成。臼后孔（postglenoid foramen, pgf）窄长裂隙状，位于关节窝的后部，其前半部自关节窝外后角斜向前内下方伸延。而后半部则转向后内方延伸，直至外耳道（见下）上方。岩鼓裂（petrotympanic fissure, petf）位于听泡的前外缘，其后外端与臼后孔内端相连。听泡（auditory bulla, ab；或称鼓泡，*bulla tympanica*）很圆隆，自侧面看，深陷于颅内，其高大于其前后径长。在听泡侧壁的前下部，大体和外耳道在同一水平上有一小孔，应为供鼓索神经通过的于吉埃氏管孔（foramen of canal of Huguier, chuf）。外耳道（external auditory meatus, eam）仅为雏形管状，很短，位于听泡后缘的中上部。乳突（mastoid process, msp）很小，为扁平板状，紧贴在外耳道之后。茎乳孔（stylomastoid foramen, styf）很小，位于外耳道和乳突之间区域的下部。项面稍向前倾斜。枕髁（occipital condyle, occ）位于项嵴（nuchal crest, nc）和听泡的后方，其下缘高于听泡的下缘。

腹面观（图 7 B, 8 B, 9 A） 在吻部保存完整的 V 16286 和 HMV 1923 标本上均无前颌骨间孔

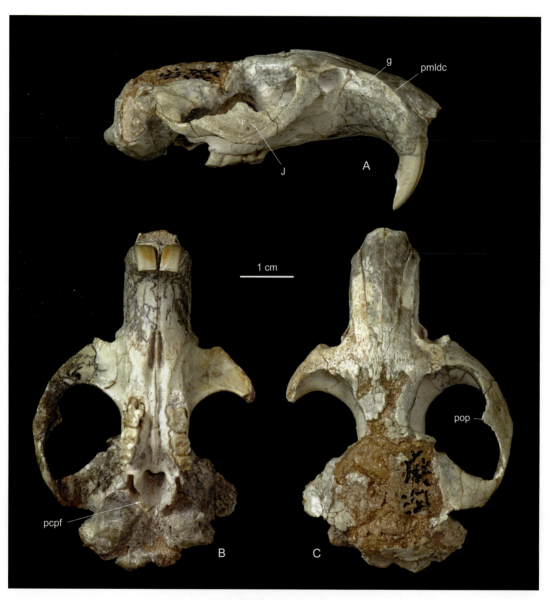

图 8 三趾马层副竹鼠头骨（HMV 1923）

Fig. 8 Skull of *Pararhizomys hipparionum* (HMV 1923)

A. 右侧面（right lateral view），B. 腹面（ventral view），C. 背面（dorsal view）

缩写（**Abbreviations**）：g. 浅纵沟（shallow longitudinal groove），J. 颧骨（jugal），pcpf. 翼管后孔（posterior foramen of pterygoid canal），pmldc. 前颌骨侧背嵴（premaxillary laterodorsal crest），pop. 眶后突（postorbital process） （interpremaxillary foramen）。门齿孔（incisive foramen, inf）窄小，位于上颌齿隙的中部，其长约为齿隙长的 1/4–1/3（见表 3）。前颌骨 - 上颌骨缝呈弱锯齿状，穿过门齿孔的中后部，向前外方延伸。颧弓板宽阔、腹面微凹，向前背外侧方延伸，并向腹外侧方倾斜，限于上颌骨内。其前缘有明显的弧形前缘嵴（araml）。外层咬肌区的弧形前缘嵴的最前点与门齿孔的前端约在同一横线上或稍后。颧弓板弧形后缘的最前点的位置在门齿孔与 M1 之间。咬肌结节（masseteric tubercle, mt）不为突出的结节状，而为卵圆形弱粗糙凹面，位于颧弓板的前部下缘，是表层咬肌（*m. masseter superficialis*）的附着处。它的位置与门齿孔后端约在同一横线上，距门齿孔较近，而距眶下孔较远。左、右颊齿列彼此近于平行或稍向前靠近。沿上颌骨中缝和腭骨中缝有矢状棱从门齿孔开始向后伸达硬腭后缘（即中翼窝前缘），并形成微向后伸的后鼻棘（posterior nasal spine），使硬腭后缘约呈"ω"形。矢状棱两侧的腭沟（palatine sulci, ps）在位于 M1–2 处的两对腭后孔（posterior palatine foramina, ppf）处加深，成为很深的窄槽。两对腭后孔中，前面的一对位于上颌骨内，M1 的后部内侧；后面一对位于腭骨内，M2 前部内侧。腭骨在腭面出露很小，上颌骨 - 腭骨骨缝的近中端可能仅达前面的一对腭孔的后端，与 M1 相对。硬腭后部腭面深凹，形成一对三角形凹面，凹面内有许多小孔。同样的三角形凹陷在某些囊鼠类（geomyids）中也有，Wahlert（1985a: 5）将其称之为副翼窝（parapterygoid fossa, pptf）。这可能是某种腺体的容纳处。后上颌孔（posterior maxillary foramen, pmf）位于副翼窝的外后角，M3 之后，与硬腭后缘约在同一横线上。中翼窝（mesopterygoid fossa, mptf, = 鼻咽窝, nasopharyngeal fossa）比翼窝（pterygoid fossa, ptf）宽很多。翼窝和翼突在 HMV 1923 标本中保存稍好（图 8 B）。内翼突（internal pterygoid process, ipp）较高大，向外方倾斜，其基部向后稍趋中，使中翼窝向后稍变窄。外翼突（external pterygoid process, epp）较低而短，稍向后外方伸。咬肌神经管后孔（posterior foramen of masseteric nerve canal, mfp）和颞深神经管，以及颊肌神经管后孔（posterior foramen of buccinator nerve canal, bfp）融合为一，位于外翼突的后端基部。翼蝶管后孔（posterior foramen of alisphenoid canal, ascp）位于上述融合孔的内侧。中破裂孔附近在三件标本上都保存不好。但可以肯定的是中破裂孔较小，而且卵圆孔（*foramen ovale*, fo）可能与中破裂孔（middle lacerate foramen, mlf）相融合。

基蝶骨（basisphenoid, Bs）和基枕骨（basioccipital, Bo）在三件标本上均受压变形。从保存部分看，基蝶骨和基枕骨形成前端较窄的梯形。基结节（basilar tubercle, bt）弱。基枕骨中央有矢状棱。棱的两侧为凹面，其上偶见小的滋养孔。舌下神经孔（hypoglossal foramen, hyf）位于枕髁的前外侧。听泡（auditory bulla, ab）适度膨胀，外形为卵圆形，长轴呈前内 - 后外方向，其前内角棘突伸向前内方，并向腹方尖突。V 16286.1 的左、右听泡和 HMV 1923 的左侧听泡壁均破损，听泡内未见到隔板。于吉埃氏管孔（foramen of canal of Huguier, chuf）位于外耳道的前内方。欧氏管孔（foramen of Eustachian canal, eucf）位于听泡棘突前方。颈内动脉孔（internal carotid foramen, icf）位于听泡内缘前 1/3 处，有一条明显的沟连接颈内动脉孔和颈静脉孔。颈静脉孔（jugular foramen, juf, = 后破裂孔, posterior lacerate foramen）裂隙状，很细小。从 V 16287 腹面看，在颈静脉孔内的听泡内后侧面上，未见有供颈内动脉镫骨动脉支进入听泡的孔。枕髁腹面较宽短，左、右枕髁彼此很靠近。

后面观（图 6 C）　项面的外形约呈半圆形。枕外隆凸点（inion），即项嵴与矢状嵴交接处，稍向上凸。项面较平，无明显的枕外嵴。枕骨（Occipital, Oc）两侧缘稍稍凹入。颞骨岩乳部（petromastoid portion, Pms）位于项面的外侧下部，相对较大，约为不规则菱形，其上部的横宽明显大于枕骨大孔外侧的枕骨宽。乳突孔（mastoid foramen, msf）明显，其位置稍高于枕骨大孔上缘。副乳突在三件标本上全未保存。从保留的基部看，该突非常小，位于乳突和枕髁之间、听泡隆凸轴的后端。枕骨大孔（*foramen magnum*, fm）外形约为圆形，其高约为枕骨项面高的 1/3。枕髁的后面相当窄。

（2）下颌骨（图 9 B, C；表 3）

只有 V 16286.1 保存有下颌骨，但左、右下颌角和左下颌的冠状突和髁突均已破损。左、右下颌在下颌联合处相连。下颌骨为松鼠型，其水平支短而粗壮，舌侧面垂向稍凹，下缘在下颊齿下方较直。下颌联合部向前上方延伸，与水平支下缘约呈40°相交。在下颌联合的后端的颏突（chin process, cp）为明显的嵴形。颏突内侧有明显的长椭圆形凹面，是二腹肌（*m. digastricus*）、下颌横肌（*m. transverse mandibularis*）和颏舌骨肌（*m. geniohyoideus*）等肌肉的附着处。我们称该凹面为二腹肌窝（digastric fovea, df）。下颌齿隙长度约与下颊齿列长度相近。从侧面看，下颌齿隙略呈"S"形，其前部稍凸，后部深凹，在m1之前的齿隙的后坡较陡。咬肌窝（masseteric fossa, msf）向前伸至m1/m2的下方。其咬肌嵴（masseteric ridge, mr）很发达，向侧方张开。颏孔（mental foramen, mtf）位于m1前缘的下方，约位于水平支的下1/3处，明显低于咬肌窝的前端。

　　下颌骨上升支相当长。冠状突（coronoid process, crp）的前缘约起于m2前部的外方。其前缘与下臼齿齿槽缘的夹角约为105°–110°。冠状突上部较明显向后弯，后缘凹入较明显。其内侧面下部有明显的凹面，可能供颞肌附着，被称为颞肌窝（temporal fovea, tf）。下颌孔（mandibular foramen, mdf）小，位于冠状突后缘的下方，与下臼齿的齿槽缘约在同一水平上，在m3的后外侧，距m3的距离较短（3 mm，4 mm）。下颌切迹的深度小于下颌高的1/3。髁突（condyloid process, cdp）较冠状突（crp）稍低，向内后上方延伸。内侧供翼外肌附着的凹面不明显。髁突前缘长约8 mm。下颌髁上的关节面约成椭圆形半球面，长大于宽，内缘的界线明显，而外缘无明显的界线。翼内肌窝（internal pterygoid fovea, iptf）大而深，其前端伸达冠状突的下方，其下缘向内卷突。可惜其后部大部分破损缺失。下门齿后端在上升支的颊侧，在下颌切迹的下方形成明显的隆凸，我们称其为i2后端隆突（bulge formed by i2, i2b）。

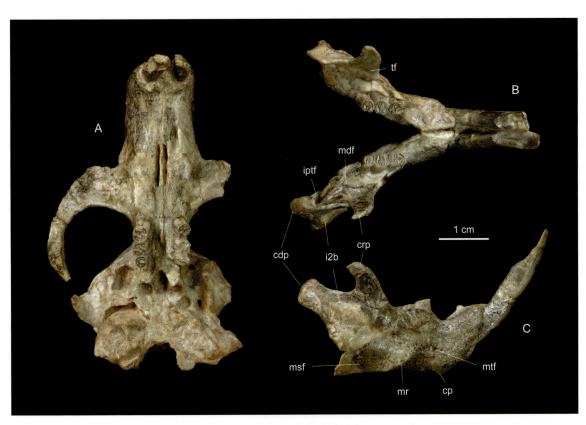

图9　三趾马层副竹鼠头骨含下颌骨（V 16286.1）

Fig. 9　Skull with mandible of *Pararhizomys hipparionum* (V 16286.1)

A. 头骨腹面（ventral view of skull），B. 下颌骨冠面（crown view of mandible），C. 右下颌骨颊侧面（buccal view of right hemimandible）

缩写（Abbreviations）：cdp. 髁突（condyloid process），cp. 颏突（chin process），crp. 冠状突（coronoid process），i2b. i2后端隆突（bulge formed by i2），iptf. 翼内肌窝（internal pterygoid fovea），mdf. 下颌孔（mandibular foramen），mr. 咬肌嵴（masseteric ridge），msf. 咬肌窝（masseteric fossa），mtf. 颏孔（mental foramen），tf. 颞肌窝（temporal fovea）

表3 临夏盆地副竹鼠属头骨和下颌骨测量和比较（单位：mm）

Table 3　Measurements and comparison of skulls and mandibles of *Pararhizomys* of the Linxia Basin (in mm)

测量项 (Parameter)*	三趾马层副竹鼠 *Pa. hipparionum*			华夏副竹鼠（新种）*Pa. huaxiaensis* sp. nov.			陇副竹鼠（新种）*Pa. longensis* sp. nov.		
	归入标本 Referred specimens			正模 Holotype	副模 Paratypes		正模 Holotype	副模 Paratypes	
	V 16286.1	HMV 1923	V 16287	HMV 1413	HMV 1924	V 16291	V 16292.1	V 16292.2	V 16292.3
1. 枕鼻长（CNL）		63		65.3	61.7				
2. 颅基长（CBL）	62.8	60.3		67	60.5				
3. 头骨前部长（LAS）	31.5	29.4		32.5	30		24.5		
4. 头骨后部长（LPS）	31.7	31	30.5	35	31				
5. 腭长（PL）		42.2		?	41.9				
6. 上颌齿隙长（LMXD）	26, 26.7	24.6, 24.1		26.2, 27.2	22.5, 23	22.1	19, 19**		
7. 腭桥长（LPB）		19	16.6	?	19				
8. 腭桥前宽（AWPB）	4	4	4.4	3.8	3.3	3.5	2.7		
9. 腭桥后宽（PWPB）	5	7	5.8	4.5	5.5	5.4	4.3		
10. 门齿孔长（LIF）	8.1	6.3	4.8	5.8	6.2	5.07	5.8	4.4	5
11. 吻部长（RL）	17.8	16.8		20.2	18.2		19, 18.5	19, 18.5	17
12. 吻部前宽（AWR）	14.2	12.9		14	13.6		11.8	14.2	13.8
13. 吻部后宽（PWR）	14.3	13.1	10.9	13.6	13.5	12.6	10.2	13.9	14.5
14. 吻部高（RH）	10.5	11.8		12.3	11.8	10.4	9.4	10.5	9.9
15. 头骨中部高（HMS）	15.3+	17.6	19.1	20	18.6	17.5	17.6		
(15/3)/%	48.6+	60		62.5	62		72		
16. 鼻骨长（NL）		27.5		28.6	26		23.5		
17. 鼻骨前宽（AWN）		9.4		9.3	8.9		6.6	~8.3	8
18. 鼻骨后宽（PWN）		0.9	1	1.7	0.9	0.9	0.5	0.9	
19. 眶间宽（IOW）	17.5	14.2	10.6	12.7	13.5	13.1	9.1	11.8	
20. 眶后收缩处宽（WPOC）	8.9	8	7.2	7.5	7.8	8.1	6.9		
21. 颧弓处宽（ZW）	46 ***	48 ***		46			42.8 ***		
22. 脑颅乳突处宽（WCM）		~26	27.8	31.2	27.2				
23. 听泡长（LAB）		12.5	11.6, 11.8	12.2	11.2, 11.6				
24. 听泡宽（WAB）		12.5	13.2, 13.7	12.5	12.2, 14.7				
25. 听泡间距（DAB）			3.1	6.2	~1.7				
26. 项面高（NH）			18.4	20.4					
27. 枕骨大孔高（HFM）		8.2	~6.5	7	7.4				
28. 枕骨大孔宽（WFM）		8.4	>7	7.8	7.5				
29. 下颌骨长（MDL）				50, 51					
30. 下颌骨高（MDH）	22.9			23.9+			20.6+		
31. 下颌髁突高（HCPM）	19.5			18.5, 19.3					
32. 下颌切迹深（DMN）	6.2						~5		
(32/30)/%	27.1						24.1		
33. 下颌齿隙长（LMD）	11.1, 11.5			14.1			10.5, 10.6		
34. 下颌齿隙处高（HMD）	10, 10.3			12.4			8.9, 9.2		
35. 下颌水平支高（HHR）	11.6, 12.3			11.5, 12			10.1, 10.2		

* 见第 11 页（see p.11）。

** 中间有错动 (specimen slightly slip-cracked)。

*** 有些颧弓处宽 = 半部宽 × 2（some zygomatic width = half width × 2）。

（3）牙齿（图10；表4）

齿式为 1·0·0·3/1·0·0·3。臼齿为前 - 后侧高冠型。在臼齿的侧面，釉质曲线（dentine tract, DT，依 Repenning, 2003）与臼齿冠面斜交。臼齿冠面有 2 或 3 条褶沟和 2 或 3 条横脊。本书中描述的三件标本都是中年或老年个体。在臼齿的褶沟中只有少数是开放的。有些褶沟因磨蚀已变成封闭的盆，而在磨蚀更深的标本中则已完全消失。

M1 是上臼齿中最大的，外形呈四边形，长大于宽，舌侧稍短于颊侧。冠面具两条颊侧沟（前边凹和中凹）和一条舌侧沟（内凹）以及 2 或 3 条横脊。中凹最长，其舌端稍向后弯。前边凹几乎呈横向延伸。内凹伸向前尖，插入到前边凹和中凹之间。M2 外形近正方形。有一颊侧褶沟（中凹）和一舌侧褶沟（内凹）。中凹长于内凹，其舌端稍向后弯。内凹在中凹之前，稍向前颊侧方向延伸，与中凹部分重叠。M3 的外形呈四边形，其后面较窄，并稍向后凸。与 M2 相似，M3 也有一条颊侧褶沟（中凹）和一条向前颊侧延伸的舌侧褶沟（内凹）。但中凹横向要短于内凹，而深度要大于后者，呈横向或稍向前舌侧延伸，与内凹相遇或部分重叠（见图 10 和表 4）。

I2 强烈弯曲，其前端并稍向后弯，属垂直伸型齿（见图 8 A）。I2 的后端起自上颌骨的前缘，门齿孔的后外方，距 M1 较远。I2 的横切面呈三角形，舌侧角圆钝，唇面几乎是平的。但在唇面中央近内

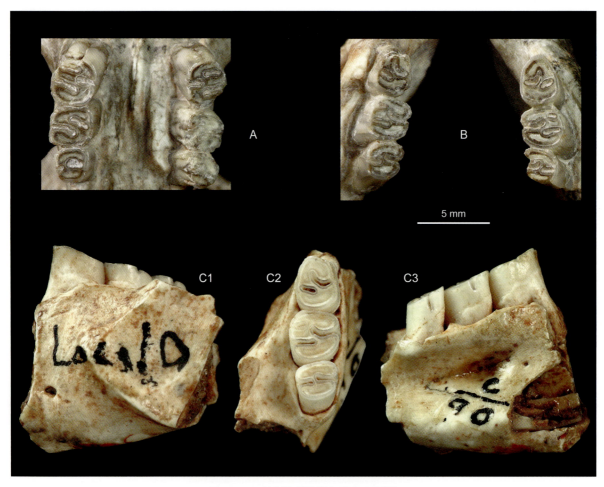

图 10　三趾马层副竹鼠臼齿和下颌骨

Fig. 10　Molars and mandible of *Pararhizomys hipparionum*

A–B. V 16286.1，臼齿冠面（occlusal view of molars）；A. 左和右上臼齿（left and right upper molars），B. 左和右下臼齿（left and right lower molars）；C. V 412（正模），左下颌骨部分水平支具 m1–3（partial horizontal ramus of left hemimandible with m1–3, holotype）：C1. 颊侧面（buccal view），C2. 冠面（occlusal view），C3. 舌侧面（lingual view）

表 4　临夏盆地副竹鼠属牙齿测量和比较（单位：mm）

Table 4　Measurements and comparison of teeth of *Pararhizomys* of the Linxia Basin (in mm)

测量项 (Parameter)*	三趾马层副竹鼠 Pa. hipparionum						华夏副竹鼠（新种）Pa. huaxiaensis sp. nov.						陇副竹鼠（新种）Pa. longensis sp. nov.					
	V 16286.1		HMV 1923		V 16287		HMV 1413		HMV 1924		V 16291		V 16292.1		V 16292.2		V 16292.3	
							正模 Holotype		副模 Paratypes				正模 Holotype		副模 Paratypes			
	Ⓛ	Ⓡ	Ⓛ	Ⓡ	Ⓛ	Ⓡ	Ⓛ	Ⓡ	Ⓛ	Ⓡ	Ⓛ	Ⓡ	Ⓛ	Ⓡ	Ⓛ	Ⓡ	Ⓛ	Ⓡ
M1–3 (长, length, L)	9.8	9.2	10.6	10.6	9.4		10.4	10.2	10.1	10.1	10	10.2	10.1	10.2				
M1 L	3.2	3.2	~3.9	~3.7	4	3.9	3.6	3.4	4.1	3.9	4.1	4	3.8	3.9				
M1 W	3.3	3.2	3.6	~3.4		3.6	3.5	3.5	3.6	3.6	~3.6		~3.6	~3.6				
M1 Hpl	3	3					2.8	2.8										
M1 Hpb1							3.2											
M1 Hpb2							4	4.1										
M1 (Hpl/Hpb2)/%							70	68										
M2 L	2.9	2.9	3.5	3.4	2.9	~2.5	3.2	3	2.9	3	3.2	3	3.2	3.2				
M2 W	3.1	3.4	3.5	3.4	3.3	3.1	3.5	3.5	3.2	3.2	3.4	~3.2	~3.5	~3.8				
M2 Hpl							2.2	2.2										
M2 Hpb2				2.5				2.6										
M2 (Hpl/Hpb2)/%							85											
M3 L	2.6	2.6	2.9	2.9			2.6	2.7	2.6	2.5			2.8	2.9				
M3 W	~2.6	2.7	2.8	2.8			2.7	2.7	2.6	2.6			~2.9	~2.8				
M3 Hpl							2	2										
M3 Hpb2				1.7	1.5	1.5	1.3	1.5										
M3 (Hpl/Hpb2)/%							154	133										
I2 L	4.8	4.8	4.4	4.2	3.7	3.7	4.8	4.8	4.6	4.8	3.8	3.8	4.1	4.1	4.4	4.4	4	4
I2 W	3.7	3.8	3.6	3.6	3.1	3.1	3.9	3.9	4	4	3.2	3.2	3.8	3.8	3.7	3.8	3.5	3.6
m1–m3 L	10.1						10.8	10.9					10.8	10.5				
m1 L	~3.3	3.5					3.7	3.8					3.9	4				
m1 W	2.8	3					3	3					3.5	3.4				
m1 Hpl							2.5	2.4										
m1 Hpb2							3.5											
m1 (Hpl/Hpb2)/%							71											
m2 L	3.2	3					3.2	3.1					3.7	3.7				
m2 W	3.4	3.2					3.5	3.5					3.8	3.8				
m2 Hpl	~2.1	2.2					2.2	1.9										
m2 Hpb2							3.1											
m2 (Hpl/Hpb2)/%							71											
m3 m3	2.8						3.3	3.1					3.1	2.8				
m3 W	2.8						3.3	3.3					3.2	3.1				
m3 Hpl	2.2						1.5	1.5										
m3 Hpb2							3.2	3										
m3 (Hpl/Hpb2)/%							47	50										
i2 L	4.9	4.8					5.1	5.2					4.4	4.4				
i2 W	3.4	3.5					3.6	3.6					3.5	3.5				

* 测量项定义及缩写详见第 12–13 页（For explanation and abbreviations of parameters please refer to pp. 12–13）。

侧处有一条很显著的纵棱（见图 8 B）。釉质层覆盖在整个唇面，只有少量覆盖在内、外侧面。

V 16286.1 的下臼齿的形态结构与正模（V 412）的很相似。m1 的外形约呈长大于宽的卵圆形，前缘较窄。V 16286.1 的 m1 冠面仅保留有一条颊侧沟（下外凹）和一条舌侧沟（下中凹）。可能它的下原凹已被磨蚀掉了。下外凹和下中凹已被磨蚀成封闭的盆。下中凹长，强烈弯曲，其颊部转向前伸。下外凹在下中凹之后稍向后舌侧延伸。m2 呈颊侧较长的四边形，冠面具两条颊侧沟（下原凹和下外凹）和一条舌侧沟（下中凹）。下原凹和下外凹均已被磨蚀成横向伸长的、封闭的盆。下原凹短于下外凹。下中凹颊部稍向前弯，弯向下原凹，其舌端向外开口。m3 外形近正方形，其后缘较窄，并稍向后圆凸。m3 只有一条舌侧褶沟和一条颊侧褶沟（下外凹和下中凹）。下中凹稍长于下外凹，其颊部稍向前弯。下外凹呈横向延伸，与下中凹相对。

i2 强壮，平缓地向前上方延伸（图 9 B, C）。在下颌骨之前伸出的部分很长，其前端的与 I2 的磨蚀面和齿槽间的距离也长。i2 后端起自下颌骨上升支。其横切面为窄的三角形。其唇面和外侧面较平，舌侧较窄而圆。其釉质层覆盖的情况与 I2 的相似：覆盖于整个唇面，在内、外仅覆盖了很少一部分。V 16286.1 左、右侧 i2 唇侧纵嵴的数量有些不同，在左 i2 的唇面只有一条纵嵴（= 内侧纵嵴），而在右 i2 唇面有两条纵嵴：内侧纵嵴较明显，外侧纵嵴弱。这表明 i2 唇侧纵嵴由两条退化成一条的现象在晚中新世的 *Pa. hipparionum* 中已出现。

（4）下门齿釉质层的微细结构（图 11）

在临夏盆地现有的 *Pa. hipparionum* 的材料中，仅 V 16286.1 下颌保存有下门齿。由于其左、右下颌联合部牢固地缝合在一起，为了尽量少破坏原标本，仅切取了其左 i2 的最前端。切取的部分保存亦不完全，仅显示出 i2 唇侧部分，舌侧大部分缺失（图 11 A）。外侧约 1/3 部分的釉质层较厚，而内侧 2/3 部分较薄（图 11 A）。造成这种现象的原因不清楚，还有待进行更多的观察和研究。

为了显示其厚度的变化，笔者对不同部位的釉质层的厚度分别进行了测量。

横切面（图 11 A, B, C）显示：*Pa. hipparionum* 的下门齿的微细结构属单系（uniseral）。未见无釉柱的最外层（prismless external layer, PLEX）。釉质层的总厚分别约为 144 μm（沿纵切面）、158 μm [横切面最薄处（图 11 A a）] 和 168 μm [横切面最厚处（图 11 A b, B）；以下测量位置与此相同]。釉质层明显分为内、外层。外层（*portio externa*, PE）很薄，其厚度分别为 26 μm、29 μm 和 38 μm。釉柱间质（interprismatic matrix, IPM）不倾斜，在纵、横切面上均与门齿唇侧表面垂直。内层（*portio interna*, PI）很厚，其厚度分别为 118 μm、129 μm 和 130 μm。这样，釉质层的外层厚与总厚之比（PE/T），分别为 18%、18% 和 23%。釉质层的外层厚与内层厚之比（PE/PI）分别为 22%、22% 和 29%。从这一测量结果看，后两点的内层的厚度差别不大，而外层的厚度差别大，这表明内侧部分较薄可能与磨损有关。

横切面显示外层的施氏明暗带（Hunter-Schreger bands, HSB）均与唇面近于垂直，而内层的施氏明暗带（HSB）均从釉 - 齿质界面（enamel dentine junction, EDJ）稍向近纵嵴方倾斜，但倾斜的角度较小。其倾斜度从舌侧往唇面方向稍有变化：近舌侧的倾斜度较大，往唇面倾斜度变小（与唇面近于垂直）。唇、舌部倾角的差异向门齿的内、外侧方向也逐渐变小，变得不明显；而向近纵嵴的方向逐渐增大。

V 16286.1 左 i2 唇侧仅有一纵嵴（= 内纵嵴，mr）。在横切面上可见纵嵴处的釉质层明显变厚，而且其内、外侧的施氏明暗带 (HSB) 均向纵嵴处汇集（图 11 A, C）。在纵嵴处，可见 HSB 的延伸的方向由舌部往唇部发生了明显的变化：在近釉 - 齿质界面（EDJ）处舌部倾角较大，两侧的 HSB 彼此约呈垂直相交，而往唇侧方向逐渐转为彼此近于平行，向唇侧延伸，并渐渐过渡到外层（PE）。

李强（Li, 2010: 57–60, Figs. 10C, 11C）曾描述了产自内蒙古宝格达乌拉的 *Pa. hipparionum* 的下门齿的微细结构。从李强所示的纵、横切面图来看，V 16286.1 的门齿的微细结构与宝格达乌拉标本（V 16306）者很接近。

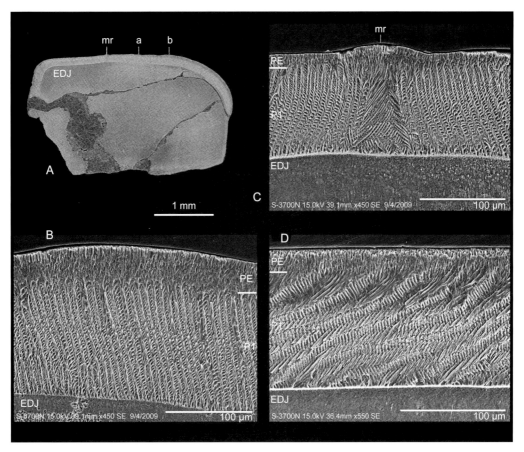

图 11   三趾马层副竹鼠的左下门齿（V 16286.1）釉质层的微细结构

Fig. 11   Enamel microstructure of left lower incisor of *Pararhizomys hipparionum* (V 16286.1)

A−C. 横切面（cross section）：A. 左 i2 前端后视（posterior view of anterior end of left i2），B. b 点处横切面
（cross section at b），C. 内侧纵嵴处横切面（cross section at mr）；D. 纵切面（longitudinal section）

缩写（**Abbreviations**）：a, b. 釉质层厚度测量处（places where the thickness of the enamel are measured），EDJ. 釉 - 齿
质界面（enamel dentine junction），mr. 内纵嵴（medial ridge），PE. 外层（*portio externa*），PI. 内层（*portio interna*）

（5）颅后骨骼

在临夏盆地采集的 *Pa. hipparionum* 标本中，只有 V 16286 标本保存了一左桡骨的下半段（V
16286.2）和一段右股骨的骨体（V 16286.3）。

桡骨（radius, Ra；图 12；表 15）   仅左桡骨的下半部保存，长 22 mm（图 12，表 15）。桡骨仍
保留有远端的骺线。骨体稍向后弯，向远端逐渐变粗。骨体前面近中部有一明显的斜沟自外上方向内
下方延伸，可能是肌腱通过处。该斜沟的内上方以明显的斜棱为界。骨体外侧下部有较发达的纵向沟棱，
可能供桡尺韧带等附着。

桡骨远端最大宽 4.7 mm，前 - 后最大径长 4.5 mm（见表 15）。远端关节面宽为 3.8 mm，该面为
不规则的四边形。关节表面大部分为圆滑的凹面，仅其后内半部前后圆凸。表面未见有分隔腕舟骨
（scaphoid）和月骨（lunar）关节面的痕迹，很可能该两腕骨已愈合为腕舟 - 月骨（scapho-lunar, Sc-lu）。
桡骨茎突（styloid process of radius, stypr）明显向下凸出。远端背面有一明显的纵沟，可能供指的伸肌
腱通过。远端外侧面稍凹而粗糙，可能是尺骨和桡骨在远端的骨间韧带附着处。远端内侧面有一明显
的纵沟，可能供伸肌腱（拇伸肌或拇展肌肌腱）通过。

股骨（femur, Fe, V 16286.3）   仅保留了骨体的中段，含第三转子和小转子的下端。第三转子很明
显向外伸出。骨体中部前面横向圆凸，后面在第三转子附近横向凹入，而在下部横向较平。骨体中部

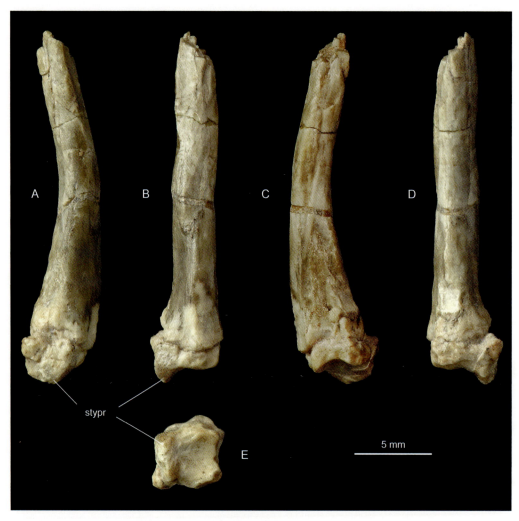

图 12 三趾马层副竹鼠左桡骨下半部（V 16286.2）

Fig. 12 Lower half of left radius of *Pararhizomys hipparionum* (V 16286.2)

A. 内侧面（medial view），B. 前面（anterior view），C. 外侧面（lateral view），D. 后面（posterior view），E. 远端面

（distal view）

缩写（Abbreviation）：stypr. 桡骨茎突（styloid process of radius）

（第三转子下端处）宽约 5 mm，前后径长约 4 mm。

　　李强（Li, 2010）报道了在内蒙古宝格达乌拉发现的几件 *Pa. hipparionum* 的颅后骨骼，称它们为左距骨（V16306.15）和 3 件蹠骨（V 16306.16–18），但未进行描述。笔者观察了上述标本，发现被他称为蹠骨的三件肢骨很可能都不是蹠骨，而是掌骨。其中的 V 16306.16 保存较完整，显然不是蹠骨，而应为右第三掌骨。V 16306.17 也保存得较好，只是近端破损。从它的整个肢骨的形态和保存的部分近端关节面来判断，V 16306.17 也不是蹠骨，而是左第四掌骨。V 16306.18 的近端虽已缺失，但其骨体和远端的形态，以及尺寸大小都与 V 16306.16 和 V 16306.17 的相近，特别与 V 16306.16（右第三掌骨）的很相似，只是掌侧纵隆嵴两侧凹槽大小不对称的情况相反，它很可能是左第三掌骨。这样，V 16306.16–18 都不是蹠骨，而应是掌骨。为了对 *Pa. hipparionum* 的颅后骨骼的特点有较多的了解，笔者在此也对在内蒙古宝格达乌拉发现的上述距骨和掌骨作简短的描述。

　　第三掌骨（McIII；图 13；表 16）　右第三掌骨（V 16306.16）保存完好，左第三掌骨（V 16306.18）保存有骨体下部和远端。近端顶视近三角形，有三个关节面。中间与头状骨（magnum）的关节面（fmg）最大，约为外宽内窄的梯形，横向明显凹入，背 - 掌向圆凸。外侧与钩骨（uniform）的关节面（fun）窄于与头状骨的关节面。该关节面为不规则的三角形，掌端较窄锐；表面较平，向外上方倾斜。该两关节面约以直角相交，并被一明显的、向上圆凸的弧形的、背 - 掌向的棱脊分开。内侧与第二掌骨（McII）的关节面（fmcII）最小，位于近端面的前内角。该面约呈三角形，表面稍圆凸，主要面向内上方。该关节面与头状骨关节面间的夹角较钝，两关节面间的棱脊也较低而短。在该面的掌侧有一小凹面。而在第三掌骨近端内侧，该小凹面掌侧的其余部分则为凹凸不平的粗糙面。第三掌骨近端外侧与第四掌骨的关节面（fmcIV）约为圆弧形，稍向外掌下方倾，与其上的与钩骨关节面有明显的掌 - 背向棱脊为界。其下方为一大而深的凹坑。近端背面较平。背侧外缘，与 McIV 关节面的前方，有一很发达的、棱嵴状的结节状隆起，表面粗糙，可能供掌骨近端骨间韧带附着。笔者暂称该棱嵴为韧带嵴（crest for ligament, cl）。

骨体（corpus, shaft）直。上部较窄，横切面约为三角形。往下骨体逐渐变宽大，约为横宽的扁圆柱形。背面垂向较平直，横向圆凸。在背面内侧上部，距头状骨关节面下方约 1 mm 处，有一很显著的、

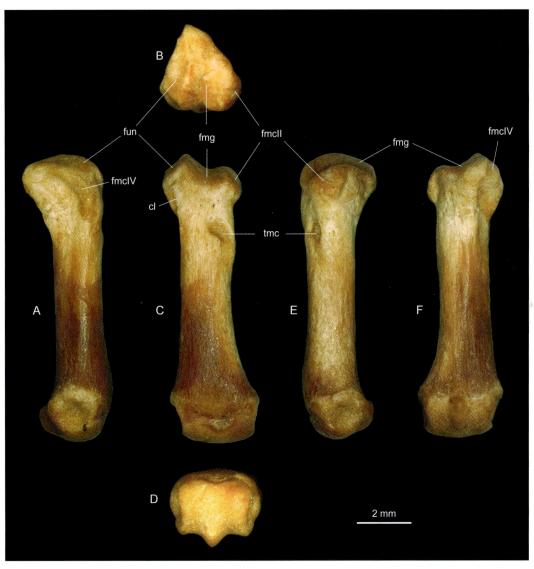

图 13　三趾马层副竹鼠右第三掌骨（V 16306.16）

Fig. 13　Right third metacarpal of *Pararhizomys hipparionum* (V 16306.16)

A. 外侧面（lateral view）、B. 近端面（proximal view）、C. 背面（dorsal view）、D. 远端面（distal view）、E. 内侧面（medial view）、F. 掌面（volar view）

缩写（**Abbreviations**）：cl. 韧带嵴（crest for ligament）、fmg. 与头状骨关节面（facet for magnum）、fmcII. 与第二掌骨关节面（facet for McII）、fmcIV. 与第四掌骨关节面（facet for McIV）、fun. 与钩骨关节面（facet for unciform）、tmc. 掌骨粗隆（tuberosity of metacarpal）

粗糙的凹坑，可能供腕桡侧伸肌短头（*m. extensor carpi radialis brevior*）附着。笔者暂称其为掌骨粗隆（tuberosity of metacarpal, tmc）。在掌面上部近内侧处有明显的滋养孔。

远端与第一指节骨的关节面的背部横向稍圆隆，无纵嵴。纵嵴仅在远端关节面的掌部存在。纵嵴两侧为横向凹入、纵向圆凸沟槽。远端两侧有深的窝，供指关节侧韧带附着。

测量见表16。

第四掌骨（McIV；图14；表16）　左第四掌骨（V 16306.17）近端破损。McIV 比 McIII 粗短。近端保留了部分与 McIII 的关节面（fmcIII）。该关节面约成三角形，表面较平，面向内上方倾斜。近端背侧近内侧有一很发达的、粗大的棱嵴一直向下约延伸到骨体的中部。这可能是与 McIII 的骨间韧带附着处，也称其为韧带嵴（cl）。从 McIII 外侧也有相应发达的韧带嵴来看，很可能在前爪用力时，McIII 和 McIV 近端连接得比较结实。

骨体粗壮而直。其上部的横切面为四边形，中部稍细，往下又变宽大，横切面变为宽扁的卵圆形。远端与 McIII 的相似。只是远端内侧比外侧前后径要大些；远端关节面的背部比 McIII 的稍窄长些；关节面掌部纵棱两侧的凹槽稍不对称，外侧者较内侧者宽，并较明显向外倾。

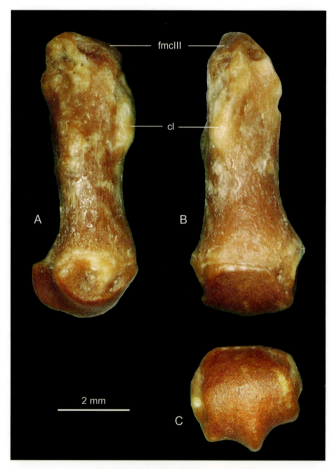

图 14　三趾马层副竹鼠左第四掌骨（V 16306.17）

Fig. 14　Left fourth metacarpal of *Pararhizomys hipparionum* (V 16306.17)

A. 内侧面（medial view），B. 背面（dorsal view），C. 远端面（distal view）

缩写（Abbreviations）：cl. 韧带嵴（crest for ligament），fmcIII. 与第三掌骨关节面（facet for McIII）

距骨（astragalus, As；图 15）　左距骨（V 16306.15）保存完好，高而窄。距骨滑车 [trochlea (tro) of astragalus] 占据了距骨近端的顶面和背面，向背外远方伸。滑车蹠侧端变窄。滑车沟很弱，位置靠近内嵴。滑车的内、外嵴不对称。滑车内嵴（medial ridge, mr）的外坡面较陡而窄，内坡面更陡，近垂直，但横向稍圆凸。滑车外嵴（lateral ridge, lr）的外坡面很陡而平直，与腓骨远端关节；内坡面较内嵴的外坡面平缓而宽。滑车前面往远端分成两支：外支短小；内支较大，往背下方延伸达距骨颈的背面，其远端变窄，并稍转向内方倾斜（可能与胫骨远端内滑车向前下方的突起相关节）。

距骨颈（neck of astragalus, nas）细长。距骨蹠面与跟骨有两个较明显的关节面：①位于近端外侧部的关节面。前人对此关节面的命名并不相同：最早被称为外关节面（ectal facet，见 Wood, 1962 和 Bleefeld et McKenna, 1985 等）；后被称为后跟关节面（posterior calcaneal facet，见 Voss, 1988 和 Rose et al., 2008 等）；而 Szalay（1985）同时称其为跟距关节面（calcaneo-astragalar，见 Szalay, 1985 和 Meng et al., 2003 等）和近端跟距关节面（proximal calcaneo-astragalar）等。综合分析啮形类的距骨与跟骨的关节面的相对位置和演化历程，笔者称此关节面为与跟骨的近端关节面，或称为近端跟骨关节面（proximal calcaneal facet, pcf）。该关节面较大，为四边形；呈内上-外下方伸长，长轴明显凹入，而横向较平直。位于近端跟骨关节面的下方的舌状面（tongue-shaped facet, tosf）很小，呈弓形，横向伸长；面向蹠下方，与近端跟骨关节面约以 90° 相交。②与跟骨载距突（sustentacular process, sup）关节的载距关节面（sustentacular facet,

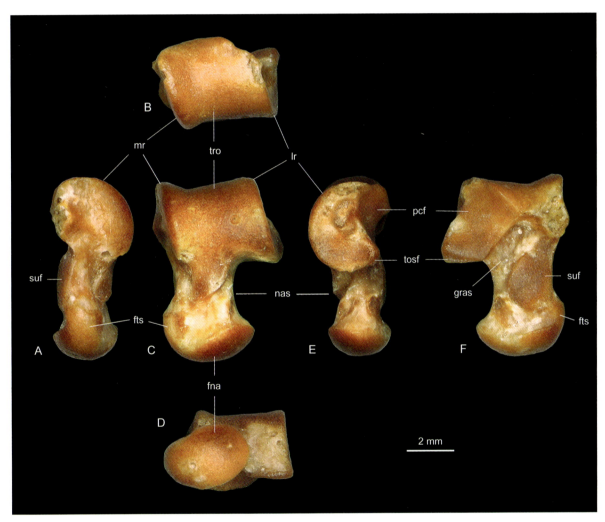

图 15　三趾马层副竹鼠左距骨（V 16306.15）

Fig. 15　Left astragalus of *Pararhizomys hipparionum* (V 16306.15)

A. 内侧面（medial view），B. 近端面（proximal view），C. 背面（dorsal view），D. 远端面（distal view），E. 外侧面（lateral view），F. 蹠面（plantar view）

缩写（Abbreviations）：fna. 与跗舟骨关节面（facet for navicular），fts. 与胫侧籽骨关节面（facet for tibial sesamoid），gras. 距骨沟（groove of astragalus），lr. 滑车外嵴（lateral ridge），mr. 滑车内嵴（medial ridge），nas. 距骨颈（neck of astragalus），pcf. 近端跟骨关节面（proximal calcaneal facet），suf. 载距关节面（sustentacular facet），tosf. 舌状面（tongue-shaped facet），tro. 滑车（trochlea） suf）位于内侧，与上述关节面间以内上 - 外下方向延伸的距骨沟（groove of astragalus, gras）分开。载距关节面为卵圆形，其长轴也略呈上内 - 下外方向。其内上端伸达滑车蹠内角下方的结节下缘；其下端与远端关节面仅在其内下角较接近，其余大部分明显分开。该关节面表面稍圆凸。在蹠面的内上角，滑车的后内下方、载距关节面的内上方的结节明显（表面稍破损），可能供内侧韧带附着。

距骨的远端与跗舟骨（navicular）相关节的关节面——与跗舟骨关节面（facet for navicular, fna）为卵圆形，横径大于背 - 蹠径。关节面表面呈球面圆凸，主要面向下方。其内侧还有一关节面，也为卵圆形，但较前者稍短小，主要面向内方。它可能与胫侧籽骨（tibial sesamoid = 内跗骨 medial tarsal）关节，被称为与胫侧籽骨关节面（facet for tibial sesamoid, fts）。其近端几乎与载距关节面相连。远端的两关节面平滑过渡，之间无明显的界线。这可能表明，*Pa. hipparionum* 有一较发育的胫侧籽骨（tibial sesamoid）。

距骨最大高 77 mm；近端最大宽 55 mm；近端最大背 - 蹠径长 36 mm；距骨颈最窄处横径为 26 mm，背 - 蹠径为 22 mm；近端跟骨关节面的长径 3.2 mm，短径 2.7 mm；载距关节面的长径 3.2 mm，短径 2 mm；远端关节面宽 39 mm；远端关节面背 - 蹠最大长 26 mm。

2. 比较

Pararhizomys hipparionum 的正模是左下颌的水平支的中段。与正模比较，临夏盆地 V 16286.1 的下颌的特征与正模几乎完全一样，如：下颌骨短粗；颏孔位于 m1 前缘的下方，位置低于咬肌窝前端；下臼齿为后侧高冠型，具齿根；m1–3 的舌侧褶沟与颊侧褶沟的深度相近；m1–2 具两颊侧褶沟和一舌侧褶沟（可惜两者的 m1 的下原凹均已磨蚀掉了！）；m1 的下中凹强烈前弯；m3 外形约为近正方形，下外凹和下中凹几乎相对；i2 唇面有一明显的纵嵴等。另外，臼齿的尺寸也与正模的相近（见表 4, 5）。

Kowalski（1968）将产自蒙古国 Altan Teli 地点的一个带下颌的头骨归入 *Pa. hipparionum*。根据我们对该标本的模型（Lindsay 提供）的观察，它与临夏盆地的标本也很相似，如：头骨较低；吻部窄，其两侧近于平行；前颌骨和鼻骨的前端约在同一垂直线上；前颌骨的侧背嵴长，与鼻骨 - 前颌骨缝前半部近于平行，前颌骨背部呈长条形（见 Kowalski, 1968: Pl XXI 1b）；项面向前倾斜微弱；I2 强烈弯曲，唇侧并具一显著的纵棱；臼齿为前 - 后侧高冠型齿；和 m1 下中凹的颊部强烈向前弯曲等。此外，两件标本在头骨和牙齿的尺寸上都很相近（比较表 3 至表 5）。显然，临夏盆地的 V 16286.1 等标本应归入 *Pa. hipparionum*。

3. 讨论

李强（Li, 2010: 55–57，Fig. 8）将当时已知的 *Pa. hipparionum* 标本按地质年代排序。他在其文章的中文摘要（Li, 2010: 48）中指出副竹鼠的进化趋势是："从早期到晚期，其颊齿的个体有从小到大、釉质曲线高度 H 值逐渐减小，即齿冠逐渐增高的趋势。"

比较了 *Pa. hipparionum* 的所有已知的（包括临夏盆地的）标本后，笔者发现本书中所建议的冠高替代指标（Hp）比李强所建议的冠高指数（index H）能更确切地反映齿冠高度的变化。例如，在 *Pa. hipparionum* 中，M2 的舌侧和颊侧的褶沟的冠高替代指标（Hpl 和 Hpb2）在产出时代约为 9.5 Ma 的 V 16286.1 中 Hpb2 为 2.5 mm，在产出时代约为 8 Ma 的 V 14178 中 Hpl 和 Hpb2 分别为 3 mm 和 2.6 mm，而在产出时代约为 6 Ma 的 V 16306.8 中则分别为 3.5 mm 和 3.2 mm（见表 4, 5）。这表明，在 *Pa. hipparionum* 的 M2 中，Hp 冠高替代指标越大，颊齿的齿冠越高。

李强（Li, 2010: 56）根据他所建议的进化趋势推测，产于陕西府谷地点 10（Loc. 10）的正模（V 412）的时代很可能介于陕西秦安（约 7 Ma）和内蒙古的宝格达乌拉（约 6 Ma）之间。然而，在他的表 4 中，正模的 m1–2 的舌侧釉质曲线的高度（LNH：m1 为 0.9 mm；m2 为 0.8 mm）明显小于陕西秦安和内蒙古的宝格达乌拉两者的 LNH 值（前者分别为 1.3 mm 和 1.5 mm；后者分别为 1 mm 和 0.9 –0.95 mm），而不是在二者之间，与其结论相悖。

表 5　临夏盆地以外地点的副竹鼠牙齿测量和比较（单位：mm）

Table 5　Measurements and comparison of teeth of *Pararhizomys* from localities other than the Linxia Basin (in mm)

测量项（Parameter）*		三趾马层副竹鼠 *Pa. hipparionum*										三趾马层? 副竹鼠 *Pa. hipparionum?*			秦副竹鼠 *Pa. qinensis*	
		V 412	V 14178	V 14179	V 16306.1	V 16306.6	V 16306.7	V 16306.8	V 16306.9	V 16306.12	MgM-V/65**	V 16306.13	V 16306.14	V 16307	V 14177.1	V 14177.2
M1–3 L											9.8					
M1	L		3.6			3.9	3.8				4.4				3.4	
	W		3.5			3.7	3.7				3.6				3.2	
	Hpl		3.6			3.7	3.7								3.6	
	Hpb1		3			3.1	3.2								3.1	
	Hpb2		3.8			4	4.1								3.8	
	(Hpl/Hpb2)/%		95			93	90								95	
M2	L		2.8					3	3.1		3			3.2		2.5
	W		3.5					3.7	3.9		3.3			3.7		3.2
	Hpl		3					3.5						2.4		2.1
	Hpb2		2.6					3.2						3.5		2.5
	(Hpl/Hpb2)/%		115					109						69		84
M3	L		2.7							2.8	2.4					
	W		2.6							2.8	2.5					
	Hpl		2.2							1.6						
	Hpb2		1.3							0.6						
	(Hpl/Hpb2)/%		169							267						
I2	L										4.3					
	W										3.6					
m1–3 L		10.4		10.5	11.6						10					
m1	L	3.8		3.9	4.1						3.8	4				
	W	3		3	3.4						3.2	3.3				
	Hpl	2.7		3	2.5							2.1				
	Hpb1											3.7				
	Hpb2											3.2				
	(Hpl/Hpb2)/%											66				
m2	L	3.4		3	3.4						3.3	3.6				
	W	3.6		3.5	3.9						3.6	4				
	Hpl	2.2		2.9								2.6				
	Hpb1											3.2				
	Hpb2	2.1		2.7+								3.4				
	(Hpl/Hpb2)/%	105		107								77				
m3	L	2.6		3.2	3.2						3					
	W	3.1		3.2	3.2						2.9					
	Hpl	2		2	2.4											
	Hpb2	2.1		2.2	2.1											
	(Hpl/Hpb2)/%	95		91	114											
i2	L	4.3									4.2					
	W	3.4									3.4					

＊ 测量项同表 4（parameters as in Tab. 4）。

＊＊ 据 Kowalski, 1968 (after Kowalski, 1968)。

李强（Li, 2010: 55）认为产自秦安的 V 14179 "个体较小和牙齿的釉质曲线高度较大"，因此怀疑该标本是否应归入 *Pa. hipparionum* 种。而应用本书建议的方法测量则表明，V 14179 的臼齿的大小仍在 *Pa. hipparionum* 的大小变异范围之内。此外，V 14179 的 m2 和 m3 的舌侧和颊侧的冠高指标（Hpl 和 Hpb2）彼此很相近（见表 5）。而这一特征也是 *Pa. hipparionum* 正模的主要鉴别特征之一。因此，V 14179 还是应该归入 *Pa. hipparionum*。

被李强归入 *Pa. hipparionum* 的产自内蒙古宝格达乌拉的单个牙齿中，有两个下臼齿 [V 16306.13（m1）和 V 16306.14（m2）]，在下中凹强烈弯曲这一特点上与 *Pa. hipparionum* 的确很相似。但是，它们的舌侧褶沟明显深于颊侧者，其舌侧与颊侧冠高指标的比值（Hpl/Hpb2）在 m1（V 16306.13）中为 66%，在 m2（V 16306.4）中为 77%。这明显与 *Pa. hipparionum* 的其他标本不同。同样，V 16307（M2）也有类似的情况。因此，这三枚牙齿是否能归入 *Pa. hipparionum* 尚存疑。在表 5 中笔者暂将它们列为三趾马层？副竹鼠（*Pararhizomys hipparionum*?），以示种存疑。

总之，*Pa. hipparionum* 目前已知的材料，除正模（V 412）外，还有两件含下颌的头骨（V 16286.1 和 Z. Pal. No. MgM-V/65）、两件不含下颌的头骨（V 16287 和 HMV 1923）、两件基本完整的下颌（V 14179 和 V 16306.1）和一些单个的牙齿及几件肢骨。其中，V 16287 头骨与另外三件头骨有些不同：它的吻部和门齿都更纤细些。很可能 V 16287 代表一雌性个体，而其余三件代表雄性个体。

华夏副竹鼠（新种） *Pararhizomys huaxiaensis* sp. nov.

（图 16–21, 75；表 1–4, 6–8, 14）

正模 HMV 1413，同一个体的完整的带下颌的头骨、前三枚颈椎、肱骨近端和肩胛骨的肩臼角，产自 LX 200041（山城村），柳树组中部上层（详见表 1, 2）。

副模 HMV 1924 和 V 16291，头骨两件：HMV 1924 产自 LX 200503（牛扎湾），V 16291 产自 LX 200030（松树沟村）；上述两件标本均产于柳树组中部上层（详见表 1, 2）。

产地与层位 甘肃临夏盆地：和政县山城村和牛扎湾，广河县松树沟村，柳树组中部上层（详见表 1, 2）；庆阳（？），上中新统（见下）。

鉴别特征 前颌骨侧面门齿上方的凹槽很浅，门齿棱平缓。颧弓板弧形后缘大体与 M1 前缘平齐，其最前点仅稍前于 M1。下颌齿隙明显长于下臼齿齿列长。颏突和咬肌嵴都更发育。下颌孔的位置较靠前，位于冠状突的正下方。臼齿齿冠较高，均具显著的前 - 后侧高冠和舌侧高冠；M1–2 和 m1–3 舌侧沟明显深于颊侧沟，而在 M3 中则浅于颊侧沟。M3 的内凹和中凹近于横向，彼此相对，不重叠。m1 的下中凹颊端仅稍前弯，伸向下原凹。

区别特征 新种与 *Pa. hipparionum* 和 *Pa. qinensis* 的共同区别是：臼齿齿冠较高，舌侧高冠显著。该种与 *Pa. hipparionum* 的区别还有：下颌齿隙较长，明显长于下臼齿齿列长；M3 的内凹和中凹彼此相对，不重叠；m1 的下中凹较短，其颊端仅稍前弯，伸向下原凹；该种与 *Pa. qinensis* 的区别还有：头骨吻部和上颌骨齿隙较长和 M2 较长。

词源 Huaxia，华夏，中国的古称；-ensis，拉丁文，表示地点的形容词阳性词尾。

1. 描述

（1）头骨（图 16, 17 A1–A5, B；表 3）

HMV 1413 标本的头骨和下颌骨均保存得很好。两者原本咬合在一起（图 16），为了观察和研究的方便，后被拆分开（图 17A）。头骨与 *Pa. hipparionum* 的很相似，具鼠型的颧 - 咬肌结构。头骨比例上较长、窄而低。吻部较窄，其两侧缘彼此近于平行。头骨的前、后部（以 M1 前缘为界）的长度相近。鼻骨为楔形，其前端与前颌骨的前端约在同一垂线上；其后端很窄，与前颌骨后缘的内端和眼眶的前缘约在同一横线上。鼻骨背面横向和纵向均圆凸。前颌骨具有很显著的侧背嵴。该侧背嵴长，与鼻骨 - 前颌骨缝前半部近于平行。前颌骨的背侧面为长条带形。前颌骨侧面由 I2 形成的纵棱较平缓，纵棱上方的凹沟较浅。前颌骨 - 上颌骨缝的背部段约为直线，纵向延伸，与上颌骨 - 额骨缝以钝角相连。眶下孔为横宽的椭圆形，具腹侧裂隙。眶下管大，其周围内壁凹入较深。眶下管内也有一骨质隔板将腹侧裂隙与大的眶下管分开，隔板的内侧有一明显的鼻泪孔。眶下管前端内缘、上缘和外缘也都以明显的棱嵴为界。在外缘附近内层咬肌附着区也明显向背侧延伸。外层咬肌起始区也仅限于上颌骨内，不伸达前颌骨。其前缘嵴为圆弧形，向前不超过眶下孔腹侧裂隙，后于门齿孔的前端。颧弓板弧形后缘大体与 M1 前缘平齐，其最前点仅稍前于 M1 的前缘。颧弓也为圆弧形，其中部较粗大，横切面也为三角形。颧骨短，前端不伸达眼眶前缘。泪骨主要位于眼眶的内侧，在背侧出露很少，与上颌骨和额骨相连。泪结节弱小。眶后收缩很明显。额嵴很微弱，向后内方延伸，左、右额嵴在眶后收缩前相遇形成明显的矢状嵴。顶骨约为不规则的六边形，长大于宽。关节窝很长，向后伸达项嵴，将外耳道挤至关节窝的后部下方。臼后孔较长大，为裂隙状，在关节窝后部，外耳道的上方稍前，沿鳞骨 - 鼓骨缝之间延伸。硬腭后缘的位置在 M3 的后缘之后。硬腭后部的副翼窝也很明显。中翼窝明显宽于翼窝。咬肌神经孔和颊肌神经孔是分开的。而颞深神经孔与咬肌神经孔则已愈合。该三条神经管的后孔与 *Pa. hipparionum* 的相似，也合并为一。翼蝶管后孔也位于该融合的三者后孔的内侧。翼管后孔（posterior foramen of pterygoid canal, pcpf）也与 *Pa. hipparionum* 相似，位于翼窝的内后侧，内翼突的后基部。卵圆孔与中破裂孔融合为一。头骨基部在形状和结构上与 *Pa. hipparionum* 的很相似，但听泡内侧的隆凸程度要比 *Pa. hipparionum* 的大些，表面有一或多或少明显的棱从棘突伸达外耳道。听泡内侧棘突之后有很清楚的颈内动脉孔。颈静脉孔细裂隙状。听泡内壁未见有连接颈内动脉孔和颈静脉孔的沟，在颈静脉孔附近也未见有供蹬骨动脉进入的孔。舌下神经孔被骨板分成 2-3 个小孔。枕髁腹面较宽短，左右枕髁彼此很靠近，之间仅以一窄沟相隔。

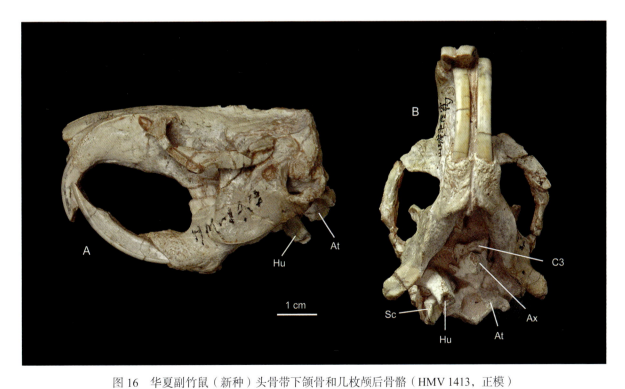

图 16　华夏副竹鼠（新种）头骨带下颌骨和几枚颅后骨骼（HMV 1413，正模）

Fig. 16　Skull with mandible and some postcranial bones of *Pararhizomys huaxiaensis* sp. nov. (HMV 1413, holotype)

A. 左侧面（left lateral view），B. 腹面（ventral view）

缩写（Abbreviations）：At. 寰椎（atlas），Ax. 枢椎（axis），C3. 第三颈椎（third cervical），Hu. 肱骨（humerus），Sc. 肩胛骨（scapula）

项面的形态结构与 *Pa. hipparionum* 的相似。外形约呈半圆形。项面也较平，并稍向前倾斜。在 HMV 1413 有弱的枕外嵴从项嵴垂向地向下伸达枕骨大孔上缘，而在 HMV 1924 未见此嵴。颞骨岩乳部为不规则的菱形，其内上角较明显的向内侧隆凸。枕骨大孔大，约为圆形。HMV 1413 的副乳突（paramastoid process, pmp）保存得很好（见图 17 A）。它们很小，约为四角锥形，由枕骨的外下角向外下方、稍向后方延伸。

（2）下颌骨（图 16，17 A6-A7；表 3）

下颌骨与 *Pa. hipparionum* 的也很相似，为松鼠型下颌骨。它的水平支较短而粗壮。下颌齿隙的后部凹入较深。颏突比 *Pa. hipparionum* 者更发达，明显较向下突，其下缘的嵴较圆凸，其内侧的二腹肌窝较大。下颌联合向前上方延伸，其下缘与臼齿下方的下颌水平支下缘的夹角约为 50°。咬肌窝向前伸至 m1 的下方。咬肌嵴很发达，比 *Pa. hipparionum* 的更发达些，呈圆弧形，明显向外下方伸张，往后一直伸达下颌骨角突下缘。颏孔位于 m1 前下方，水平支的下 1/3 处，低于咬肌窝的前端。冠状突前缘的起点和倾斜度均与 *Pa. hipparionum* 的相似，其内侧下部供颞肌附着的颞肌窝也很显著，只是冠状突的后缘不很向后弯曲。髁突比冠状突低，主要向后，仅稍向后上方延伸，其上部明显向内弯。髁突前缘近于水平，仅稍向前 - 下方延伸，较直而长（下颌髁关节面前缘距冠状突后缘的距离约 10 mm）。髁突上部内侧表面有浅的凹面，可能为翼外肌附着处（本书称之为翼外肌窝，external pterygoid fovea, eptf）。下颌髁关节面为长椭圆形，其内、外缘的界线都较明显。下颌切迹较浅，其深度约为下颌高的 1/3。下颌孔的位置较靠前，位于冠状突的中部下方，距 m3 的距离短（3 mm, 3.5 mm）。下门齿后端在下颌骨上升支的颊侧形成很显著的隆凸（i2b）。HMV 1413 左侧的隆凸表面因被磨损，露出 i2 后端的腔。HMV 1413 的下颌骨上升支的下部保存完好。翼内肌窝（iptf）大而深，为三角形的凹面。其前端起自冠状突的下方，下缘稍呈弧形，较平缓地稍向后上方伸。下颌角突（ap）很发育，约呈三角锥状，向后内上方弯突。角突后缘与下颌髁后缘约在同一垂线上。下颌骨上升支后缘在下颌髁与角突之间的凹入很深。

（3）牙齿（图 18；表 4）

牙齿与 *Pa. hipparionum* 的也很相似。齿式为 1·0·0·3/1·0·0·3。臼齿也为前 - 后侧高冠型。臼齿齿冠高于 *Pa. hipparionum* 者，而且具明显舌侧高冠。M1 和 m1–2 具两颊侧沟和一舌侧沟，M2–3 和 m3 则具一颊侧沟和一舌侧沟。M1 的内凹伸向前尖，稍插入到前边凹和中凹之间。M3 的内凹与中凹相对，彼此不重叠；冠面上内凹横向长于中凹，但侧面向下的深度则浅于后者。m3 近方形，其下外凹和下中凹均近于横向延伸，彼此相对。I2 强烈弯曲，其前部并稍向后弯伸，属垂直伸型齿，其后端起自上颌骨近门齿孔后缘处。门齿唇侧有一很显著的纵棱。下门齿强壮，向前上方伸出，纵向弯曲度较 *Pa. hipparionum* 的大，前部明显向上方弯。下门齿出露于齿槽的部分很长。左、右下门齿唇侧均只有一条明显的纵嵴。

图 17　华夏副竹鼠（新种）头骨和下颌骨

Fig. 17　Skulls and mandible of *Pararhizomys huaxiaensis* sp. nov.

A. HMV 1413，头骨含下颌骨（skull with mandible），正模（holotype）：A1–A5. 头骨（skull），A1. 背面（立体照片）（dorsal view, stereopair），A2. 腹面（立体照片）(ventral view, stereopair)，A3. 前面（anterior view），A4. 后面（posterior view），A5. 左侧面（left lateral view）；A6–A7. 下颌骨（mandible），A6. 冠面（occlusal view），A7. 左下颌骨颊侧面（buccal view of left hemimandible）。B. HMV 1924，头骨腹面（立体照片）（ventral view of skull, stereopair）

缩写（**Abbreviations**）：ap. 角突（angular process），eptf. 翼外肌窝（external pterygoid fovea），i2b. i2 后端隆突（bulge formed by i2），iptf. 翼内肌窝（internal pterygoid fovea），pmp. 副乳突（paramastoid process）

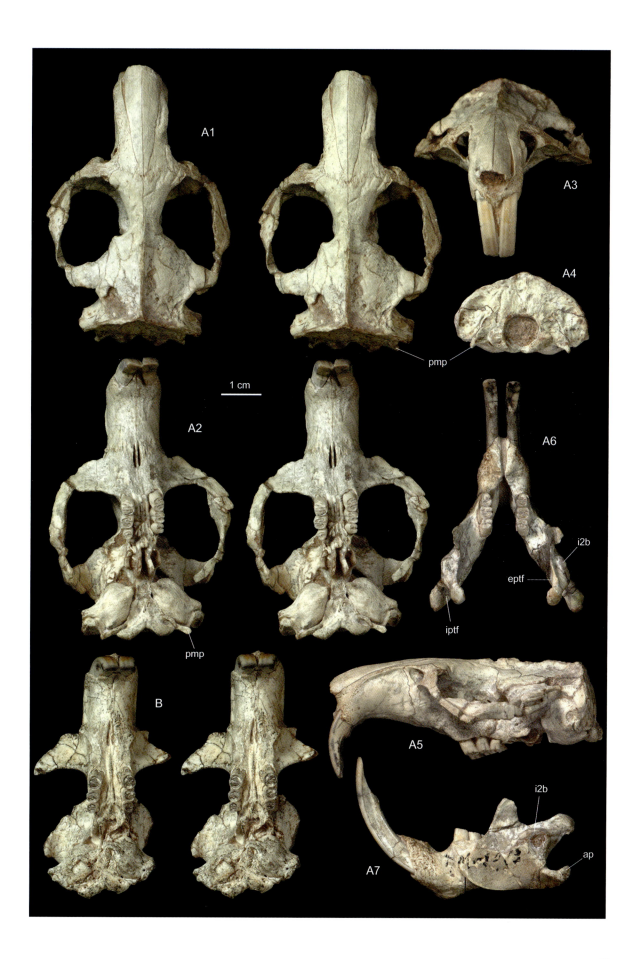

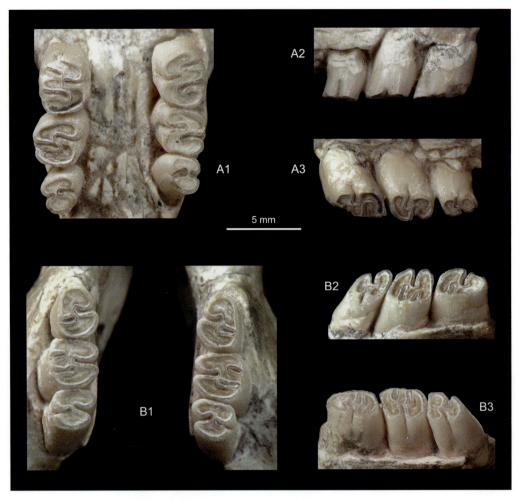

图 18　华夏副竹鼠（新种）臼齿（HMV 1413，正模）

Fig. 18　Molars of *Pararhizomys huaxiaensis* sp. nov. (HMV 1413, holotype)

A. 上臼齿（upper molars）：A1. 左和右上臼齿冠面（occlusal view of left and right upper molars），A2. 右 M1–3 颊面（buccal view of right M1–3），A3. 右 M1–3 冠舌面（occlusolingual view of right M1–3）；B. 下臼齿（lower molars）：B1. 左和右下臼齿冠面（occlusal view of left and right lower molars），B2. 右 m1–3 冠颊面（occlusobuccal view of right m1–3），B3. 右 m1–3 冠舌面（occlusolingual view of right m1–3）

（4）颅后骨骼（图 16，19–21）

　　HMV 1413 保存了几件颅后骨骼，包括寰椎、枢椎、第三颈椎、肩胛骨和肱骨的近端等。

　　寰椎（atlas, At；图 16，19；表 6）　　HMV 1413 的寰椎保存较完好，只是后关节面和寰椎翼多少有些破损。自前、后看，寰椎的整个外形约呈椭圆环形，宽大于高，前后很短。背弓（dorsal arch, da）约呈前后短而横向伸长的条带状。背结节（dorsal tubercle, dt）粗壮，位于背弓背面正中的后部。其后坡较陡，前坡稍缓。背结节前坡未见有纵棱和凹陷的痕迹。背弓腹面较平滑，纵向较平直。在背弓两侧内部有一横向延伸的供颈动脉和脊神经通过的横管（本书称该横管为寰椎侧椎管，atlas' lateral vertebral canal）。背弓背面外端的中部稍前有一小孔，内含侧椎管的外开孔（称为侧椎孔，lateral vertebral foramen, lvf，见图 31）和翼孔（*foramen alare*, fa），这里合称为侧椎 - 翼孔（lateral vertebro-alar foramen, lvaf）。侧椎管经由侧椎 - 翼孔，向内伸达椎管（vertebral canal, vc）的远中壁上的内开口，这里称之为内侧椎孔（internal lateral vertebral foramen, ilvf），而向外一直伸达寰椎翼腹侧的翼孔（fa）。内侧椎孔位于前关节凹的后内上角后方。腹弓（ventral arch, va）比背弓稍厚而前后较短。腹结节（ventral tubercle, vt）向后尖突，其两侧未见有凹陷的痕迹。腹弓的背面保存不好，齿窝（odontoid fossa, odf）

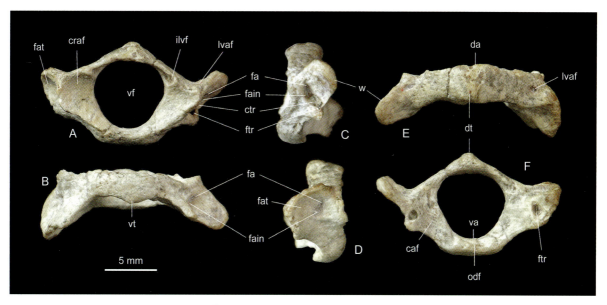

图 19　华夏副竹鼠（新种）寰椎（HMV 1413，正模）

Fig. 19　Atlas of *Pararhizomys huaxiaensis* sp. nov. (HMV 1413, holotype)

A. 前面（anterior view），B. 腹面（ventral view），C. 左侧面（left lateral view），D. 右侧面（right lateral view），E. 背面（dorsal view），F. 后面（posterior view）

缩写（Abbreviations）：caf. 后关节面（caudal articular facet），craf. 前关节凹（cranial articular fovea），ctr. 横突管（*canalis transversarius*），da. 背弓（dorsal arch），dt. 背结节（dorsal tubercle），fa. 翼孔（*for. alare*），fain. 下翼孔（*for. alare inferior*），fat. 寰椎窝（*fossa atlantis*），ftr. 横突孔（*for. transversarium*），ilvf. 内侧椎孔（internal lateral vertebral for.），lvaf. 侧椎 - 翼孔（lateral vertebro-alar for.），odf. 齿窝（odontoid fossa），va. 腹弓（ventral arch），vf. 椎孔（vertebral for.），vt. 腹结节（ventral tubercle），w. 寰椎翼（wing）

表 6　临夏盆地副竹鼠类寰椎（正模）测量和比较（单位：mm）

Table 6　Measurements and comparison of atlases (holotypes) of pararhizomyines of the Linxia Basin (in mm)

测量项（Parameter）	华夏副竹鼠（新种） *Pa. huaxiaensis* sp. nov.	土著假竹鼠（新属、新种） *Ps. indigenus* gen. et sp. nov.	甘肃假竹鼠（新属、新种） *Ps. gansuensis* gen. et sp. nov.
	HMV 1413	V 16293	HMV 1942
1. 最大长（max L）	6	6	7.7+
2. 背弓中长（dorsal sagittal L）	3.7	3.2	3.5
3. 腹弓中长（ventral sagittal L）（包括腹棘突）	2.4 (2.9)	2.1 (2.7+)	2.7 (3)
4. 最大宽（max W）	18.1	17.4 (= 8.7 × 2)	
5. 两前关节凹宽（two cranial articular foveae W）	13.9	14	14.7
6. 两后关节面宽（two caudal articular facets W）		10	10.6
比值（ratio)/%			
1）1：4	33	34	
2）3：4	13	12	
3）5：4	77	80	

的界线不清楚。椎孔 (vertebral foramen, vf) 近卵圆形，下部稍窄于上部。

　　寰 - 枕髁关节的前关节凹（cranial articular fovea, craf）约呈"肾形"，其上缘内部与背弓前缘约在同一弧线上。前关节凹上部较宽，明显凹入，两关节面分得较开；关节凹的下部逐渐变窄，左、右两关节凹彼此靠近，但不相连，中间由切迹分开。后面与枢椎关节的面——后关节面（caudal articular facet, caf）也是上宽下窄，横向稍凹。左、右后关节面的上部分得较开，下部逐渐靠近。后关节面表面横向微凹。

相对于椎孔而言，寰椎翼（wing, w）较短小，自前后看，斜向外上方伸展；自侧面看，为一自前上斜向后下方伸展的板状；翼的外缘约呈圆弧形，较粗糙。外缘前端距前关节凹的前缘约 1 mm。外缘后端伸达后关节面的后外角。寰椎翼的后背侧面的中部增厚，似形成一三角横棱。横突孔（foramen transversarium, ftr）大，位于该横棱的下方与后关节面上缘之间。翼的腹面微凹。HMV 1413 右侧的寰椎窝（fossa atlantis, fat）保存较好，窝大而浅。寰椎窝的前部有一呈前上 - 后下斜向延伸的短沟。该沟的两端有两个孔：前上方的较小，为椎间管腹外侧的开口——翼孔（foramen alare, fa）；后下方的较大，为横突管的前开口，被 Gromova（1959: 90）称为下翼孔（foramen alare inferior, fain）。翼孔与下翼孔间的距离很短（约 1 mm）。HMV 1413 的左寰椎窝已破损，显出横突孔与下翼孔相通的横突管（canalis tranversarius, ctr），该横突管较大而长。

枢椎（axis, Ax；图 16, 20；表 7）　HMV 1413 的枢椎保存较好，只是棘突和右横突部分破损。枢椎明显长于寰椎。椎体（vertebral body, vb）较宽扁。其背面较平。齿突（odontoid process, odpr）为前端钝的扁圆椎形，背面和前端稍磨损，而腹面为纵向直、横向圆凸的关节面，与寰椎的齿窝关节（facet for odontoid fossa of atlas, fodf）。枢椎前关节面（anterior articular facet, aaf）位于齿突两侧。左、右关节面均为表面稍圆凸的三角形，面向前外方。其外上缘和下缘均为稍圆凸的弧形。其内下角与齿突侧面相连。在该连接处的下方，有明显的切迹将前关节面与齿突关节面分开。椎体腹面约成横宽的扁圆形。

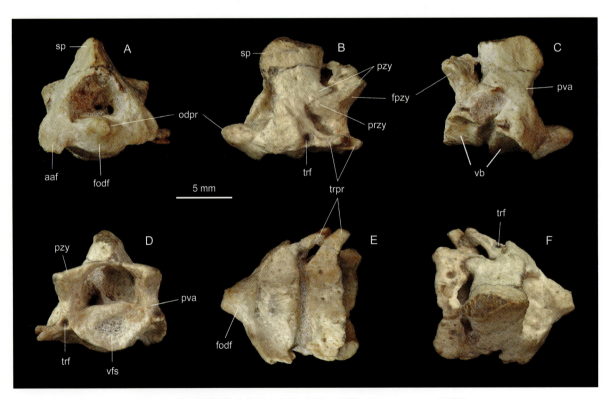

图 20　华夏副竹鼠（新种）枢椎和第三颈椎（HMV 1413，正模）

Fig. 20　Axis and third cervical vertebra of *Pararhizomys huaxiaensis* sp. nov. (HMV 1413, holotype)

A. 前面（anterior view），B. 左侧面（left leteral view），C. 右侧面（right lateral view），D. 后面（posterior view），E. 腹面（ventral view），F. 背面（dorsal view）

缩写（Abbreviations）：aaf. 前关节面（anterior articular facet），fodf. 齿突与寰椎齿窝的关节面（facet for odontoid fossa of the atlas），fpzy. 后关节突上关节面（facet on postzygapophysis），odpr. 齿突（odontoid process），przy. 前关节突（prezygapophysis），pva. 椎弓根（pedicle of vertebral arch），pzy. 后关节突（postzygapophysis），sp. 棘突（spinous process），trf. 横突孔（transverse for.），trpr. 横突（transverse process），vb. 椎体（vertebral body）；vfs. 椎窝（vertebral fossa）

表7 临夏盆地副竹鼠类枢椎（正模）测量和比较（单位：mm）

Table 7 Measurements and comparison of axes (holotypes) of pararhizomyines of the Linxia Basin (in mm)

测量项（Parameter）	华夏副竹鼠（新种） *Pa. huaxiaensis* sp. nov.	土著假竹鼠（新属、新种） *Ps. indigenus* gen. et sp. nov.	甘肃假竹鼠（新属、新种） *Ps. gansuensis* gen. et sp. nov.
	HMV 1413	V 16293	HMV 1942
1. 最大长（max L）	6.7	7.1	8
2. 背弓中长（dorsal sagittal L）	5.7		
3. 腹弓中长（ventral sagittal L）（包括腹棘突）	4	4	4.3
4. 最大宽（max W）	15 (= 7.5 × 2)	11.4	13.7+
5. 两前关节凹宽（two cranial articular foveae W）	10.6	10.1	10.1
6. 两后关节面宽（two caudal articular facets W）	8		9.2
7. 椎孔高 × 宽（vert. for. H × W）	5 × 3.8	− × 4.6	4.6 × 4.6
比值（ratio）/%			
1）4：6（6：4）	188 (53)		149 (67)
2）5：6（6：5）	133 (75)		110 (91)
3）3：2	70		
4）7H：7W	132		100

其前缘以由齿突后下缘和前关节面的下缘共同形成的前凸的圆弧形崤为界，只是在齿突关节面与前关节面间有切迹分开。椎体腹面无明显的腹棘（ventral spine），只是腹面近中部分稍隆凸，该隆凸往后变得较宽大明显。腹面两侧部分为较宽的凹面。椎窝（vertebral fossa, vfs）与第三颈椎的椎体连在一起，未能修理开。椎孔为腹面较平的圆形，横径稍大于高。椎弓根（pedicle of vertebral arch, pva）较强壮。

棘突（spinous process, sp）上部破损，只保存了基部。棘突基部横切面为中部增厚的扁豆状，长大于宽。棘突前缘较窄，向后上方斜伸。棘突由前往后逐渐变宽，在后约2/3处最宽，再往后又逐渐变窄。因此棘突上部后缘也应为窄脊形。后关节突（postzygapophysis, pzy）由椎弓根后缘向后伸出。与第三颈椎前关节突关节的关节面面向下后方，与水平面的夹角仅20º左右。

横突（transverse process, trpr）约为窄的三角形板，由椎体和椎弓根伸出，基部由背、腹两侧根组成：背侧根大于腹侧根。横突远中端向后外方延伸。横突背面稍向前外方倾斜。横突孔（transverse foramen, trf）较大，位于横突基部背、腹根之间。横突管很短。

第三颈椎（third cervical vertebra, C3；图16，20；表8） HMV 1413的第三颈椎保存较好，只是其椎弓背部后缘和右横突稍破损。椎体腹面的长度稍大于寰椎腹弓的长，而其椎弓背部的前后长度稍短于寰椎背弓的长。椎体为短而宽的扁圆柱形。椎体背面较平直。腹面约成梯形，前缘窄于后缘。腹面近中部分稍隆凸，往后变得较宽，无明显的腹棘（medial keel）。腹面两侧也为较宽的凹面。椎窝（vertebral fossa, vfs）为横宽的卵圆形，明显凹入。椎孔约呈半圆形，宽大于高。

椎弓背部为前后短、横向延伸的条带状，稍向背侧圆凸。棘突较低窄。在椎弓背部背面，棘突两侧各有一明显的孔。椎弓根较粗壮。前关节突（przy）较明显，从椎弓根前缘向前伸。与枢椎的后关节突关节的关节面位于其前背侧。后关节突位于椎弓根后缘，位置比前关节突稍高，但向后伸出得很少，几乎不超过椎弓背部后缘。后关节突上与第四颈椎前关节突的关节面（fpzy）为横向较窄的长卵圆形，表面较平直，面向后下外方，与前关节突上的关节面近于平行。前、后关节面间最短距离约为1.1 mm。

第三颈椎的横突主要向外后方伸出，也约呈窄的三角锥形，但比枢椎横突要粗大。横突由两根组成，背侧根由椎弓根起始，腹侧根由椎体侧面起始；腹侧根明显大于背侧根。横突远中端在枢椎横突的后下方向外后方延伸，与后者部分重叠。横突孔（trf）位于横突基部背侧和腹侧两根之间。横突管较枢椎的长。

肩胛骨（scapula, Sc；图16，21 B） HMV 1413只保存有右肩胛骨的关节角（glenoid angle）。肩臼（glenoid cavity, glc）为卵圆形的凹面。其径长（约7 mm）大于肩胛骨颈宽（肩胛颈最小宽为

表8　临夏盆地副竹鼠类第三至第七颈椎（正模）测量和比较（单位：mm）

Table 8　Measurements and comparison of 3rd–7th cervical vertebrae (C3–C7, holotypes) of pararhizomyines of the Linxia Basin (in mm)

测量项（Parameter）	华夏副竹鼠（新种） *Pa. huaxiaensis* sp. nov.	土著假竹鼠（新属、新种） *Ps. indigenus* gen. et sp. nov.					甘肃假竹鼠（新属、新种） *Ps. gansuensis* gen. et sp. nov.	
	HMV 1413	V 16293					HMV 1942	
	C3	C3	C4	C5	C6	C7	C6	C7
1. 椎体长（max L）	3.3	2.7	2.5	2.5	2.6	2.9	3	~2.9
2. 椎头宽（caput W）		5.1					6	~6
3. 椎头高（caput H）		3.1					3.5	
4. 前关节突外宽（prezyg. W）	8.5	9	9.5	10.5	10.5	10.5		11.2+
5. 后关节突外宽（postzyg. W）	9.5	~9		10.4	10.4	10.6		10.6
6. 前、后关节突最大长（max L of pre- & post-zyg.）	5.1, 5.5	5	5, 4.9	5	5, 4.9	4.6		~5, 5
7. 椎孔高（vert. for. H）	3.5	3.7						3
8. 椎孔宽（vert. for. W）	4.5	5						5
9. 横突处最大宽（transverse proc. max W）	13 (= 6.5 × 2)	13 (= 6.5 × 2)	14.4	14.5	15.4	15		13.5+
10. 椎窝宽（vertebral fossa W）		5					6	5.7
11. 椎窝高（vertebral fossa H）		3.1					3.7	3.9
12. 背弓中长（dorsal sagittal L）		~2.1	~2	2	2	1.6		2

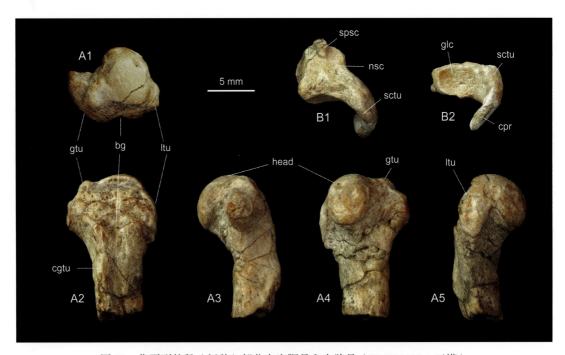

图 21　华夏副竹鼠（新种）部分右肩胛骨和右肱骨（HMV 1413，正模）

Fig. 21　Partial right scapula and humerus of *Pararhizomys huaxiaensis* sp. nov.

(HMV 1413, holotype)

A. 右肱骨上部（Upper part of right humerus）：A1. 近端面（proximal view），A2. 前面（anterior view），A3. 外侧面（lateral view），A4. 后面（posterior view），A5. 内侧面（medial view）；B. 右肩胛骨的关节角（glenoid angle of right scapula）：B1. 外侧面（lateral view），B2. 肩臼端面（glenoid cavity view）

缩写（Abbreviations）：bg. 臂二头肌沟（bicipital groove），cgtu. 大结节嵴（crest of greater tuberosity），cpr. 喙突（coracoid process），glc. 肩臼（glenoid cavity），gtu. 大结节（greater tuberosity），head. 肱骨头（head of humerus），ltu. 小结节（lesser tuberosity），nsc. 肩胛颈（neck of scapula），sctu. 肩胛结节（scapular tuber），spsc. 肩胛冈（spine of scapula）

5.5 mm）。肩胛冈（spine of scapula, spsc）只保留了下端。其起始点约在肩胛颈的最窄处，距肩臼外侧缘约 4 mm。肩胛结节（scapular tuber, sctu）低。喙突（coracoid process, cpr）很发育，明显向内下方尖突。

肱骨（humerus, Hu；图 16, 21 A；表 14）　HMV 1413 只保存了右肱骨的近端部分，而且其前面已破损。从保存部分看，近端的宽大于其前后径长（见表 14）。肱骨头（head of humerus, head）稍向后伸，与肩臼的关节面约呈横向较窄的卵圆形球面，其纵向和横向都很隆凸，使关节面几近半圆球形。大结节（greater tuberosity, gtu）较粗大，主要向远中侧伸，其顶端稍低于肱骨头的最高点。大结节被一明显的凹区分为内、外两部分。内部（可能是冈上肌的附着处）较小而高，其顶端内侧以一小的浅凹与肱骨头的近端关节面相隔。外部较内部大，但明显较低。冈下肌可能附着在大结节的外部和内、外部间的凹区中。小结节（lesser tuberosity, ltu）较大结节低小。其上有垂向的浅沟，将其分为前、后两部分。前部前面破损。骨体上部为三角柱形，宽大于前后径。结节间沟（intertubercular sulcus, 又被称为臂二头肌沟，bicipital groove, bg）及其外侧的大结节嵴（crest of greater tuberosity, cgtu）都很明显。

2. 比较

由上面的描述可以明显地看出，HMV 1413 等标本无论是在头骨和下颌骨的形态上，还是在牙齿的结构上都与 *Pa. hipparionum* 很接近，应被归入 *Pararhizomys* 属。*Pararhizomys* 属目前已知包括三趾马层副竹鼠（*Pa. hipparionum*）和秦副竹鼠（*Pa. qinensis*）两个种。与上述二种比较，HMV 1413 等标本还具有一些不同于它们的特点。

HMV 1413 等标本与 *Pa. hipparionum* 的区别是（*Pa. hipparionum* 的特点被列在括号内）：下颌齿隙明显长于下齿列（两者长度相近，见表 3, 4）；M1–2 和 m1–3 为舌侧高冠，其舌侧褶沟明显深于颊侧褶沟，其 Hpl/Hpb2 值在 M2 中为 85%，在 M1 和 m1–2 中 ≤ 71%（无明显的舌侧高冠现象，其舌侧褶沟与颊侧褶沟的深度相近，上述比值分别为 109%–115% 和 ≥ 90%，见表 4, 5）；M3 的舌侧褶沟与颊侧褶沟在冠面上通常彼此相对，不重叠（两褶沟部分重叠）；m1 的下中凹较短，稍弯，其颊部伸向下原凹（下中凹较长，较强烈弯曲，其颊部向前伸超过下原凹）等。

上述标本与 *Pa. qinensis* 的区别是（*Pa. qinensis* 的特点被列在括号内）：头骨具较长的吻部和上颌齿隙，它们分别为 22.1 mm 和 27.2 mm（两者均较短，上颌齿隙为 16.5 mm；依 Zhang et al., 2005: 4）；M1 的内凹深于前边凹和中凹，其 Hpl/Hpb2 值为 68% 和 70%（颊侧和舌侧褶沟的深度相近，其 Hpl/Hpb2 值为 95%）等。

显然，上述的标本应代表 *Pararhizomys* 属中一新种，被称为华夏副竹鼠（*Pararhizomys huaxiaensis*）。

Pararhizomys huaxiaensis 目前仅知三件标本。其中 HMV 1413 和 HMV 1924 的头骨结构比 V 16291 者更粗壮些，并具有较宽而高的吻部和较强大的门齿。很可能前两件标本代表雄性个体，而 V 16291 代表雌性个体。

3. 讨论

2011 年李强曾口头告知笔者，他认为，德日进 1942 年在研究与桑氏原鼢鼠（*Prosiphneus licenti*）模式标本一起发现而后来修理出来的一批被归入该种的材料中，一件左 M1–2 标本（Teilhard de Chardin, 1942, Fig. 32，现编号为 RV 42023），很可能是副竹鼠类的右 m1–2。笔者赞同李强的看法。因该标本的"M2"冠面较长的褶沟明显向前弯。这与副竹鼠类的 m1 的下中凹很相似，而与原鼢鼠的 M2 的内凹不同。后者的内凹通常较短，不向前弯，而是伸向前尖。此外，Teilhard de Chardin（1942）的 Fig. 32 显示的该标本的"M1"的颊、舌侧都有两褶沟。据笔者观察，该"M1"仅一侧有两褶沟，而另一侧仅有一褶沟。而在该褶沟前只有一破损的裂缝，笔者未观察到褶沟的痕迹。如果该侧只有一褶沟，这也与副竹鼠的 m2 相似，而与鼢鼠类的 M1 不同。有意义的是，该"M1"和"M2"的齿冠较高，不但有前 - 后侧高冠现象，而且还是舌侧高冠，其舌侧褶沟明显深于颊侧褶沟。这些特征也与 *Pa. huaxiaensis* 的很相似。如果上面的分析合理的话，这件标本就应该是 *Pa. huaxiaensis* 的右 m1–2 了。这样，在甘肃庆阳地区也有副竹鼠类化石。这还表明，*Pa. huaxiaensis* 有可能延续到晚中新世晚期，即保德期。只可惜该标本保存不好，而且为较幼年的个体。需要发现更多更好的材料予以确认。

陇副竹鼠（新种） *Pararhizomys longensis* sp. nov.
（图 22，23，75；表 1–4）

正模 V 16292.1，一件不完整含下颌的头骨，产自甘肃省东乡族自治县那勒寺乡郭泥沟东北（60°）约 300 m（LX 200042）；柳树组下部，晚中新世早灞河期（详见表 1，2）。

副模 V 16292.2 和 V 16292.3，两件头骨前部；产地和层位同正模。

鉴别特征 前颌骨侧面的纵棱很显著，其上方的纵向凹槽较宽而深。颧弓板的位置较后，其弧形后缘的最前点约位于 M1 中部。下颌齿隙与下颊齿齿列的长度相近。臼齿齿冠较低，前 - 后侧高冠仅在 M1–2 和 m3 较明显，在 m1–2 中弱，在 M3 中不明显。M3 的向前外方斜伸的内凹和向前内方斜伸的中凹彼此部分重叠。m1 的下中凹颊端稍前弯。下门齿短而纵向弯曲度较大。

区别特征 与 *Pararhizomys hipparionum*，*Pa. huaxiaensis* sp. nov. 和 *Pa. qinensis* 的共同区别是：前颌骨侧面的纵棱更明显，纵棱上方的凹槽宽大而深；颧弓板的位置较靠后，其弧形后缘的最前点约位于 M1 的中部；臼齿齿冠较低。

与 *Pa. hipparionum* 和 *Pa. huaxiaensis* sp. nov. 的区别还在于下门齿较短，并较强烈弯曲；与 *Pa. hipparionum* 的区别还在于 m1 的下中凹较少弯曲；与 *Pa. huaxiaensis* sp. nov. 的区别还在于下颌齿隙较短，其长约与下臼齿列长相近和 M3 内凹和中凹斜向前延伸并部分重叠；与 *Pa. qinensis* 的区别还在于具有较长的上颌齿隙。

词源 Long，陇，为甘肃省的简称。

1. 描述

（1）头骨（图 22 A，B1 - B4；表 3）

V 16292.1 头骨（图 22 B1–B4）仅保存了吻部、I2、上颊齿、部分左颧弓和右颧弓的前根，以及脑颅的前部。该头骨并因受压稍变形。V 16292.2 和 V 16292.3 均只保存有吻部或面部。郭泥沟标本的头骨的基本结构与 *Pa. hipparionum* 和 *Pa. huaxiaensis* sp. nov. 的一致：具鼠型颧 - 咬肌结构。吻部两侧缘彼此近于平行，但吻部腹面前端向下弯曲更显著。鼻骨为楔形，其前端与前颌骨前端约在同一垂线上，鼻骨的后端窄，与前颌骨的后端和眼眶的前缘约在同一横线上。前颌骨的侧背嵴很明显，与鼻骨 - 前颌骨缝近于平行。前颌骨背面为窄长条形，外侧面上部有明显的纵棱和纵向凹槽等。但该凹槽明显较宽大而深，受压后都或多或少凹陷。眶下孔也为横宽的椭圆形，也有腹侧裂隙。眶下管内壁凹入也较深，其前端由棱嵴形成的界线也很明显。颧弓约成圆弧形。颧骨短，其前端不伸达眼眶前缘。颧弓板限于上颌骨内。外层咬肌起始区明显凹入，其前缘以明显的棱嵴为界。颧弓板的前缘约位于门齿孔的中部水平。颧弓板弧形后缘的位置较靠后，其最前点约位于 M1 的中部。泪结节很弱小。眼眶很小。额嵴很弱，向后内方延伸，左、右额嵴在眶后收缩附近愈合形成明显的矢状嵴。前颌骨 - 上颌骨缝从门齿孔的中部穿过。腭桥部分破损，腭沟和腭后孔情况不清。颞深神经孔和咬肌神经孔也愈合为一裂缝，颊肌神经孔与该两孔分开。左、右两臼齿列彼此近于平行。

（2）下颌骨（图 22 B5，B6；表 3）

V 16292.1 的下颌骨保存得较完整，只是上升支部分破损，缺角突及左冠状突和髁突。下颌骨的形态结构与 *Pararhizomys* 的很相似，属松鼠型下颌骨。下颌骨的水平支较短而粗壮。下颌联合较陡地向前上方延伸，与水平支下缘的夹角约为 45°。颏突和二腹肌窝都明显。下颌齿隙的后部凹入较深。颏孔位于 m1 的前下方，水平支的下 1/3 处，低于咬肌窝前端。咬肌窝前端伸至 m1 的下方。咬肌嵴很发达，并向外伸张。下颌切迹浅，其深度约为下颌骨高的 1/4。下门齿齿槽后端在上升支的颊侧形成显著的隆凸。

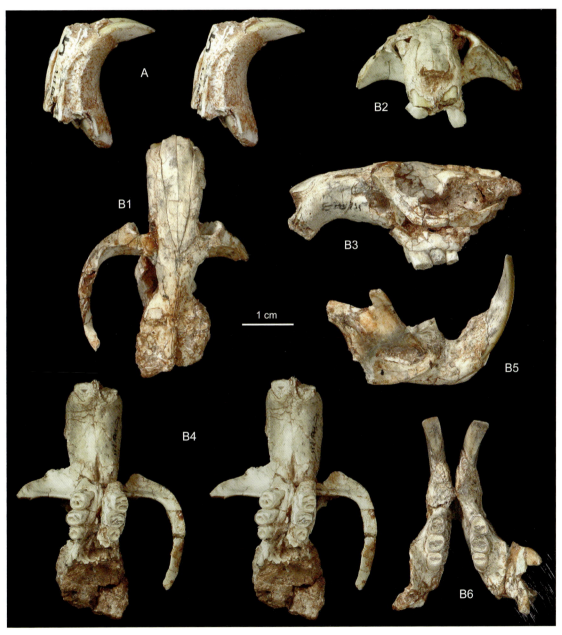

图 22　陇副竹鼠（新种）头骨和下颌骨

Fig. 22　Skulls and mandible of *Pararhizomys longensis* sp. nov.

A. V 16292.3 吻部右侧面（立体照片）（right lateral view of muzzle, stereopair）。B. V 16292.1（正模），头骨含下颌骨（skull with mandible, holotype）：B1–B4. 头骨（skull），B1. 背面（dorsal view），B2. 前面（anterior view），B3. 左侧面（left lateral view），B4. 腹面（立体照片）（ventral view, stereopair）；B5–B6. 下颌骨（mandible），B5. 右下颌骨颊侧面（buccal view of right hemimandible），B6. 冠面（crown view）

（3）牙齿（图 23；表 4）

齿式为 1·0·0·3/1·0·0·3。臼齿主要为前 - 后侧高冠型。V 16292.1 是一老年个体。上、下臼齿冠面的褶沟均被磨蚀成封闭的环形。左 m2 的下原凹甚至被磨蚀成圆形的坑。左、右 m1 和右 m2 的冠面的前颊部分被磨蚀成凹面。该处未见下原凹的痕迹。很可能它们的下原凹已被完全磨蚀掉了。很可能在较少磨蚀时 m1 和右 m2 也和左 m2 一样都有下原凹和下外凹两个颊侧褶沟存在。如果这种推测有道理的话，V 16292.1 的臼齿的冠面结构也与 *Pa. hipparionum* 和 *Pa. huaxiaensis* sp. nov. 者相似：M1 和 m1–2 均具有两颊侧褶沟和一舌侧褶沟，M2–3 和 m3 均只有一颊侧褶沟和一舌侧褶沟。M3 与 *Pa. hipparionum* 的相似：内凹和中凹斜向前延伸并部分重叠；m3 的外形为近方形，后缘稍窄于前缘，下外凹和下中凹彼此相对。

I2 强烈弯曲，其唇侧具很发达的纵棱。下门齿粗壮，其齿槽前伸出的部分短，纵向的弯曲度与 *Pa. huaxiaensis* sp. nov. 者相近，而较 *Pa. hipparionum* 者大，前部明显向上弯伸。左 i2 的唇侧有一条纵嵴，右 i2 有两条纵嵴。这似乎表明 i2 唇侧由两条纵嵴向一条纵嵴的退化过程也已出现在郭泥沟的标本中。

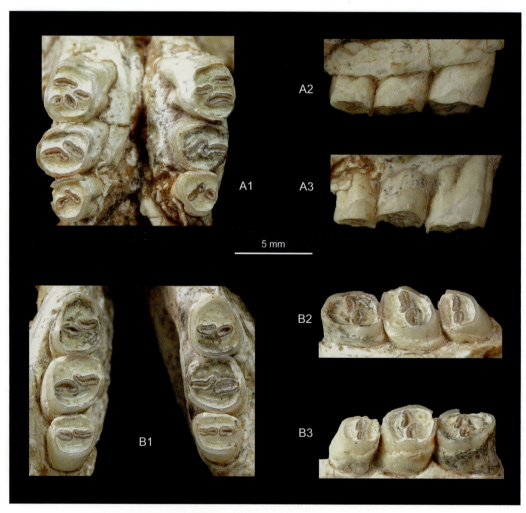

图 23　陇副竹鼠（新种）臼齿（V 16292.1, 正模）

Fig. 23　Molars of *Pararhizomys longensis* sp. nov. (V 16292.1, holotype)

A. 上臼齿（upper molars）：A1. 左和右上臼齿冠面（occlusal view of left and right upper molars），A2. 右 M1–3 颊面（buccal view of right M1–3），A3. 左 M1–3 舌面（lingual view of left M1–3）；B. 下臼齿（lower molars）：B1. 左和右下臼齿冠面（occlusal view of left and right lower molars），B2. 左 m1–3 冠颊面（occlusobuccal view of left m1–3），B3. 左 m1–3 冠舌面（occlusolingual view of left m1–3）

2. 比较

前面的描述表明，郭泥沟的标本在头骨、下颌骨和牙齿的形态结构上都符合 *Pararhizomys* 属的特征，应归入该属。然而，与 *Pararhizomys* 属的三种（*Pa. hipparionum*，*Pa. qinensis* 和 *Pa. huaxiaensis* sp. nov.）比较，郭泥沟的标本还有一些不同的特征。它们与上述三种不同的是：前颌骨侧面的纵棱较明显，纵棱上方的凹槽较宽而深（见图 22 A, B3）；颧弓板弧形后缘的最前点的位置较靠后，大约位于 M1 的中部水平；臼齿齿冠较低，前 - 后侧高冠仅在 M1–2 和 m3 较明显，在 m1–2 较弱，而在 M3 不明显（见图 23）。

此外，郭泥沟的标本与 *Pa. hipparionum* 和 *Pa. huaxiaensis* sp. nov. 的区别还在于下门齿较短；不同于 *Pa. hipparionum* 还在于下门齿弯曲度较大和 m1 的下中凹较少弯曲；不同于 *Pa. huaxiaensis* sp. nov. 在于 M3 具斜向延伸并部分重叠的内凹和中凹，以及下颌齿隙的长度与下齿列的相近（见表 3, 4）；与 *Pa. qinensis* 的区别还在于具有较长的上颌齿隙 [前者为 19 mm（见表 3）；后者为 16 mm（依 Zhang et al., 2005）]。

郭泥沟的标本显然代表不同于 *Pararhizomys* 上述各种的一个新种，这里称之为陇副竹鼠（*Pararhizomys longensis*）。

假竹鼠属（新属） *Pseudorhizomys* gen. nov.

模式种　*Pseudorhizomys indigenus* gen. et sp. nov.。

其他包括种　*Pseudorhizomys gansuensis* gen. et sp. nov.，*Pseudorhizomys planus* gen. et sp. nov.，*Pseudorhizomys pristinus* gen. et sp. nov.，可能还有 *Pseudorhizomys? hehoensis* (Zheng, 1980)。

分布与时代　中国甘肃，可能还有西藏；晚中新世，中—晚灞河期（~10–7 Ma）。

鉴别特征　前颌骨前端向前伸，超过鼻骨的前端。鼻骨背面纵向平直。前颌骨背面为狭窄的三角形。前颌骨侧背嵴短，前端稍斜向前内侧伸，与鼻骨 - 前颌骨缝汇合。前颌骨 - 上颌骨缝在背侧呈弧形向后外方弯，与上颌骨 - 额骨缝相联，共同形成一圆弧形。眶下孔近圆形。颧弓前根和颧弓板的位置均更靠前：眼眶前缘明显前于鼻骨和前颌骨的后端；颧弓板弧形后缘的最前点与门齿孔后端几乎在同一横线上。外层咬肌附着区的前缘为圆弧形或 S 形。颧骨长，其前端伸达眼眶前缘。顶骨约为短宽的梯形。副翼窝小而浅。硬腭后缘（= 中翼窝前缘）约与 M3 的后部位置相对。中翼窝与翼窝的宽度相近或前者稍宽于后者。左、右内翼突不倾斜，而与腹面近于垂直、彼此近于平行地纵向延伸。颞深神经孔与咬肌神经孔和颊肌神经孔一起合并为裂隙状，但颞深神经管的后孔则与咬肌神经管和颊肌神经管的后孔分开。项面下部相对较宽，颞骨岩乳部分相对较小。下颌齿隙的后部凹入较浅。下颌骨水平支下缘向下圆凸。颏孔的位置较高，约位于水平支的中部，与咬肌窝的前端约在同一水平上。下颌切迹较深，其深度约为下颌骨高的 2/5。

新属的臼齿齿冠较 *Pararhizomys* 的低，前 - 后侧高冠仅在 m3 明显，在 M1 有或无，而在 M2–3 和 m1–2 无。M2–3 具 1–2 条颊侧褶沟（前边凹或有或无），一条舌侧褶沟，而下臼齿具 1 或 2 条舌侧褶沟（下后边凹或有或无）。m3 为长卵圆形，长大于宽，下中凹向前外侧斜伸，与下外凹部分重叠，通常具下原凹。I2 纵向弯曲程度较 *Pararhizomys* 者小，其前端向前下方伸，而不向后弯，属前伸型齿（proodont）；其后端始自上颌骨中部，靠近 M1；唇面无显著的纵棱。下门齿唇侧通常具两条纵嵴。

区别特征　*Pseudorhizomys* gen. nov. 与 *Pararhizomys* 的区别见表 9。

词源　希腊文 pseudes，假的；rhizomys，竹鼠，现生的鼠类。属名含并非真正竹鼠之意。

评注　郑绍华（1980）描述的黑河低冠竹鼠（*Brachyrhizomys hehoensis*）很可能属于假竹鼠属（新属）。但因有关材料较少，暂时有疑问地将其归入 *Pseudorhizomys* gen. nov.，被称为黑河假竹鼠（属存疑）（*Pseudorhizomys? hehoensis*）（详见下）。

表 9　副竹鼠与假竹鼠（新属）的主要区别特征

Table 9　Major distinguishing characters of *Pararhizomys* and *Pseudorhizomys* gen. nov.

	特征（Character）	副竹鼠 *Pararhizomys*	假竹鼠（新属）*Pseudorhizomys* gen. nov.
1	鼻骨背面 dorsal side of nasal	纵向较圆凸 convex longitudinally	纵向较平直 straight longitudinally
2	前颌骨和鼻骨的前端 anterior ends of nasal and premaxilla	在同一垂线上 in the same vertical line	鼻骨前端后于前颌骨前端 that of nasal posteriorly to that of premaxilla
3	前颌骨侧背棱 dorsolateral crest of premaxilla	长，与鼻骨 - 前颌骨缝平行 long, parallel to naso-premaxillary suture	短，往前与鼻骨 - 前颌骨缝汇合 short, anteriorly joining with naso-premaxillary suture
4	前颌骨背部 dorsal part of premaxilla	长条形 long, band-form	窄的三角形 narrow, triangular
5	前颌 - 上颌骨缝背侧部分 dorsal part of premaxillo-maxillary suture	纵向直，纵向延伸，与上颌 - 额骨缝形成钝角 straight and longitudinal, forming an obtuse angle with maxillo-frontal suture	向后外方弯曲，与上颌 - 额骨缝共同形成圆弧 curved postero-laterally, forming an arch with maxillo-frontal suture
6	眶下孔的形状 outline of infraorbital foramen	横宽的椭圆形 wide oblate	圆形 circular
7	颧弓前根 anterior root of zygomatic arch	位置靠后，眼眶前缘的最前点与鼻 - 额骨缝在同一横线上 posteriorly positioned, anterior border of orbit aligned with naso-frontal suture	位置靠前，眼眶前缘的最前点前于鼻 - 额骨缝 anteriorly positioned, anterior border of orbit anterior to naso-frontal suture
8	颧弓板 zygomatic plate	位置靠后，其弧形后缘的最前点后于门齿孔 posteriorly positioned, anteriormost point of its arched posterior border posterior to incisor foramen	位置靠前，其弧形后缘的最前点与门齿孔的后端在同一横线上 anteriorly positioned, anteriormost point of its arched posterior border aligned with posterior end of incisive foramen
9	颧骨 jugal	短，颧骨前突不达眼眶前缘 short, anterior end does not reach to the anterior border of orbit	长，颧骨前突伸达眼眶前缘，通常与泪骨相连 long, anterior end extends to anterior border of orbit, usually meeting lachrimal
10	顶骨形状 shape of parietal	不规则的长方形，长大于宽 irregularly rectangular, longer than wide	约为六边形，宽大于长 hexagonal, wider than long
11	副翼窝 parapterygoid fossa	较大而深 larger and deeper	较小而浅 smaller and shallower
12	硬腭后缘的位置 position of posterior border of palate	位于 M3 之后 posterior to M3	与 M3 后缘约在同一横线上 aligned with posterior border of M3
13	中翼窝的宽度 width of mesopterygoid fossa	明显宽于翼窝 much wider than pterygoid fossa	与翼窝的宽度相近 subequal to pterygoid fossa
14	内翼突 internal pterygoid processes	向外倾斜，往后彼此靠近 slanting laterally and convergent posteriorly	不倾斜，彼此近于平行地纵向延伸 vertical and parallel with each other
15	咬肌 - 颞深神经孔和颊肌神经孔 masticatory-deep temporal & buccinator f.	分开 separated	合并为一 confluent
16	岩乳骨在项面上出露 petromastoid exposed on nuchal side	较大 larger	较小 smaller
17	下颌骨齿隙的后部 posterior part of mandibular diastema	凹入较深，具较陡的后坡 deeply concave with steep posterior slope	凹入较浅，具较缓的后坡 slightly concave, with gentle posterior slope
18	下颌骨水平支的下缘 lower border of horizontal ramus	较平直 straight	向下圆凸 convex
19	颏孔的位置 position of mental foramen	低于咬肌窝的前端 lower than the anterior end of masseteric fossa	与咬肌窝的前端约在同一水平上 at same level as the anterior end of masseteric fossa
20	下颌切迹深度 depth of mandibular notch	较浅，不达下颌骨高的 1/3 shallower, less than 1/3 mandibular height	较深，约为下颌骨高的 2/5 deeper, ~ 2/5 of mandibular height

	特征（Character）	副竹鼠 *Pararhizomys*	假竹鼠（新属）*Pseudorhizomys* gen. nov.
21	I2	强烈弯曲，前端向下后方弯，其后端起自上颌骨的前缘近门齿孔处 strongly curved, anterior end curved posteriorly, posterior end originated from anterior margin of maxillary near posterior end of incisor foramen	较少弯曲，前端向下弯，但不向后弯，其后端起自上颌骨中部 M1 之前 less curved, anterior end not bending posteriorly, posterior end originated from middle of maxillary in front of M1
22	I2 唇侧纵棱 longitudinal crest of I2	很发达 well-developed	弱或无 absent or very weak
23	臼齿齿冠 molar crown	较高，白齿均为前 - 后侧高冠 higher crowned, all molars mesiodistal hypsodont	较低，前 - 后侧高冠仅在 m3 明显 lower crowned, mesiodistal hypsodont distinct only on m3
24	M2–M3 的颊侧褶沟 buccal reentrant of M2–3	仅具一条颊侧褶沟，无前边凹 only one reentrant (mesosinus), without anterosinus	具 1 或 2 条颊侧褶沟（通常具前边凹） with one or two reentrants (usually with anterosinus)
25	m3	长宽相近的方形，后缘较少圆凸 subguadrate, subequally long and wide, posterior side less convex	长大于宽的卵圆形，后缘较窄而圆凸 oval, longer than wide, posterior side more convex
26	m3 的下中凹和下外凹 mesosinusid and sinusid of m3	通常横向延伸，彼此相对 usually transverse and directly opposite to each other	下中凹向前外方伸，与下外凹部分重叠 mesosinusid extending antero-laterally, partially overlapping sinusid
27	m3 下原凹 protosinusid on m3	无 absent	通常有 usually present

土著假竹鼠（新属、新种）*Pseudorhizomys indigenus* gen. et sp. nov.
（图 24–55, 76；表 1, 6–8, 10–17, 21）

正模 Ｖ 16293，同一年轻个体的含下颌骨的头骨和部分颅后骨骼，产于 LX 200004（杨家山）；柳树组中部上层（详见表 1, 10）。

副模 Ｖ 16294，一中年个体的含下颌骨的头骨，产于 LX 200020（南面沟村）；柳树组中部上层（详见表 1, 10）。

鉴别特征 吻部和上颌齿隙长；外层咬肌附着区的前缘约为"S"形，其弧形的下部向前凸，超过眶下孔的腹侧裂隙。颞嵴后部变平缓。听泡前面内侧有明显的凹面。下颌齿隙后部凹入部分相对较短。冠状突较长，较向后倾；其上部向后弯曲，后缘稍凹入。下颌髁关节面约为纺锤形。

M1 和 m3 为明显的前 - 后侧高冠型齿。M1 颊侧釉质曲线有明显折曲。上白齿均具两颊侧沟（前边凹总是存在）；内凹横向短，不与前边凹重叠。M3 内凹与中凹均约横向延伸，彼此相对，或相通；中凹的深度明显大于内凹的深度。M2–3 的前边凹和 m3 的下原凹均退化为封闭的盆。m1–3 无下后边凹，下外凹不分叉。m3 下外凹的位置较靠前，与下中凹的舌部约在同一横线上。I2 唇面横向圆凸，无纵棱。

第二掌骨的掌骨粗隆为结节状隆凸，位于骨体中部背面。第五掌骨近端与钩骨的关节面为卵圆形，表面圆凸；骨体中部为扁圆柱形。拇指的爪指骨远端缘薄锐，无锯齿。

词源 Indigenus，拉丁文，本地的，土生的；意指该类动物可能为东亚土生种类。

1. 描述

（1）头骨（图 24 A, 25；表 11）

正模 （Ｖ 16293）为一幼年个体头骨，保存不如 Ｖ 16294 者好，但其基本形态和结构仍基本保存。Ｖ 16294 为成年个体，头骨保存较好，只是左颧弓和左枕髁缺失，右颧弓保存不全。

背面观（图 24 A2, 25 D） 吻部较长而窄，其两侧缘向前彼此稍靠近。鼻骨呈楔形，其后端很狭窄，

表10 临夏盆地产假竹鼠（新属）化石的地点和层位

Table 10 Localities and horizons yielding fossil *Pseudorhizomys* gen. nov. in the Linxia Basin

新种 New species		标本 Specimens	化石编号 IVPP Catal.	野外地点号 IVPP Loc.	产出层位 Horizon	地方动物群 Local fauna
土著假竹鼠 *Ps. indigenus*	正模 holotype	头骨带下颌骨和部分颅后骨骼 skull with mandible and partial postcranial bones	V 16293	LX 200004	柳树组中部上层 upper part of middle Liushu Fm.	杨家山 Yangjiashan
	副模 paratype	带下颌骨的头骨 skull with mandible	V 16294	LX 200020		
甘肃假竹鼠 *Ps. gansuensis*	正模 holotype	头骨带下颌骨和部分颅后骨骼 skull with mandible and partial postcranial bones	HMV 1942	LX 200502	柳树组中部下层 lower part of middle Liushu Fm.	大深沟 Dashengou
	副模 paratypes	头骨带下颌骨 skull with mandible	V 16297	LX 200007		
		头骨 skull	V 16298	LX 200011		
		部分头骨带左半下颌支 partial skull with left hemimandible	V 16299	LX 200046		
		头骨前部 anterior part of skull	V 16300	LX 200037		
	归入标本 referred specimens	头骨带下颌骨 skull with mandible	V 16296	LX 200019	柳树组中部上层 upper part of middle Liushu Fm.	杨家山 Yangjiashan
		部分头骨带右半下颌支 partial skull with right hemimandible	V 16301	无 no	层位不清 uncertain	不确定 uncertain
		部分头骨 partial skull	V 16302			
		头骨前部 anterior part of skull	V 16303			
平齿假竹鼠 *Ps. planus*	正模 holotype	头骨前部 anterior part of skull	V 16304	无 no	层位不清 uncertain	不确定 uncertain
原始假竹鼠 *Ps. pristinus*	正模 holotype	头骨带下颌骨 skull with mandible	V 16305	LX 201001	柳树组中部下层 lower part of middle Liushu Fm.	大深沟 Dashengou

与前颌骨-额骨缝的最后端约在同一横线上。鼻骨在纵、横向上不像*Pararhizomys*的那样圆凸，而较平直。左、右鼻骨后部稍向近中方向倾斜，形成"V"形浅中纵谷。左、右鼻骨-前颌骨缝除稍弯曲的前端外，绝大部分较直，向后彼此靠近。前颌骨有一明显的侧背棱（pmldc）。与*Pararhizomys*不同的是，该棱嵴很短，不与鼻骨-前颌骨缝平行，而是在前部与后者汇合。这样，前颌骨的背面形成尖端向前的窄三角形。锯齿状的前颌骨-额骨缝不像*Pararhizomys*那样直着斜向前外方延伸，而是外部稍弯向外方延伸。前颌骨-上颌骨缝在头骨背侧不是呈纵向延伸的直线，而是向后外方弯曲，其后端与上颌骨-额骨缝的内端相连，共同形成外弯的圆弧形。眶下孔（iof）也限于上颌骨内，但比*Pararhizomys*的大，近圆形，其背缘圆弧形的弯曲度显然要比*Pararhizomys*的大。从前面看，可见眶下孔也有明显的腹侧裂隙，与眶下管以隔板分开，但该隔板不很向上延伸。隔板内侧有鼻泪孔（nlf）。在背面没有看见泪骨，也未见明显的泪结节。左侧顶骨和鳞骨保存好。顶骨宽短，侧缘在后半部斜向前侧方，其前半部后段急剧向前趋中，使顶骨的前端变得很窄，插入额骨内。两顶骨约组成一宽短的六边形，大体形态可能和*Ps. gansuensis*（新属、新种）者相近（见后）。鳞骨的前内角向前伸展很远，使额骨后端变得很窄。

V 16293仅保存了左颧弓前部的一部分，V 16294则仅保留了右颧弓前部。从保留的部分看，颧骨很长，伸达眼眶前缘。颧弓中部相当粗壮。颧弓前根的上后缘凹入相当深。其弧形上后缘（即眼眶的前缘）的最前点明显位于鼻骨和前颌骨后端之前。眶后收缩显著。额嵴较*Pararhizomys*者稍更明显，左、右两额嵴向后内方靠近，在眶后收缩附近会合成矢状嵴。矢状嵴很明显。颞窝很大。颞嵴（temporal crest）短，向侧面弯曲，凹入较深。颞嵴后部变得较低缓。

侧面观（图24 A3, 25 B） 头骨背缘的轮廓在眶后收缩处稍凸，其前的部分向前下方延伸。可惜V 16293和V 16294的矢状嵴均或多或少破损，使其颅部背缘的轮廓不太清楚。前颌骨前端向前伸，超

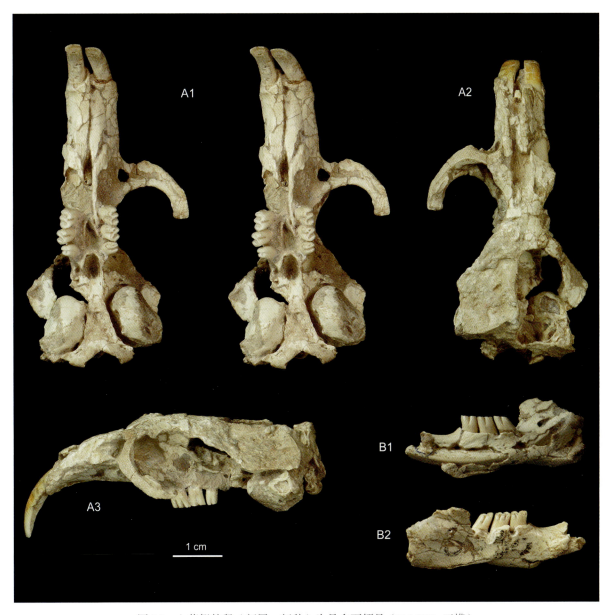

图 24 土著假竹鼠（新属、新种）头骨含下颌骨（V 16293，正模）

Fig. 24 Skull with mandible of *Pseudorhizomys indigenus* gen. et sp. nov. (V 16293, holotype)

A. 头骨（skull）：A1. 腹面立体照片（ventral view, stereopair），A2. 背面（dorsal view），A3. 左侧面（left lateral view）；

B. 右下颌骨（right hemimandible）：B1. 舌侧面（lingual view），B2. 颊侧面（buccal view）

过鼻骨的前端。上颌齿隙很长，比上颊齿列长很多（见表 11, 12）。M1 以前的头骨前部伸长，其长与自M1 前缘以后的头骨后部的长相近（见表 11）。前颌骨的侧面也有一由上门齿（I2）的齿槽形成的纵向弧形隆凸（V 16294 的左侧在眶下孔内侧可见 I2 出露，表明 I2 是由眶下孔内侧通过的）。在该隆凸上方的前颌骨侧面约呈三角形，表面稍凹。外层咬肌附着区也限于上颌骨内，向前不伸达前颌骨。其前缘（araml）不是简单的圆弧形，而约成"S"形：其上部以眶下孔腹侧裂隙的外侧缘（outer rim of ventral slit of infraorbital foramen）为界，明显向后凹入；而其下部则明显向前圆凸，向前超过眶下孔腹侧裂隙。表层咬肌起点，即咬肌结节（mt），为一圆形小坑，位于门齿孔后端外侧稍前，离眶下孔较远。在V 16294 头骨右侧上颌骨的上颧弓基部之下有一垂向窄条形的破裂部，可能与 Wahlert（1985a: figs. 2, 3）所描述的上颌骨与泪骨间的未骨化区相当。

眶部诸孔的位置与 *Pararhizomys* 者相近，但似乎靠得更紧一些。蝶腭孔（spf）相当大，位于 M1 的上方。后上齿槽孔（psaf）明显，位于蝶腭孔的前外下方，M1 前缘之上方。在后上齿槽孔的后下方，蝶腭孔的后外下方，M1 和 M2 之间的上方，有一小孔，向后通过一条沟槽进入前翼裂（aafi）的下端。这个小孔，按其位置判断，应该是 Wahlert 所说的背腭孔（dpf），应为通过颈内动脉翼腭动脉的腭部（或称眶下动脉）的腭降动脉小支的开孔（通向腭后孔）；也是 Greene（1935）所说的通过三叉神经上颌支（V2）中的后上颌齿槽神经（posterior superior alveolar nerve）的开孔。筛孔（etf）保存好，在额骨内，狭长椭圆形，与视神经孔大小相近。视神经孔（opf）位于筛孔后下方，M1 的后部上方。两者都比蝶腭孔小很多。在头骨左侧（图 25 B），在该两孔的后方见有一很大的孔，实际是一破孔，在右侧未见此孔。前翼裂（aafi）大，分不出上、下两部分，位于蝶腭孔的正后方。后两者一起形成眶窝的最深处。在前翼裂外壁的下部有一小孔，三叉神经的颧骨神经支可能由此孔通过，被 Evan 和 Christensen（1979）称为小翼孔（*foramen alare parvum*, fap）。与 *Pararhizomys* 不同，头骨上的颞深神经孔、咬肌神经孔和颊肌神经孔三孔合一，形成一裂隙。关节窝（glf）很长，向后一直伸达项嵴，听泡的外侧部也参与了关节窝的组成，并将外耳道（eam）挤至其后部下方。臼后孔（pgf）位于关节窝后部，外耳道的前上方的鳞 - 鼓骨缝处，可能与岩鼓裂相愈合。其后缘稍前于外耳道的前缘，距项嵴水平距离约为 4 mm。乳突（msp）明显。茎乳孔（stylomastoid foramen, styf）位于外耳道和乳突间的下裂凹中。项面稍向前倾斜。项面向前上方倾斜的程度比副竹鼠属者更强。听泡下缘比枕髁下缘低很多。

腹面观（图 24 A1, 25 A）　门齿孔约位于上颌齿隙的中部，其长约为上颌齿隙长的 1/4。前颌骨 - 上颌骨缝仅近中部向后延伸，穿过门齿孔的中部。其两侧的主要部分（约位于门齿孔前端两侧）呈强烈锯齿状，总体横向平伸。颧弓板宽阔，向前上方延伸，腹侧面微凹。其前后位置大体与门齿孔相同：其弓形前缘的最前点约与门齿孔前端相对，其弧形后缘的最前点约与门齿孔的后端对齐。附着表层咬肌的咬肌结节（mt）为坑状，位于门齿孔后端外侧稍前。I2 起自上颌骨中部，其后端靠近 M1，在 M1 之前形成一明显的隆凸。门齿孔的侧缘形成的纵棱向后延伸，在门齿后端形成的隆凸前分出一支斜向外后方，伸至 M1 的前外角。左、右上颊齿列向后稍分开。腭沟（ps）深，在腭沟中可见两对腭后孔（ppf），

图 25　土著假竹鼠（新属、新种）头骨（Ⅴ 16294）

Fig. 25　Skull of *Pseudorhizomys indigenus* gen. et sp. nov. (V 16294)

A. 腹面，立体照片（ventral view, stereopair），B. 左侧面，立体照片（left lateral view, stereopair），C. 前面（anterior view），D. 背面（dorsal view），E. 后面（posterior view）

缩写（Abbreviations）

头骨上的孔和其他有关结构（foramina and other structures）：aafi. 前翼裂（anterior alar fissure），ab. 听泡（auditory bulla），araml. 外层咬肌附着区前缘（anterior rim of area for *m. masseter lateralis*），ascp. 翼蝶管后孔（posterior for. of alisphenoid canal），bf. 颊肌神经孔（buccinator for.），bt. 基结节（basilar tubercles），chuf. 于吉埃氏管孔（for. of canal of Huguier），dpf. 背腭孔（dorsal palatine for.），dtf. 颞深神经孔（deep temporal for.），eam. 外耳道（external auditory meatus），eoc. 外枕嵴（external occipital crest），epp. 外翼突（external pterygoid process），etf. 筛孔（ethmoidal for.），eucf. 欧氏管孔（for. of Eustachian canal），fap. 小翼孔（*foramen alare parvum*），fm. 枕骨大孔（*for. magnum*），fo. 卵圆孔（*for. ovale*），glf. 关节窝（glenoid fossa），hyf. 舌下神经孔（hypoglossal for.），I2. 第二上门齿（second upper incisor），icf. 颈内动脉孔（internal carotid for.），inf. 门齿孔（incisive for.），iof. 眶下孔（infraorbital for.），ipp. 内翼突（internal pterygoid process），juf. 颈静脉孔（jugular for.），mf. 咬肌神经孔（masticatory for.），mlf. 中破裂孔（middle lacerate for.），mptf. 中翼窝（mesopterygoid fossa），msf. 乳突孔（mastoid for.），msp. 乳突（mastoid process），mt. 咬肌结节（masseteric tubercle），nlf. 鼻泪孔（nasolacrimal for.），occ. 枕髁（occipital condyle），opf. 视神经孔（optic for.），pcpf. 翼管后孔（posterior for. of pterygoid canal），pgf. 臼后孔（postglenoid for.），pmf. 后上颌孔（posterior maxillary for.），pmldc. 前颌骨侧后背嵴（premaxillary laterodorsal crest），pmp. 副乳突（paramastoid process），ppf. 腭后孔（posterior palatine for.），pptf. 副翼窝（parapterygoid fossa），ps. 腭沟（palatine sulcus），psaf. 后上齿槽孔（posterior superior alveolar for.），ptf. 翼窝（pterygoid fossa），spf. 蝶腭孔（sphenopalatine for.），styf. 茎乳孔（stylomastoid for.）

骨骼（bones）：As. 翼蝶骨（alisphenoid），Bo. 基枕骨（basioccipital），Bs. 基蝶骨（basisphenoid），F. 额骨（frontal），J. 颧骨（jugal），M. 上颌骨（maxilla），N. 鼻骨（nasal），Oc. 枕骨（occipital），P. 顶骨（parietal），Pl. 腭骨（palatine），Pm. 前颌骨（premaxilla），Pms. 岩乳部（petromastoid portion），Pt. 翼骨（pterygoid），Sq. 鳞骨（squamosal）

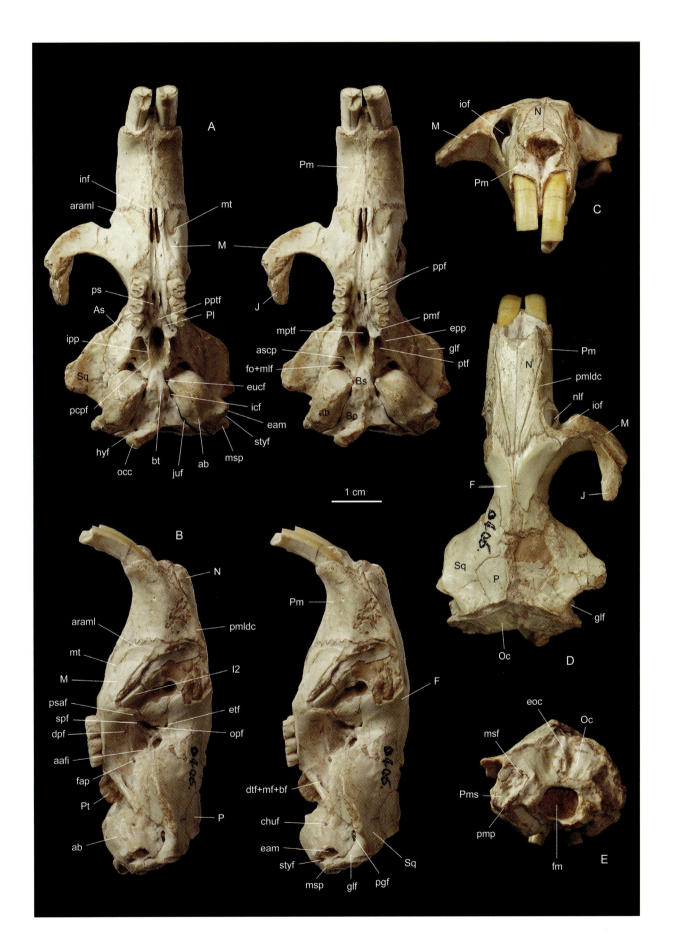

位于 M1 和 M2 之间的内侧，前面的一对较大，后面的一对很小。后上颌孔（pmf）位于 M3 后内方。副翼窝（pptf）较浅，其上仍有一系列小滋养孔。中翼窝（mptf）相对较窄，其宽度与翼窝（ptf）相近。左、右两内翼突（ipp）不倾斜，而是与腹面近于垂直、彼此近于平行地向后延伸。外翼突（epp）为较低而短的纵嵴，稍向后外方伸。硬腭后缘（即中翼窝前缘）与 M3 的后缘约在同一横线上。卵圆孔与中破裂孔合并为一个大孔（fo-mlf）。咬肌神经管和颊肌神经管的后孔（mfp 和 bfp）与翼蝶管后孔愈合，而颞深神经管后孔则开孔于卵圆孔＋中破裂孔内。翼蝶管后孔位于愈合的卵圆孔＋中破裂孔的内前下方。翼管后孔（pcpf）位于翼窝的后部内侧，靠近内翼突的基部处，与翼蝶管后孔（ascp）约在同一横线上。该孔左侧为单孔，而右侧为双孔，应该为翼腭动脉之翼支（pterygoid a.）伸向脑内的开孔。

V 16293 和 V 16294 的基枕部都保存较好，只是前者的基蝶骨和基枕骨的腹面稍破损。基蝶骨（Bs）和基枕骨（Bo）的腹面共同形成前端较窄的梯形，基蝶骨 - 基枕骨缝已完全消失。有一对很发育的基结节（bt）。两结节以中央纵沟相隔。基枕骨中央的纵棱明显，纵棱两侧为宽阔的凹面，其上未见滋养孔。舌下神经孔（hyf）大而单一，位于枕髁前外侧。V 16293 的一对枕髁保存较好，纵向短，横向宽，表面圆凸。两枕髁前端彼此不相连，中间有明显的切迹分开。听泡（ab）明显膨胀，卵圆形，长轴呈前内 - 后外方向。棘突很发育，伸向前内下方。在听泡表面，有一微弱的凸棱，自棘突向后外方伸至听泡的最高隆凸点。在该棱的外侧、听泡的前面内侧有一明显的凹面，向翼窝方向倾斜。该凹面的前内侧凹入较深，在 V 16293 还形成一纵沟。于吉埃氏管孔（chuf）在外耳道的前内方开孔。欧氏管孔（eucf）在听泡棘突前内方开孔。颈内动脉孔（icf）大，位于棘突后方。在听泡内侧，由该孔伸出一明显的稍向下后方斜的沟。颈静脉孔（juf）在较年幼的正模中较大，在 V 16294 中呈裂隙状。在听泡内侧由颈静脉孔也伸出一稍向前下方斜伸的沟，但该沟不与由颈内动脉孔伸出的沟相连。V 16293 的左侧听泡和 V 16294 的左、右侧听泡均保存较好，在颈静脉孔附近的听泡内后侧壁上均未见有供蹬骨动脉进入听泡的孔。但在 V 16294 的听泡腹面后端有一通入听泡的小孔，其作用不清。

后面观（图 25 E）　头骨项面在外形上与 *Pararhizomys* 的很相似，也为半圆形，表面也相当平。但具稍明显的呈垂向延伸的外枕嵴（eoc）。枕骨下部相对较宽。颞骨岩乳部（Pms）相对较小，其上部宽仅稍宽于枕骨大孔两侧的枕骨面。乳突孔（msf）位于颞骨的岩乳部上缘与枕骨缝的近内端处。副乳突（pmp）很小，向外下方延伸。枕骨大孔也为卵圆形，其宽大于高。左、右枕髁的后面彼此分得很开。

（2）下颌骨（图 24 B，26；表 11）

V 16293 的左、右下颌水平支保存尚好，但上升支破损严重，髁突仅在左下颌骨上保存。V 16294 的左、右下颌均保存较好，仅角突和左下颌冠状突破损。

下颌骨为松鼠型。下颌水平支短而粗壮。下缘在臼齿下方稍圆凸。下颌联合向前上方延伸，其下缘与水平支的下缘约成 40° 的夹角。颏突（cp）和其内侧的二腹肌窝（df）都很明显。下颌齿隙长长于下颊齿列的长。从侧面看，下颌齿隙也约呈前凸后凹的"S"形，但其后部凹入较 *Pararhizomys* 者浅。咬肌窝（msf）向前伸至 m1 中部的下方。咬肌嵴（mr）很发达，并向外伸张。颏孔（mtf）单一，位于 m1 前缘的下方，水平部的中部，与咬肌窝的前端约在同一水平上。

下颌上升支长。冠状突（crp）的前缘虽也起自 m2 颊侧，但冠状突向后的倾斜度较 *Pararhizomys* 的稍大，其前缘与下臼齿齿槽缘的夹角约为 120°，其上部稍向后弯，其后缘仅稍凹。冠状突内侧下面供颞肌附着的颞肌窝（tf）大而明显。下颌切迹较深，其深约为下颌高的 2/5。下颌孔（mdf）大，位于冠状突后缘的下方，与颊齿列的冠面约在同一水平上，距 m3 的纵向距离较 *Pararhizomys* 的要远（7–7.5 mm）。髁突（cdp）较冠状突低而长，也主要向后上方延伸。其上部稍向近中方弯曲。髁突内侧上部有一明显的凹面供翼外肌附着，被称为翼外肌窝（external pterygoid fovea，eptf）。下颌髁与关节窝的关节面约为半纺锤形，表面圆凸，前端较尖窄，后端较浑圆，表面稍向后外方倾斜。附着翼内肌的翼内肌窝（iptf）大而深，约成三角形，其前端伸达冠状突的下方。下门齿后端在上升支颊侧形成很显著的隆凸（i2b）。下颌角突（ap）未保存，其大小特征不清楚。

表 11　假竹鼠（新属）头骨和下颌骨测量和比较（单位：mm）

Table 11　Measurements and comparison of skulls and mandibles of *Pseudorhizomys* gen. nov. (in mm)

测量项（Parameter）*	土著假竹鼠（新属、新种）*Ps. indigenus* gen. et sp. nov.		甘肃假竹鼠（新属、新种）*Ps. gansuensis* gen. et sp. nov.		平齿假竹鼠（新属、新种）*Ps. planus* gen. et sp. nov.	原始假竹鼠（新属、新种）*Ps. pristinus* gen. et sp. nov.
	正模 Holotype	副模 Paratype	变异范围 Range	个数 N	正模 Holotype	
	V 16293	V 16294			V 16304	V 16305
1. 枕鼻长（CNL）			65.9	1		
2. 颅基长（CBL）		66.7	64–67.1	2		52.6, 54.3
3. 头骨前部长（LAS）		35	25.6–35	5		27.3
4. 头骨后部长（LPS）	31.9	33.2	30.9–35.6	4		26.6
5. 腭长（PL）		44.5	36.1–45	4		34.7
6. 上颌齿隙长（LMXD）	24.4	30.7	21–34	14	18, 18.2	23, 24.4
7. 腭桥长（LPB）		17	13–16.8	6		
8. 腭桥前宽（AWPB）	3.6	3.6	2.8–4.2	8	3.4	
9. 腭桥后宽（PWPB）	~4.7	4.9	4–6.6	7	3.8	
10. 门齿孔长（LIF）		7.3	6–9.3	5	5	
11. 吻部长（RL）		20	14–20.7	6	~11.5	17.4
12. 吻部前宽（AWR）	10.9	13	9.7–13.4	7	9.5	10.1
13. 吻部后宽（PWR）		15	10.6–15.1	8	10.6	10.3+
14. 吻部高（RH）	8.5	12	8.3–11.7	5	9.3	9.9
15. 头骨中部高（HMS）	16+	23.8	16.2–21.8	8	15.2	18.5
(15/3)/%		68	60–68	5		68
16. 鼻骨长（NL）		26.8	19.8–23.5	5	~18	
17. 鼻骨前宽（AWN）	6.6	9.8	5.3–10	5	6.2	6
18. 鼻骨后宽（PWN）		1.2	1–1.6	5		0.7
19. 眶间宽（IOW）			10.2–14.8	7	~12.1	
20. 眶后收缩处宽（WPOC）	7.6	9.2	7–12.8	7		6.6
21. 颧弓处宽（ZW）			36.2–53.6	5		39.4
22. 脑颅乳突处宽（WCM）		28	27.7–31.6	4		21.4
23. 听泡长（LAB）	12.5	12.2	10–12.1	10		12.2
24. 听泡宽（WAB）		12.9	10.3–13.3	8		8.5
25. 听泡间距（DAB）		4.2	2.6–5.7	5		4
26. 项面高（NH）		18.5	16.1–19.5	2		17.8
27. 枕骨大孔高（HFM）	5.5	5.8	6.2	1		7.7
28. 枕骨大孔宽（WFM）	7.5	7.5	6.5–8.3	3		6.4
29. 下颌骨长（MDL）		50.8**, 52.5**	40.9–49.2	3		38.7**, 39.1**
30. 下颌骨高（MDH）		27.1	23.1–28.6	4		23.6
31. 下颌髁突高（HCPM）		20.8	17.5–21.2	6		19.7, 19.5
32. 下颌切迹深（DMN）		10+	9.6–11.1	3		8.6
(32/30)/%		36.9+	36.9–38.8	3		36.4
33. 下颌齿隙长（LMD）		13.4, 13.6	11.5–16.3	8		12.3, 10
34. 下颌齿隙处高（HMD）	9.4, 9.2	11.1, 11.2	7.9–10.6	8		6.8, 8.6
35. 下颌水平支高（HHR）	10.3, 10.1	12.1, 12	10.1–13.2	8		10.1, 10.2

* 测量项同表 3（parameters of measurements as in Tab. 3）。

** 下颌骨前端—髁突后缘（anterior end of mandible–posterior border of condyloid process）。

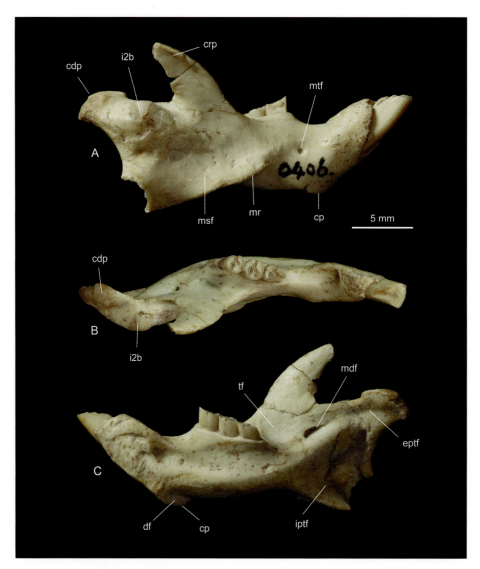

图 26　土著假竹鼠（新属、新种）右下颌骨（V 16294）

Fig. 26　Right hemimandible of *Pseudorhizomys indigenus* gen. et sp. nov. (V 16294)

A. 颊侧面（buccal view），B. 冠面（crown view），C. 舌侧面（lingual view）

缩写（**Abbreviations**）：cdp. 髁突（condyloid process），cp. 颏突（chin process），crp. 冠状突（coronoid process），df. 二腹肌窝（digastric fovea），eptf. 翼外肌窝（external pterygoid fovea），i2b. i2 后端隆突（bulge formed by i2），iptf. 翼内肌窝（internal pterygoid fovea），mdf. 下颌孔（mandibular for.），mr. 咬肌嵴（masseteric ridge），msf. 咬肌窝（masseteric fossa），mtf. 颏孔（mental for.），tf. 颞肌窝（temporal fovae）

（3）牙齿（图 27；表 12）

齿式 1·0·0·3/1·0·0·3。臼齿的基本形态与 *Pararhizomys* 的相似：脊形齿，齿冠中等高度，具齿根。臼齿的尺寸从前往后变小。臼齿齿冠比 *Pararhizomys* 者相对较低，前-后侧高冠只在 M1 和 m3 中明显存在。上臼齿的颊侧褶沟深于舌侧褶沟。而下臼齿则相反，其舌侧褶沟深于颊侧褶沟。

V 16293 和 V 16294 的上臼齿均保存较好。M1 冠面约为四边形，长大于宽，前缘稍宽于后缘，颊侧长于舌侧。齿冠为前侧高冠型，釉质曲线仅在颊侧有明显折曲。M1 的冠面具两颊侧褶沟（前边凹和中凹）和一舌侧褶沟（内凹）。两条颊侧褶沟主要呈横向延伸，而且在 V 16293 和 V 16294 均向颊侧开放。但中凹横向比前边凹长，其舌端明显向后弯曲。舌侧的内凹在 V 16293 仍向舌侧开放，但在 V 16294 已磨蚀成封闭的盆。内凹横向比两颊侧褶沟都短，呈横向向两颊侧褶沟之间的前尖方向延伸。

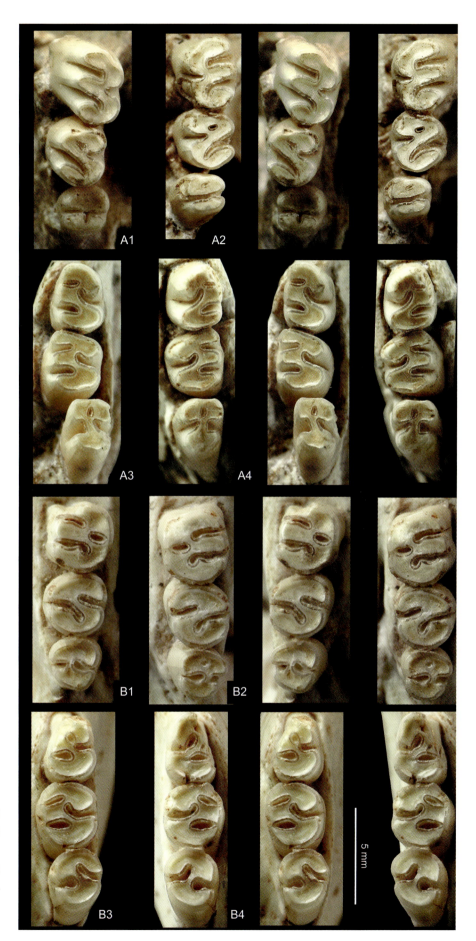

图27 土著假竹鼠（新属、
新种）臼齿冠面（立体照片）

Fig. 27 Occlusal view
(stereopair) of molars of
Pseudorhizomys indigenus gen.
et sp. nov.

A. V 16293（正模，holotype）：
A1. 右 M1–3（right M1–3），
A2. 左 M1–3（left M1–3），A3. 左
m1–3（left m1–3），A4. 右 m1–3
（right m1–3）；B. V 16294（副模，
paratype）：B1. 右 M1–3（right
M1–3），B2. 左 M1–3（left M1–
3），B3. 左 m1–3（left m1–3），
B4. 右 m1–3（right m1–3）

表 12　假竹鼠属（新属）牙齿测量和比较（单位：mm）

Table 12　Measurements and comparison of teeth of *Pseudorhizomys* gen. nov. (in mm)

测量项 (Parameter)*		土著假竹鼠（新属、新种）*Ps. indigenus* gen. et sp. nov. 正模 Holotype V 16293 Ⓛ	Ⓡ	副模 Paratype V 16294 Ⓛ	Ⓡ	甘肃假竹鼠（新属、新种）*Ps. gansuensis* gen. et sp. nov. 变异范围 Range	个数 N	平齿假竹鼠（新属、新种）*Ps. planus* gen. et sp. nov. 正模 Holotype V 16304 Ⓛ	Ⓡ	原始假竹鼠（新属、新种）*Ps. pristinus* gen. et sp. nov. 正模 Holotype V 16305 Ⓛ	Ⓡ
M1–3 L		10	9.8	9.8	9.5	8.4–10	13		9.8	8.1	8.4
M1	L	3.8	3.9	3.8	3.8	3.6–4	15	3.8	3.8	3.4	3.6
	W	3.3	3.4	3.5		3.1–3.7	15	3.4	3.5	3	3
	Hpl	2.4	2.4			2.3 2.4	5			0.9	0.9
	Hpb1	2		2.1		1.9–2.8	7			1	1
	Hpb2	2.3		2		1.9–3	7			1.3	1.3
	(Hpl/Hpb2)/%	104.3				80–100	4			69	69
M2	L	3.1	3	2.9	3.1	2.6–2.9	15	2.9	2.8	3	3
	W	3.2	3	3.1	3.1	2.8–3.5	15	3	3.1	2.7	2.7
	Hpl	2.1	2.1	1.9	1.9	1.8–2.2	5			0.9	0.9
	Hpb1									1	1
	Hpb2	1.6	1.6	1.5	1.6	1–2.2	6			0.6	0.7
	(Hpl/Hpb2)/%	131	131	127	119	95–200	6			150	129
M3	L	2.5	2.7	2.6	2.7	2.3–2.8	14		2.5	2.2	2
	W	2.7	2.6	2.4	2.6	2.1–2.8	14	2.5	2.5	2.3	2.2
	Hpl	1.8	1.8	2.1	1.8	1.2–1.9	5	2.2	1.9		
	Hpb1										
	Hpb2	1.6	1.3	1.3	1.3	0.8–1.3	8		1.5		
	(Hpl/Hpb2)/%	113	138	152	138	142–150	3		127		
I2	L	3.3	3.3	4.5	4.3	3.4–4.7	16	2.3	2.3	3.5	3.4
	W	3.5	3.5	4.5	4.3	3.5–4.5	16	2.5	2.5	3.4	3.4
m1–m3 L		11.6	11.7	10.8	10.7	9.6–11.4	8			9.4	9.4
m1	L	3.5	3.7	3.6	3.7	3.5–4	8			3.5	3.4
	W	2.9	2.8	2.9		2.7–2.9	8			2.3	2.4
	Hpl	1.8	1.7	1.7	1.7	1.6	1			1.3	1.3
	Hpb1	3									
	Hpb2	2.3	2.5								
	(Hpl/Hpb2)/%	78	68								
m2	L	3.5	3.5	3.3	3.4	3–3.5	8			3.1	3
	W	3.2	3.2	3.3	3.2	2.9–3.5	7			2.6	2.6
	Hpl	1.8	1.9	2	2	1.8	1			1.3	1.2
	Hpb1	3.1	3							1	1.3
	Hpb2	2.7	2.6							1	1.2
	(Hpl/Hpb2)/%	67	73							130	100
m3	L	3.4	3.4	3.3	3.2	2.9–3.4	7			2.6	2.6
	W		2.7	2.9	2.9	2.3–3	6			2.4	2.3
	Hpl	1.8	1.8	1.7	1.7	1.8–2.1	6			1.5	1.6
	Hpb2	1.5	1.5	1.4	1.4	1.4–1.5	4				1.5
	(Hpl/Hpb2)/%	120	120	121	121	120–129	4				107
i2	L	3.8		4.8	4.9	3.6–5.2	6			3.7	3.3
	W	3.5	3.6	4	4	3.5–4.3	7			3.5	3

* 测量项同表 4（parameters of measurements as in Tab. 4）。

其颊侧端仅伸达前边凹的舌端水平，不与前边凹重叠，只与中凹的舌端重叠。原凹在 V 16293 明显，但在 V 16294 很弱。

M2 冠面约为卵圆形，长稍大于宽。前侧高冠不明显，釉质曲线在颊侧的折曲弱或无。冠面也具两颊侧褶沟和一舌侧褶沟。在幼年的 V 16293 中，臼齿磨蚀程度较轻，其 M2 的前边凹仍存在，但已成封闭的釉质坑。而在 V 16294 的 M2，其前边凹仅留有极微弱的残余痕迹。这似乎表明此种的 M2 的前边沟已退化变小、变封闭，很容易被磨蚀掉。中凹横向很长，呈向后舌方弯伸的弧形。内凹在中凹之前，主要为横向，或稍向前颊方斜伸，与中凹部分重叠。M2 的内凹比 M1 的横向稍长些，但其颊端仍仅达前边凹舌端水平，并不超过后者，即不与后者重叠。

M3 冠面为卵圆形，具稍窄的后缘。未见前侧高冠和釉质曲线折曲现象。冠面也具两颊侧褶沟和一舌侧褶沟。在 V 16293 的左、右 M3 和 V 16294 的左 M3，前边凹都变成了封闭的釉质坑，而在 V 16294 的右 M3 则未见有此釉质坑。很可能后者的前边凹由于磨蚀较深而被磨蚀掉了。显然，M3 的前边凹可能与 M2 的一样，也退化变小、变封闭，容易经磨蚀而消失。中凹和内凹都是横向延伸，彼此相对。中凹短于内凹，其深度显然比内凹的深。M3 的冠面的结构有一些变异。如 V 16294 的中凹和内凹彼此只相对，并不相通。而在 V 16293，中凹和内凹相连，形成一条横沟，将 M3 分成前、后两部分。此外，在 V 16293 的右 M3，从该横沟的中部还有一向后延伸的短沟（见图 27A1）；而在其左 M3 未见上述短沟。

I2 的纵向弯曲度较 Pararhizomys 的少，其前端主要向前、向下延伸，不向后弯，属前伸型齿。I2 的横切面为三角形。其唇面横向稍圆凸；舌侧角较圆缓，表面有一明显的纵沟。釉质层主要覆盖在门齿的唇面，仅稍稍伸达两侧面。唇面与远中面圆缓过渡，之间无明显的界线。在唇面与近中面间有一很细的纵棱作为界线。唇面平滑，上无明显的纵棱。

m1 的冠面为后缘较宽的卵圆形，长大于宽。未见后侧高冠现象。冠面也具两颊侧褶沟（下原凹和下外凹）和一舌侧褶沟（下中凹）。V 16293 的 m1 保存较好。两颊侧褶沟均是横向延伸，并向颊侧开口。下外凹横向长于下原凹。舌侧褶沟（下中凹）是褶沟中最长者，向前弯曲。其颊部向前颊方、朝向下原凹延伸，但不伸达下原凹。V 16294 的左 m1 和右 m1 的前颊部均已破损，下原凹的情况不清。它们的下中凹较长，舌端仍向外开口。在右 m1，下中凹向前弯；而在左 m1，由于磨蚀，下中凹几乎被分成两部分：横向开口的舌部和颊侧的、向前颊方延伸的封闭的盆。

m2 较 m1 短而宽。冠面为椭圆形，长稍大于或等于宽，颊侧长于舌侧。未见后侧高冠现象。m2 的冠面形态结构与 m1 的相似：具两颊侧褶沟和一舌侧褶沟。该两颊侧褶沟均近于横向延伸，下原凹比下外凹短。下中凹是褶沟中最长者，向前弯曲；其颊部朝向下原凹延伸，但不伸达下原凹。在较少磨蚀的 V 16293，三条褶沟都向外开口。而在 V 16294，下原凹和下外凹已被磨蚀成封闭的盆；而由于下中凹的深度比颊侧的褶沟深，其舌侧仍是开口的。

m3 的冠面约呈后缘较窄的卵圆形，长大于宽，为后侧高冠齿。冠面也具两颊侧褶沟和一舌侧褶沟。下原凹在较少磨蚀的 V 16293 的 m3 中已退化成封闭的坑，而在 V 16294 的 m3 中已完全消失了。在 V 16294 下中凹横向长于下外凹，其舌端向前弯曲；在 V 16293 下中凹短，其舌端伸向下原凹，但不伸达后者。下外凹通常为横向，伸向下中凹，与下中凹的舌部约在同一横向上。

i2 纵向的弯曲度与 I2 的相近，主要向前和向上方延伸。与 I2 相似，i2 的横切面也成等边三角形，具稍圆凸的唇面和浑圆的舌侧角。但比例上较 I2 窄，其前后长明显大于横宽。釉质层主要覆盖在唇面，稍延至两侧面。唇面与外侧面彼此圆滑过渡，之间无明显的界线。唇面与内侧面间有明显的界线。在 V 16294 的 i2 的唇面具两条细的纵嵴，而在 V 16293 则只有一条明显的纵嵴。

（4）颅后骨骼（图 28-55）

V 16293 保存有颅后骨骼前半部的大部分（包括七枚颈椎和七枚胸椎，几枚肋骨，部分左、右前肢）和一枚后肢的趾节骨。V 16293 的脊椎椎骨系列基本上是原位保存，只是肢骨部分或多或少离开原有的位置，较分散（图 28-30）。因 V 16293 的骨架原保存时的部分骨骼是挤压在一起的，有的骨骼被埋在其他骨骼之间或围岩中（见图 28, 30），为了较全面地了解有关骨骼的形态特征和被埋在下面的骨骼的情况，我们将部分骨骼修理后取出，显露出其下面被埋的部分骨骼（见图 29）。

颈椎（cervical vertebrae, C's）

Pseudorhizomys indigenus gen. et sp. nov. 的颈部较短，七枚颈椎紧密排列。V 16293 的七枚颈椎背侧（从寰椎背弓前缘至第七颈椎背弓后缘）曲线总长约 30 mm，腹侧（从寰椎腹弓前缘至第七颈椎椎体后下缘）曲线总长约为 21 mm。

寰椎（atlas, At；图 28-31；表 6） V 16293 的寰椎保存较好，只是寰椎翼稍有破损。该寰椎在大小和总体形态上都和 *Pa. huaxiaensis* sp. nov. 者很接近（见前）。其外形约为椭圆的环形，宽大于高，前后很短。

背弓（da）约为前后短、横向伸长的条带形。背结节（dt）位于背弓背面正中，很粗大，主要向上隆凸，表面较粗糙。其前坡中央有一明显的窝，可能是与关节囊混合在一起的含有许多弹性纤维的寰枕背侧

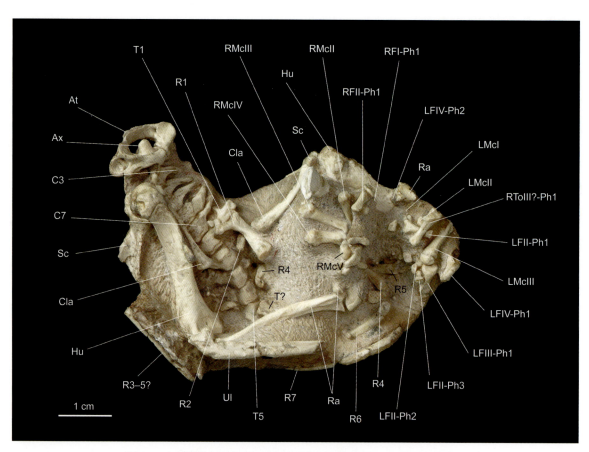

图 28 土著假竹鼠（新属、新种）部分骨骼腹面（V 16293, 正模）

Fig. 28 Ventral view of partial skeleton of *Pseudorhizomys indigenus* gen. et sp. nov. (V 16293, holotype)

缩写（Abbreviations）：At. 寰椎（atlas），Ax. 枢椎（axis），C3. 第三颈椎（3rd cervical vertebra），C7. 第七颈椎（7th cervical vertebra），Cla. 锁骨（clavicle），Hu. 肱骨（humerus），LMcI–LMcIII. 左第一至第三掌骨（left 1st–3rd metacarpals），LFII-Ph1–Ph3. 左第二指第一至第三指节骨（1st–3rd phalanges of left 2nd finger），LFIII-Ph1. 左第三指第一指节骨（1st phalanx of left 3rd finger），LFIV-Ph1–Ph2. 左第四指第一至第二指节骨（1st–2nd phalanges of left 4th finger），Ra. 桡骨（radius），R1–R7. 第一至第七肋骨（1st–7th ribs），RFI-Ph1. 右第一指第一指节骨（1st phalanx of right 1st finger），RFII-Ph1. 右第二指第一指节骨（1st phalanx of right 2nd finger），RMcII–RMcV. 右第二至第五掌骨（right 2nd–5th metacarpals），RToIII?-Ph1. 右第三趾（？）第一趾节骨（1st phalanx of right 3rd toe？），Sc. 肩胛骨（scapula），T1. 第一胸椎（1st thoracic vertebra），T5. 第五胸椎（5th thoracic vertebra），T?. 后部胸椎位置未定（uncertain posterior thoracic vertebra），Ul. 尺骨（ulna）

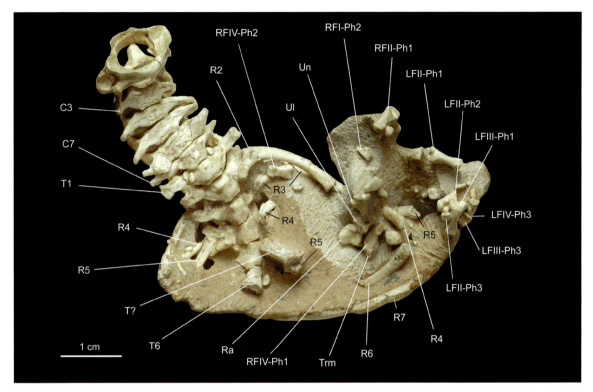

图 29　土著假竹鼠（新属、新种）部分骨骼腹面（V 16293，正模）（部分骨骼已被取出）

Fig. 29　Ventral view of partial skeleton of *Pseudorhizomys indigenus* gen. et sp. nov. (V 16293, holotype) (Some bones have been removed)

缩写（Abbreviations）：LFIII-Ph3. 左第三指第三指节骨（3rd phalanx of left 3rd finger），LFIV-Ph3. 左第四指第三指节骨（3rd phalanx of left 4th finger），RFI-Ph2. 右第一指第二指节骨（2nd phalanx of right 1st finger），RFIV-Ph1–Ph2. 右第四指第一至第二指节骨（1st–2nd phalanges of right 4th finger），T6. 第六胸椎（6th thoracic vertebra），Trm. 大多角骨（trapezium），Un. 钩骨（unciform），其他缩写同图 28（other abbreviations same as in Fig. 28）

膜（*membrana atlanto-occipitalis dorsalis*）附着处。侧椎管（lateral vertebral canal）的外开孔（侧椎孔，lateral vertebral foramen, lvf）和翼孔（*foramen alare*, fa）在背侧的开口相连，形成较大的侧椎 - 翼孔（lateral vertebro-alar foramen, lvaf），位于背弓背面两侧部近前缘 1/3 处。背弓腹面光滑，纵向平直。内侧椎孔（internal lateral vertebral foramen, ilvf）位于背弓腹面两侧，前关节凹后内上角之后。背弓侧部后面基部有一明显的横向凹槽，与后关节面突出的上缘共同形成椎动脉沟（groove for vertebral artery, gva），供椎动脉脊髓支（spinal branch）通过。沟中还有两个小孔，可能为营养孔或神经孔（？）。

腹弓（ventral arch, va）比背弓前后更短、上下更厚。腹弓背面与枢椎齿突关节的齿窝（odontoid fossa, odf）前后较长而平。腹弓腹面中央部分稍圆凸，两侧部分为宽缓的凹面。凹面的外后方有一对约呈三角形的凹槽，凹槽的外端似有小孔（功能不知）。腹结节（ventral tubercle, vt）大，明显向后突出。其左、右侧面基部各有一小孔（见图 31 B，可能为滋养孔？）。在腹结节的两侧，各有一粗糙的凹面。

椎孔（vertebral foramen, vf）约为上下稍扁的圆形，可分为上、下两部分：上部较宽，供脊髓通过；下部较窄，用以容纳齿突。在椎孔的侧壁上，在上、下部之间有较大的粗糙面，应是寰椎横韧带（transverse ligament）的附着处。

寰椎的前关节凹（cranial articular fovea, craf）约为肾形，其弧形上缘稍圆凸，与背弓前缘约在同一弧线上。前关节凹的上部较宽大，明显凹入；下部较窄，稍平。左、右前关节凹在上部分得较开，往下彼此靠近，但不相连，以切迹分开。寰椎的后关节面（caudal articular facet, caf）也是上宽下窄的肾形，表面微凹。其上部周缘都较薄锐。左、右关节面的上部分得较开，下部彼此逐渐靠近，与齿窝相连。

寰椎翼（w）的外下部分缺失，其整个外形不清楚。保留的部分显示寰椎翼为前上 - 后下方斜伸的

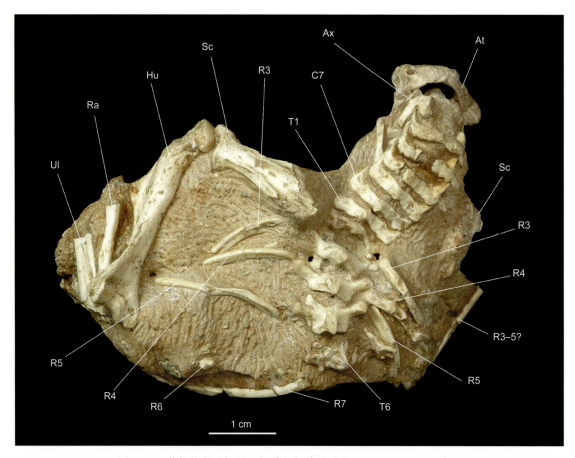

图 30　土著假竹鼠（新属、新种）部分骨骼背面（V 16293，正模）

Fig. 30　Dorsal view of partial skeleton of *Pseudorhizomys indigenus* gen. et sp. nov. (V 16293, holotype)

缩写（Abbreviations）：同图 28 和图 29（as in Figs. 28 and 29）

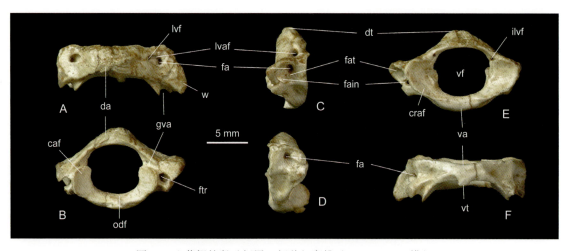

图 31　土著假竹鼠（新属、新种）寰椎（V 16293，正模）

Fig. 31　Atlas of *Pseudorhizomys indigenus* gen. et sp. nov. (V 16293, holotype)

A. 背面（dorsal view），B. 后面（posterior view），C. 右侧面（right lateral view），D. 左侧面（left lateral view），
E. 前面（anterior view），F. 腹面（ventral view）

缩写（Abbreviations）：caf. 后关节面（caudal articular facet），craf. 前关节凹（cranial articular fovea），da. 背弓（dorsal arch），dt. 背结节（dorsal tubercle），fa. 翼孔（*for. alare*），fain. 下翼孔（*for. alare inferior*），fat. 寰椎窝（*fossa atlantis*），ftr. 横突孔（*for. transversarium*），gva. 椎动脉沟（groove for vertebral artery），ilvf. 内侧椎孔（internal lateral vertebral for.），lvaf. 侧椎 - 翼孔（lateral vertebro-alar for.），lvf. 侧椎孔（lateral vertebral for.），odf. 齿窝（odontoid fossa），va. 腹弓（ventral arch），vf. 椎孔（vertebral for.），vt. 腹结节（ventral tubercle），w. 寰椎翼（wing）

弯板，外缘的上部约呈圆弧形。外缘前端稍后于前关节凹的前缘，两者相距约 1 mm。寰椎翼的后面有一横向延伸的棱，往远中侧变粗。寰椎窝（fat）大而凹入深。在寰椎窝中，翼孔（fa）的腹面开口和下翼孔（fain）都很大，两孔彼此相距较远，约 2 mm。横突管（ctr）很短，其后开孔 [= 横突孔（ftr）] 很大，位于寰椎翼后面下部，后关节面外上角的外侧，椎动脉沟的后外下方。

　　V 16293 的寰椎与 *Pa. huaxiaensis* sp. nov. 者有明显的区别：背弓背侧背结节前有明显的纵沟，背侧的侧椎 - 翼孔（lvaf）明显较大；寰椎窝较深，窝内连接翼孔和下翼孔间的沟较长，较深；腹弓后缘腹结节两侧有明显的凹面等。

　　枢椎（axis, Ax；图 28–30, 32；表 7）　V 16293 的枢椎主要保存了椎体、背弓侧部及横突部分，背弓背部和棘突完全破损。

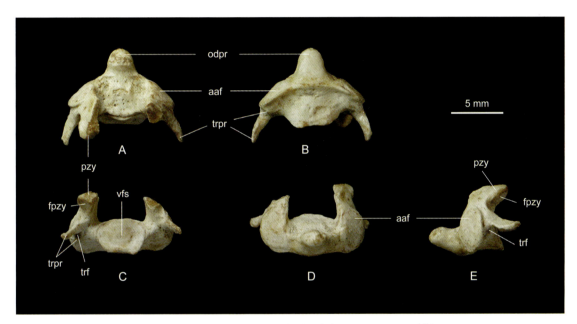

图 32　土著假竹鼠（新属、新种）枢椎（V 16293，正模）

Fig. 32　Axis of *Pseudorhizomys indigenus* gen. et sp. nov. (V 16293, holotype)

A. 背面（dorsal view），B. 腹面（ventral view），C. 后面（posterior view），D. 前面（anterior view），E. 左侧面（left lateral view）

缩写（Abbreviations）：aaf. 前关节面（anterior articular facet），fpzy. 后关节突上关节面（facet on postzygapophysis），odpr. 齿突（odontoid process），pzy. 后关节突（postzygapophysis），trf. 横突孔（transverse for.），trpr. 横突（transverse process），vfs. 椎窝（vertebral fossa）

　　枢椎明显长于寰椎。椎体为横宽的扁圆柱形。齿突（odpr）为横向稍宽而前端圆凸的扁圆锥体。齿突前端有供齿突尖韧带（apical dental ligament）附着的粗糙面。齿突背面有横向延伸的宽缓的凹槽，其横向稍圆凸，是寰椎横韧带横过的磨蚀面。齿突腹面与寰椎齿窝的关节面，横向圆凸，纵向平直。枢椎椎体背面中央部分较平直，两侧稍凹入成浅的纵沟。该纵沟的前半部表面较粗糙，可能是寰枢关节的内韧带的附着处。椎体腹面约呈扁圆形。其前缘以圆弧形棱为界，由齿突的腹缘与两前关节面明显向下突出的腹缘一起组成。椎体腹面中央前 1/3 处纵向稍凹入，横向稍圆凸，往后隆凸度增加。腹面两侧为宽的凹面。无明显的腹棘。在椎体腹面后缘有左、右对称的两个小结节。椎窝（vfs）为横宽的椭圆形，表面稍凹。椎窝与椎体间的骺线（epiphysial line）在背侧很明显，在两侧和腹缘变弱。

　　枢椎的前关节面（aaf）约为卵圆形，高大于宽。表面稍圆凸，主要面向前外方。前关节面的外侧缘和腹缘共同形成半圆弧形，向外突出；其内侧缘下端与齿突相连，只是在连接处的下方以小的切迹分开；其背缘延伸达椎弓根。

椎弓根较短。后关节突（pzy）由椎弓根后缘向后伸出。其上与第三颈椎前关节突的关节面（facet on postzygapophysis, fpzy）面向外后下方，与椎体背面的夹角约为 25°。该后关节面约为卵圆形，横轴较短，长轴呈前腹 - 后背方向，表面平直。椎弓根后缘，在后关节突与横突之间有一明显的后椎切迹（caudal vertebral notch）。

V 16293 的枢椎仅保留了部分横突（trpr）。从保留的部分看，横突由椎弓根和椎体连接处伸出。由背、腹两侧根组成：背侧根约呈窄的三角板形，向外后方伸出；腹侧根大部分断失，仅剩残根。横突孔（trf）大，从横突背、腹侧根间穿过。

第三颈椎（third cervical vertebra, C3；图 28–30, 33；表 8） V 16293 的第三颈椎保存较好，只是椎弓背部稍破损。因 V 16293 的第三颈椎至第一胸椎的诸椎体均连在一起，无法完全修开，椎体的形态未完全显露出来，只能根据前面、腹面和两侧出露的部分进行描述。从前面看，第三颈椎约为宽大于高的不规则梯形。从腹面看，第三颈椎的椎体明显短于枢椎，而长于寰椎的腹弓长。椎体约为短而宽扁的圆柱体，宽明显大于长和高。椎头和椎窝均约为宽大于高的椭圆形。椎头（caput, cap）表面较平，而椎窝表面可能稍凹。椎体腹面中央部分较平，近于水平延伸，未见有明显的腹棘；腹面两侧部分逐渐转为稍向远中方倾斜。腹面近前、后边缘处都有髁线存在。在腹面后缘也有一对左、右对称的小结节，但两者分开的距离比枢椎的稍大些。

椎弓背部呈横向伸长并向背方圆凸的条带状，前后很短（中央部分显得更短些，可能与破损有关），比椎体短很多。棘突（sp）很低弱。椎弓根（pva）较短。其前、后缘在前、后关节突之下各有一椎切迹（vertebral notch, vn）。前、后椎切迹分别与枢椎相应的后椎切迹和第四颈椎的前椎切迹形成椎间孔（intervertebral foramen, invf），供脊神经和椎动脉的脊髓支通过。前关节突（przy）从椎弓根上半部向前伸至枢椎后关节突的下方。前关节突上与枢椎后关节面关节的关节面（fprzy）约为前腹 - 后背方向伸长的卵圆形，向前内上方倾，表面平直。后关节突（pzy）在椎弓上的位置高于前关节突，向后延伸几乎不超过椎弓背部后缘。后关节突上的关节面（fpzy）也为前腹 - 后背方向伸长的卵圆形，比前关节面稍短，其表面平直，面向后外下方。后关节面的外缘与前关节面的外缘近于平行，两关节面间的最短距离约 0.8 mm。后关节面的前端稍高于前关节面的前端，而低于前关节面的后端，但其后端明显高于前关节面的后端。

横突（trpr）由两根起始：背侧根起自椎弓根，位置显然比前、后关节突低；腹侧根起自椎体的侧部。背侧根前后的长度要短于腹侧根长。横突主要向远中侧，并稍向后方伸出。横突的背、腹两根在横突孔外侧合并为扁的板状骨，并稍向后弯，其远中端逐渐变尖。横突两根间有大的横突孔（trf）穿过。横突管短。

第四颈椎（C4；图 28–30, 33；表 8） V 16293 的 C4 也保存得较好，只是椎弓背部的中央部分缺失。其基本形态结构与 C3 相似。所不同的是它的椎体要比 C3 稍短些。后关节面的前部与前关节面后部间的高差要稍大些。前、后两关节面间的最短距离加大（约为 1.1 mm）。后关节面的前、后端分别与前关节面的前、后端大约位于同一水平。横突与前、后关节突间的垂向距离稍增大。横突较粗壮，横突远中部不为扁骨板，而为较粗的棒状，横突远中端较钝。横突孔和横突管也稍大些。

第五颈椎（C5；图 28–30, 33；表 8） V 16293 的 C5 保存完好，只是椎弓背部左侧稍断开。其基本形态结构与 C4 相似。只是横突远中部稍细些，但仍比 C3 者稍粗。横突孔和横突管都较大。它的椎弓背部保存较好，可见有明显较低的棘突（sp）。该棘突顶端前部稍磨损，形态不清楚，但未见有向后伸突的现象。

第六颈椎（C6；图 28–30, 33；表 8） V 16293 的 C6 保存完好，仅两横突外端稍破损。其基本形态特征与 C4 和 C5 都很相似。但与 C5 不同的是，C6 棘突前部较低弱，往后逐渐升高（高约 1 mm），明显向后上方伸出。C6 的横突比枢椎和 C3–5 的都发达。横突腹侧根为宽大的板面，前后明显长于背侧根。横突末端分成两支。腹支被称为第六颈椎横突腹板（*lamina ventralis vertebrae cervicalis* VI, lvvc），或称横突下板（inferior lamella of transverse process），为前后伸长的板状骨，往下方弯；背支被称为第六颈椎横突背支（*lamina dorsalis vertebrae cervicalis* VI, ldvc），呈细棒状，向外前方伸出，其远端和 C3–C5 者伸达同一水平。横突孔大。

第七颈椎（C7；图 28–30，33；表 8） V 16293 的 C7 保存较好。C7 的椎弓背部与 C6 的相似，为短而横宽的条带状，棘突较发达（高约 1 mm），明显向后突出。椎弓根和前、后关节突，以及关节面的形态结构都与 C4–C6 的相似。尽管 C7 的横突的粗壮程度与 C5 的相近，但在形态结构上与 C7 之前

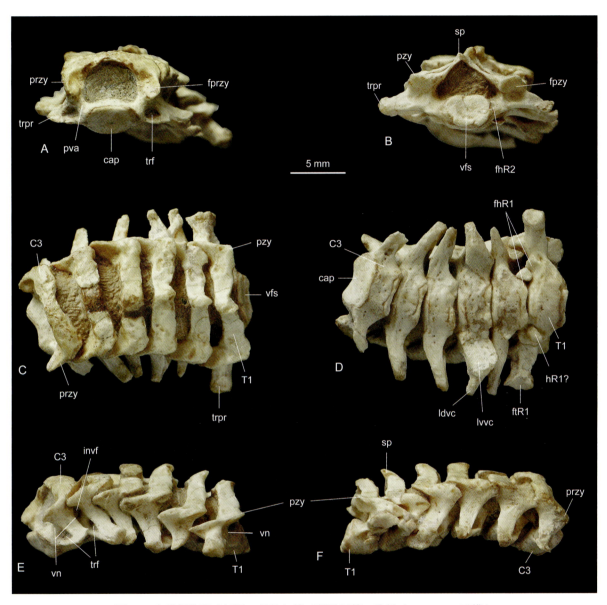

图 33 土著假竹鼠（新属、新种）第三颈椎至第一胸椎（V 16293，正模）

Fig. 33 3rd cervical vertebra–1st thoracic vertebra (C3–T1) of *Pseudorhizomys indigenus* gen. et sp. nov. (V 16293, holotype)

A. C3 前面（anterior view of C3），B. T1 后面（posterior view of T1），C. C3–T1 背面（dorsal view of C3–T1），D. C3–T1 腹面（ventral view of C3–T1），E. C3–T1 左侧面（left lateral view of C3–T1），F. C3–T1 右侧面（right lateral view of C3–T1）

缩写（Abbreviations）： cap. 椎头（caput），fhR1. 与第一肋骨头关节面（facet for head of 1st rib），fhR2. 与第二肋骨头关节面（facet for head of 2nd rib），fprzy. 前关节突上关节面（facet on prezygapophysis），fpzy. 后关节突上关节面（facet on postzygapophysis），ftR1. 与第一肋骨结节关节面（facet for tubercle of 1st rib），hR1?. 第一肋骨头的残留部分（?）（partial head？ of R1），invf. 椎间孔（intervertebral for.），ldvc. 第六颈椎横突背支（*lamina dorsalis vertebrae cervicalis* VI），lvvc. 第六颈椎横突腹板（*lamina ventralis vertebrae cervicalis* VI），przy. 前关节突（prezygapophysis），pva. 椎弓根（pedicle of vertebral arch），pzy. 后关节突（postzygapophysis），sp. 棘突（spinous process），trf. 横突孔（transverse for.），trpr. 横突（transverse process），vfs. 椎窝（vertebral fossa），vn. 椎切迹（vertebral notch）；C3. 第三颈椎（3rd cervical vertebra），T1. 第一胸椎（1st thoracic vertebra）

的所有的颈椎都有明显的区别：它的起始点较宽大，单一，不分背、腹根；横突基部无横突孔穿通；但横突基部腹面较光滑，并与较大的后椎切迹相通；横突外部不向后弯，而是转向前外方延伸；横突与前、后关节突的垂向的距离明显短于 C6 者。椎体腹面横向稍圆凸。腹面后缘也有一对左、右对称的小结节。近前、后缘处也有明显的骺线。椎体腹面后缘横向明显短于前缘，因为椎窝两侧有与小的第一肋骨头关节的关节面（facet for head of first rib，fhR1）。

综观颈椎的结构形态，除了上面各颈椎各自的特点外，从 C3 到 C7 还有一些规律性的变化：棘突由前往后逐渐变得较明显，在 C6 和 C7 开始向后延伸；横突由向后外斜伸（C5 之前）变为横向延伸（C6），至稍向前外方伸（C7），横突与前、后关节突间的垂向距离由 C3 至 C6 逐渐增大，而至 C7 又明显缩短等。

胸椎（thoracic vertebrae，T's；图 28–30, 33–34；表 13）

V 16293 保存有第一至第六胸椎和一后部胸椎。其中第一至第六胸椎均为原位保存，大多数都连在一起，未能完全修理开。

第一胸椎（T1；图 28–30, 33；表 13） V 16293 的第一胸椎保存较好，仅棘突稍破损。椎体约呈横宽的卵圆柱形，但椎体要稍长于 C7 者，而窄于颈椎椎体的宽。椎窝（vfs）表面微凹。椎窝两侧各有一约呈三角形向后外方倾斜的与第二肋骨头关节的关节面（fhR2）。椎体腹面横向稍圆凸。在椎体腹面两侧的前部各有一大的凹面，面向前下方，是与第一肋骨头的关节面（fhR1）。在椎体腹面两侧，第一与第二肋骨头关节面间有明显的凹陷区分开。在 V 16293 标本上，在左、右侧与第一肋骨头关节面的附近各保留有一小骨，这可能是第一肋骨头的残留部分（hR1?）。与 C3–C7 相似，在椎体腹面后缘也有一对对称的小结节。腹面的前、后缘附近也有骺线。

T1 的椎弓背部与 C3–C7 相似，也为短宽的横条带状。但椎弓背部中央部分向背方隆凸得较 C3–C7 者更明显些，背弓两侧向外的倾斜度要大些。棘突也比 C3–C7 的更发达些。可惜棘突顶端已破损，无

表 13　临夏盆地假竹鼠（新属）部分胸椎测量和比较（单位：mm）

Table 13　Measurements and comparison of some thoracic vertebrae of *Pseudorhizomys* gen. nov. of the Linxia Basin（in mm）

测量项（Parameter）	假竹鼠（新属）*Pseudorhizomys* gen. nov.										
	土著假竹鼠（新种）*Ps. indigenus* sp. nov.							甘肃假竹鼠（新种）*Ps. gansuensis* sp. nov.			
	V 16293（正模，Holotype）							HMV 1942（正模，Holotype）			
	第一胸椎	第二胸椎	第三胸椎	第四胸椎	第五胸椎	第六胸椎	后部胸椎	第一胸椎	第二胸椎	第三胸椎	第四胸椎
	T1	T2	T3	T4	T5	T6	T?	T1	T2	T3	T4
1. 椎体长（max L）	3.4	~3	3.5	3.7	3.8	3.7	7.2	3.2		3.6	
2. 椎头宽（caput W）		3.5								~3.7	
3. 椎头高（caput H）		3.2						3.6		3.7	
4. 前关节突外宽（prezyg. W）	11.4	11.1		7.3	6.6+			11.5		~8	
5. 后关节突外宽（postzyg. W）	10.3		7.3	6.8	6.2						
6. 前、后关节突最大长（max L of pre- & post-zyg）	5.3, 5.0			5.2				6			
7. 椎孔高（vert. for. H）	3.2	3.2									
8. 椎孔宽（vert. for. W）	5.6	4.9						5		4	
9. 横突处最大宽（transverse proc. max W）	16.2	14.2	13	12.3	11.5						
10. 椎窝宽（vertebral fossa W）	4.2					~4		4.5		4	
11. 椎窝高（vertebral fossa H）	3					2.2 (2.5)		3.7		3.5	
12. 背弓中长（dorsal sagittal L）	1.6		2.6	2.8	3.4					3	
13. 棘突高（spinous proc. H）									5.5		11

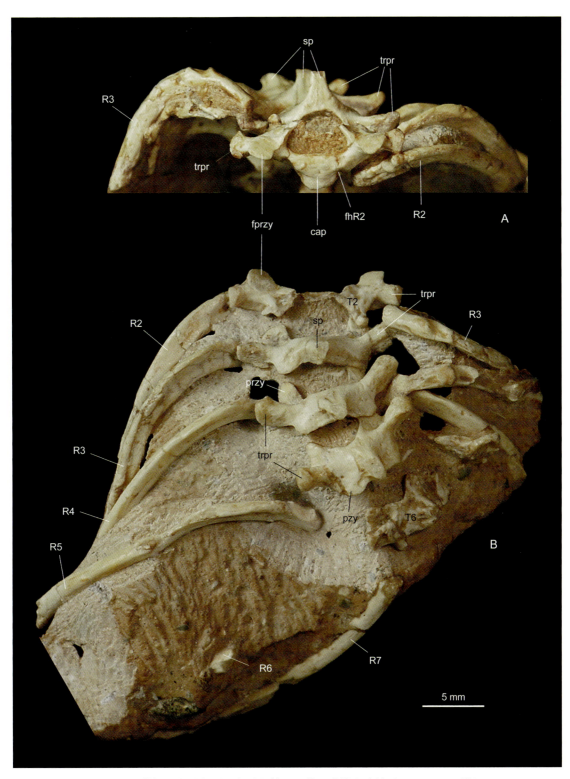

图 34　土著假竹鼠（新属、新种）第二至第六胸椎和肋骨（V 16293，正模）

Fig. 34　2nd–6th thoracic vertebrae and ribs of *Pseudorhizomys indigenus*

gen. et sp. nov. (V 16293, holotype)

A. 前面（anterior view of T2–T6）, B. 背面 (dorsal view of T2–T6)

缩写（**Abbreviations**）：cap. 第二胸椎椎头（caput of T2）, fhR2. 与第二肋骨头关节面（facet for head of 2nd rib）, fprzy. 前关节突上关节面(facet on prezygapophysis）, przy. 前关节突（prezygapophysis）, pzy. 后关节突(postzygapophysis）, R2–R7. 第二肋骨—第七肋骨（2nd rib–7th rib）, sp. 棘突（spinous process）, T2. 第二胸椎（2nd thoracic vertebra）, T6. 第六胸椎（6th thoracic vertebra）, trpr. 横突（transverse process）

法测量其高度。T1 的椎弓根较粗短。

T1 的前关节突约呈顶端向前的三角椎形,由椎弓根向前伸出。其腹侧纵棱圆缓,纵棱的内、外侧都有小的凹面。前关节突的外侧缘很薄锐。前关节面为前 - 后长大于横宽的椭圆形,表面平直,面向前上内方。后关节突从椎弓根向后,并稍向上方延伸,其后端很少超过椎弓背部后缘。从侧面看,后关节突的外侧缘也很薄,与前关节突的外侧缘几乎在同一直线上,共同形成一向前下方延伸的斜棱。此斜棱与椎体腹面约以 10° 角相交。后关节面(fpzy)为宽稍大于长的卵圆形,表面平直(左)或稍凹(右),主要面向下、稍向外方。椎孔约呈圆角三角形,宽大于高。

T1 的横突从椎体和椎弓根的两侧向外伸出,位置与前关节突约在同一水平上;其背面低于前、后关节突外侧缘形成的斜棱。T1 的横突较 C7 的发达,基部较宽大,约成四棱柱形,腹面较平。后面,在后关节突的下方,为较宽大的凹面。横突外部稍变细,约成圆柱形。横突外端钝而粗糙,表面约为稍凹的圆形,稍向下倾,与第一肋骨结节相关节(ftR1)。无横突孔。

第二胸椎(T2;图 28–30, 34;表 13) V 16293 的 T2 的椎弓背部中央和棘突,以及后关节突已破损,其余部分保存较好。椎体也约为横宽的椭圆柱形,但其宽度仅稍大于高度,而窄于 T1 者。椎体背面较平。椎头约为上宽下窄的梯形,表面稍圆凸。椎体腹面中央纵向较直、横向稍凸,腹面两侧明显向外倾斜。在腹面两侧前、后部各有一对凹面,前者为与第二肋骨头关节的关节面(fhR2),后者为与第三肋骨头关节的关节面。椎弓根较粗短。椎孔与 T1 的相似,约为圆角三角形,宽大于高。前关节突和前关节面与 T1 的形态基本相似,只是 T2 的前关节突稍短些,关节突下方纵棱的内侧无小凹,而外侧的凹面较大而深;前关节面也较短,近圆形,表面较平(左)或稍圆凸(右),主要面向前上内方。前关节突的外侧缘向后并不伸达后关节突。T2 横突的位置比 T1 的相对较高,其背缘与前关节突的侧缘约在同一水平上,将前、后关节突分开。横突横向延伸,比 T1 者粗短,为粗短柱状,往外逐渐增大。横突的外端面向外下方,与第二肋骨结节相关节。

第三胸椎(T3;图 28–30, 34;表 13) V 16293 的 T3 保存较好,仅棘突上部和横突外端破损。椎体较 T2 者更窄长,腹面两侧倾斜度更大些。腹面外侧边缘已破损,与肋骨头关节面的情况不清楚。椎弓背部也为短而宽的横条带状,但比 C7 和 T1 的都长些。棘突很发达,约呈三角锥形,向背方伸出。棘突前面较宽平,稍向后上方斜伸,表面有两条垂向的浅沟。棘突后缘基部稍宽,下有切口通达椎孔;后缘上部变窄锐,较陡直,无垂向沟。棘突两侧面主要向后外方,较平直,只是在近后缘处有极弱的垂向沟。该沟并与沿椎弓背侧后缘的一对横向的沟相连。前关节突从椎弓根伸出的位置较 T2 者低:其背侧面明显低于横突背侧缘,其腹面与横突腹面约在同一平面上。前关节突上的关节面,约呈圆形,其表面也较平,但不像 T2 那样面向前内方,而是面向正背方。后关节突的背面与横突的背侧面约在同一水平上或稍低。其上的后关节面面向内后下方。横突约成前缘较厚的板状,横向宽于 T2 者;从椎体和椎弓根伸出,向外,并稍向前方延伸。

第四胸椎(T4;图 28–30, 34;表 13) V 16293 的 C4 保存完好,仅棘突上部缺失。其基本形态结构与 T3 相似。所不同的是:T4 的椎体和椎弓背部较 T3 者稍长;棘突更粗壮,前面也有两条近于垂向的纵沟;前、后关节突的间距稍增长,两侧关节突外缘的最大外宽均较小;前关节突上的关节面表面稍圆凸,并面向外上方。横突较 T3 者粗短,明显向前斜伸,外端粗大,与第四肋骨结节的关节面面向下方。横突的位置相对更高,其背缘明显高于前关节突,也稍高于后关节突。

第五胸椎(T5;图 28–30, 34;表 13) V 16293 的 T5 保存完好,仅椎弓断裂和棘突破损。其基本结构虽与 T4 相似,但比 T4 更窄长,椎弓纵向比 T4 长许多。棘突主要位于椎弓背部后部。椎弓背部前缘和前关节突的内侧缘共同形成明显凹入的圆弧形。前、后关节突的间距更大,关节突两侧外缘的最大宽更窄。横突更粗短,其外端并稍向上翘,与第五肋骨肋骨结节的关节面为卵圆形的平面。

第六胸椎(T6;图 28–30, 34;表 13) V 16293 的 T6 只保存有部分椎体和椎弓。从保存的部分看,其椎体和椎弓的长度与 T5 相近,形态也相似。

后部胸椎(T?;图 28–29;表 13) V 16293 还保存有一枚胸椎,但该胸椎只保存了椎体部分,它不与 T6 相连,而是位于 T5 和 T6 的腹面偏左侧,相距有一段距离。该椎体长约 7.2 mm,比 T6 的长很多。因它的长度与 T6 的差距太大，它不可能是 T7，而可能是更后部的胸椎。

综观前面的六枚胸椎，它们的形态从前往后的规律性的变化是：椎体往后变窄长，椎弓往后逐渐变长；前、后关节突的间距逐渐加长，两侧关节突外侧的最大宽度往后逐渐变窄；前、后关节面的倾斜的方向均逐渐变化：前关节面的形状由长大于宽的椭圆形（T1）变短为圆形，其表面由主要面向内上方（T1 和 T2），变为面向背方（T3），后转为面向外上方（T4 和 T5）；后关节面的方向则相应地由主要向外下方（T1），变为面向腹面（T2），后变为面向内下方 (T3–T5)。横突往后变粗短。横突延伸的方向也逐渐变化：在 T1 和 T2 主要为横向外方延伸，从 T3 开始逐渐转为向前外方斜伸，而从 T5 开始其外端并向背方翘。横突与前、后关节突的相对位置逐渐变高：在 T1 横突的位置低于前、后关节突；在 T2 横突背缘变为与前关节突的侧缘约在同一水平上；在 T3 横突背缘高于前关节突的侧缘，而与后关节突的背缘约在同一水平上；至 T4 和 T5 横突背侧缘明显高于前、后关节突。棘突由前往后逐渐变长大；从近于垂直，变为向后倾，其倾斜度由弱（T3），往后逐渐增大；棘突在椎弓上的位置在 T1–T3 位于背部正中央，从 T4 到 T6 则逐渐移向后部等。

肋骨（ribs, R's；图 28–30, 35）

V 16293 保存有前面六对肋骨和左侧第七肋骨。其中第一对肋骨，左侧的第二至第七肋骨基本上是原位保存；而右侧的肋骨的位置多少有些错动，有的并挤压在一起。

第一肋骨（first rib, R1；图 28, 35） V 16293 的左、右 R1 都保存得相当完好。肋骨体（shaft of rib）为长且稍向内弯曲的扁骨。外侧面较宽，横向和纵向均圆凸，上有弱的肋骨沟（costal groove）。内侧面横向较平，纵向明显凹入。骨体上部较细，向下稍变粗大，并稍向后外方扭曲。末端为与肋软骨（costal cartilage）相连接的椭圆形粗糙面。肋骨的椎骨端较细长，肋角（angle of rib）较明显。肋骨颈（costal neck, cn）长而直，将肋骨小头（costal head, ch）和肋骨结节（tubercle of rib, tur）分得较开。肋骨颈的背侧有一条明显的棱，从肋骨小头斜伸到肋骨结节的外侧。该棱的前、后都有附着肌肉的粗糙面。肋骨小头大于肋骨结节。肋骨小头顶端有两个面，彼此成一定的角度相交，可惜其上与椎骨的关节面均已破损 [可能脱落在 C7 和 T1 的与第一肋骨头的关节面附近（见上述 T1 和图 29, 33: hR1?）]。肋骨结节顶端内侧有一卵圆形的球面，为与 T1 横突的关节面。

V 16293 的左、右第一肋骨最大直线长约 11.3 mm 和 11.1 mm；肋骨椎骨端（= 近端）长（从肋骨小头到肋骨结节后缘的距离）：6.2 mm，6.6 mm；肋骨骨体直线长（从肋骨结节至远端的直线距离）：

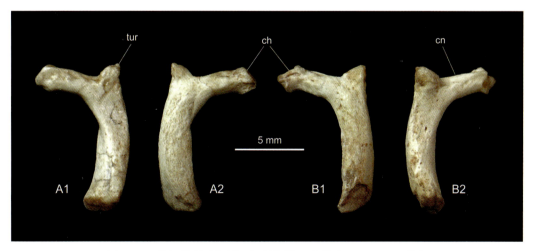

图 35 土著假竹鼠（新属、新种）第一肋骨（V 16293，正模）

Fig. 35 First rib (R1) of *Pseudorhizomys indigenus* gen. et sp. nov. (V 16293, holotype)

A. 右 R1 (right R1)：A1. 内侧面（medial view），A2. 外侧面（lateral view）；B. 左 R1 (left R1)：B1. 外侧面（lateral view），B2. 内侧面（medial view）

缩写（Abbreviations）：ch. 肋骨小头（costal head），cn. 肋骨颈（costal neck），tur. 肋骨结节（tubercle of rib）

10.5 mm，10.4 mm；肋骨后面骨体近端宽 × 厚：2.6 mm × 1.5 mm，2.6 mm × 1.5 mm；远端横宽：1.9 mm，1.8 mm；远端前后长：3.5 mm，3.5 mm。

Pseudorhizomys 的第一肋骨比鼢鼠的要窄长得多，而骨体的弯曲度则比竹鼠者要小，相对较直。

第二肋骨（second rib, R2；图 28–29, 34） V 16293 仅保存了左侧的 R2。该肋骨保存得相当完好。R2 比 R1 长而细扁。肋骨体为长而弯的扁骨，其弯曲度明显较 R1 大。肋骨体上部前后很扁，往下逐渐加厚成扁圆柱形。骨体远端为与肋软骨连接的卵圆形粗糙面。肋骨颈也较长而直，但比 R1 者短。肋骨小头稍大于肋骨结节。肋骨小头上的关节面保存不好。肋骨结节内侧有一卵圆形的关节面，表面呈鞍形，前后向圆隆，横向稍凹。

R2 的最大直线长 16.3 mm；肋骨椎骨端长 5.8 mm；肋骨骨体直线长 16.4 mm；骨体近端（肋骨结节后）宽 × 厚 2.6 mm × 0.9 mm；远端长径 2.6 mm；横径 1.8 mm。

第三肋骨和第四肋骨（R3–R4；图 28–30, 34） V 16293 的 R3 和 R4 的肋骨小头和远端均断缺。其骨体的形态与 R2 的相似，但比 R2 长很多。骨体后面的肋沟虽然很浅，但明显较 R2 者长，可伸达骨体的下部。肋骨结节的高度较 R1 和 R2 的低，仅稍高于骨体，其后面以明显的切迹与骨体分开。肋骨结节与横突的关节面为卵圆形，表面稍圆凸。

R3 和 R4 的保存部分的最大直线长（两端缺失）分别为 19 mm 和 25 mm。骨体近端（在肋骨结节附近）的宽 × 厚，在 R3 为 2.6 mm × 0.7 mm，在 R4 为 2.5 mm × 1.1 mm。

第五—第七肋骨（R5–R7；图 28–30, 34） V 16293 的左、右 R5 仅远端缺失。左 R5 保存得较右侧者稍好。R5 骨体的形态与 R2–R4 的相似，但比 R1–R4 的骨体都长。R5 的肋骨小头上有两个关节面分别与 T4 和 T5 椎体相关节。两关节面彼此呈钝角相交，之间有弱沟相隔。左 R5 保存部分的最大直线长（两端缺失）为 25.5 mm。

R6 只保存了左肋骨下部一小部分，保存部分长为 12.3 mm。R7 也仅保存有左 R7。其两端均缺失，保存部分比 R6 及以前的肋骨都长，骨体也较细些。保存部分最大直线长为 26 mm。

前肢（forelimb）

V 16293 保存的前肢有：肩带（包括锁骨和肩胛骨）、肱骨、桡骨、尺骨和部分前足骨。

锁骨（clavicle, Cla；图 28, 36） V 16293 保存有部分左、右锁骨，两者的肩峰端（acromial end）均缺失。左锁骨保存得较右锁骨好，保存有锁骨的大部分。锁骨为前、后面稍宽的片状长骨，细长，稍扭曲。

图 36　土著假竹鼠（新属、新种）锁骨（V 16293, 正模）

Fig. 36　Clavicles of *Pseudorhizomys indigenus* gen. et sp. nov. (V 16293, holotype)

A. 右锁骨（right clavicle）：A1. 后面（posterior view），A2. 前面（anterior view）；B. 左锁骨（left clavicle）：B1. 前面（anterior view），B2. 后面（posterior view）

胸骨端（sternal end）粗大。胸骨关节面近圆形，表面粗糙。骨体往肩峰端方向逐渐变细，但肩峰端的情况不明。

　　锁骨保存的长度：20 mm（左），15.6 mm（右），胸骨端（长 × 宽）：3.4 mm × 3.3 mm（左），3.4 mm × 3.1 mm（右）。

　　肩胛骨（scapula, Sc；图 28, 30, 37）　V 16293 保留了左、右肩胛骨的下半部分。肩臼（glenoid cavity, glc）为长卵圆形，其长径明显大于肩胛颈（neck of scapula, nsc）之宽，未见明显的肩臼切迹。肩臼长径的方向与肩胛冈（spine of scapula, spsc）的方向斜交。肩胛结节（scapular tuber, sctu）粗厚，自肩臼前缘向前下方伸出。喙突（coracoid process, cpr）很发达，向前下内方伸得很长，与肩胛结节间以切迹分开（肩胛结节和喙突与 Rhizomys 的很相似），主要供喙臂肌、臂二头肌和胸深肌等肌肉附着。肩胛冈（spsc）只保存了基部。其下端起始点接近肩胛颈，距肩臼外缘约 5 mm。肩峰（acromion）已完全破损。肩胛骨前缘大部分破损。冈上窝（supraspinous fossa, sspf）保存的部分约为窄的扇形，其下部很窄，宽仅为 1.5 mm。冈下窝（infraspinous fossa, ispf）保存得较多，也为狭窄的扇形。其下部的宽明显大于冈上窝者，往上逐渐变宽。肩胛骨后缘下部纵向凹入；在肩胛颈处横向变宽、变圆，往上渐变为脊形。后缘附近有几条供冈下肌和小圆肌附着的肌线（linea muscularis, lm）。在冈下窝的后部有一很发达的肌线，与后缘近于平行，向下几乎伸达肩胛窝的后缘。冈下窝在该肌线与肩胛冈之间的部分横向明显凹入，表面光滑。而在该肌线之后的部分较平或稍凹，向后方倾斜，其上部还有一细的肌线，与该粗的肌线平行，向上与后缘脊汇合。肩胛下窝（subscapular fossa, sscf）为窄长的扇形，下窄上宽。肩胛下窝的表面有一圆凸的纵棱与冈下窝的凹面相对应。在肩胛下窝的近后缘处也有一发达的肌线，往上逐渐与后缘的脊汇合，往下伸至肩胛颈。该肌线的下部与后缘之间的部分明显凹入。但在其下方，

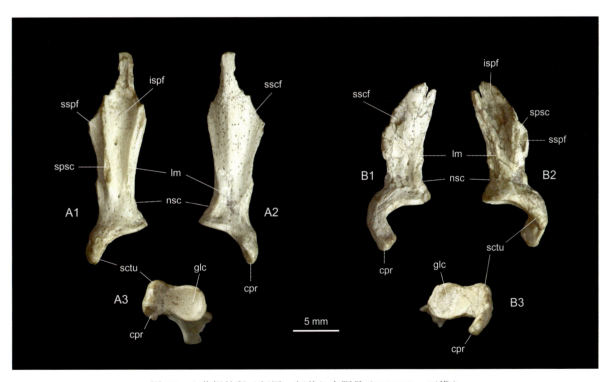

图 37　土著假竹鼠（新属、新种）肩胛骨（V 16293，正模）

Fig. 37　Scapulae of *Pseudorhizomys indigenus* gen. et sp. nov.（V 16293, holotype）

A. 部分左肩胛骨（part of left scapula）：A1. 外侧面（lateral view），A2. 内侧面（medial view），A3. 远端面（distal view）；B. 部分右肩胛骨：B1. 内侧面（medial view），B2. 外侧面（lateral view），B3. 远端面（distal view）

缩写（Abbreviations）：cpr. 喙突（coracoid process），glc. 肩臼（glenoid cavity），ispf. 冈下窝（infraspinous fossa），lm. 肌线（*linea muscularis*），nsc. 肩胛颈（neck of scapula），sctu. 肩胛结节（scapular tuber），spsc. 肩胛冈（spine of scapula），sscf. 肩胛下窝（subscapular fossa），sspf. 冈上窝（supraspinous fossa）

即肩胛颈的后面则为较粗糙隆起。该肌线和粗隆可能供臂三头肌长头和小圆肌附着。

肩胛颈最小宽：4 mm（左），4 mm（右）；肩臼（径长 × 宽）：7 mm × 3.6 mm（左），7 mm × 3.8 mm（右）。

肱骨（humerus, Hu；图 28, 30, 38；表 14）　V 16293 的左、右肱骨都保存较完整，只是左肱骨头稍破损。V 16293 为一较幼年的个体，肱骨近端可见有明显的骺线的痕迹。

肱骨的形态结构与竹鼠和山河狸的都较相似。近端宽厚，宽稍大于前后厚。肱骨头（head of humerus, head）向后上方延伸，悬于骨体的后上方。与肩臼的关节面为卵圆的半球面，明显大于肩臼。肱骨颈（neck of humerus, nhu）在肱骨头的后下方较明显。大结节（greater tuberosity, gtu）位于肱骨头的前外侧，较粗大，但不明显向上隆凸，其顶端稍低于肱骨头关节面的顶点。大结节顶端的凹面及其附近的粗糙面供冈上肌附着，其外下方的隆凸粗糙面供臂三头肌外头（lateral head of *m. triceps*

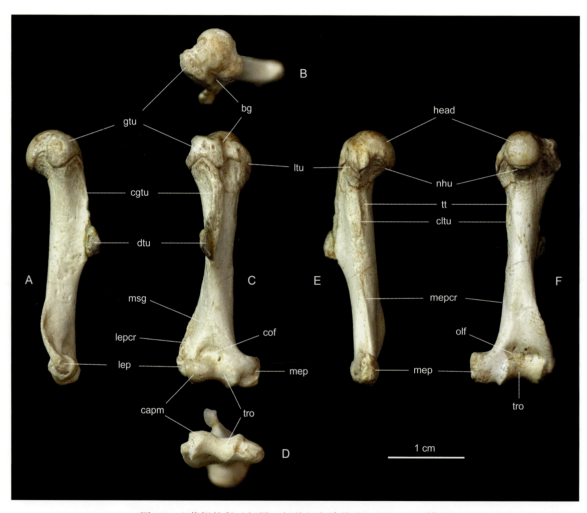

图 38　土著假竹鼠（新属、新种）右肱骨（V 16293，正模）

Fig. 38　Right humerus of *Pseudorhizomys indigenus* gen. et sp. nov.　（V 16293, holotype）

A. 外侧面（leteral view），B. 近端面 (proximal view)，C. 前面（anterior view），D. 远端面（distal view），E. 侧面（medial view），F. 后面（posterior view）

缩写（Abbreviations）： bg. 臂二头肌沟（bicipital groove），capm. 肱骨小头（*capitulum* of humerus），cgtu. 大结节嵴（crest of greater tuberosity），cltu. 小结节嵴（crest of lesser tuberosity），cof. 冠状窝（coronoid fossa），dtu. 三角肌粗隆（deltoid tuberosity），gtu. 大结节（greater tuberosity），head. 肱骨头（head of humerus），lep. 外上髁（lateral epicondyle），lepcr. 外上髁嵴（lateral epicondylar crest），ltu. 小结节（lesser tuberosity），mep. 内上髁（medial epicondyle），mepcr. 内上髁嵴（medial epicondylar crest），msg. 臂肌沟（musculo-spiral groove），nhu. 肱骨颈（neck of humerus），olf. 肘窝（olecranon fossa），tro. 滑车（trochlea），tt. 圆肌隆起（teres tuberosity）

测量项（Parameter）	假竹鼠（新属）Pseudorhizomys gen. nov.			华夏副竹鼠（新种）Pa. huaxiaensis sp. nov.	银星竹鼠 Rhizomys pruinosus		平原囊鼠 Geomys bursarius		阿尔泰鼢鼠 Myospalax myospalax		包氏囊鼠 Thomomys bottae	
	土著假竹鼠（新种）Ps. indigenus sp. nov.		甘肃假竹鼠（新种）Ps. gansuensis sp. nov.									
	正模（Holotype）											
	V 16293		HMV 1942	HMV 1413	OV 1055		OV 432		OV 478		OV 446	
	Ⓛ	Ⓡ	Ⓡ	Ⓡ	Ⓛ	Ⓡ	Ⓛ	Ⓡ	Ⓛ	Ⓡ	Ⓛ	Ⓡ
肱骨全长（total L）	32.5	32.6			42	46	24.8	24.7	27.4	27.5	21	20.9
肱骨近端最大宽 (max W of PE)		8.5	10	9.9	11.3	13	8.1	8.1	8.4	8.2	5.1	4.9
肱骨近端径长 (max APD of PE)		7.3	6.8	7.4+	9	9.2	6.2	6.1	6.5	6.3	4.7	4.4
肱骨头宽 (caput W)		5.5	5.3	6	6	6	4.2	4.3	4.7	4.8	3.2	3.2
肱骨头径长 (caput APD)	6.4	6.5	7	7	8.5	8.5	5.6	5.5	6.1	6.2	4.2	4
骨体中部最小宽 (shaft min W)	3.7	3.6			4.2	4.4	3	3	5.2	5.6	1.8	1.6
骨体中部最小宽处厚 (shaft APD at min W)	4.7	4.8			5.4	5.3	3.6	3.3	3.2	3.4	1.8	1.8
肱骨上部长（L of upper part, UL)*	16.7	16.7			23	24	14.4	14.1	17.9	17.8	11.4	11
(UL/L)/ %	51.4	51.2			54.8	52.2	58.1	57.1	65.3	64.7	54.3	52.6
远端最大宽 (max W of DE, MWD)	11.2	11.2			14.3	14.3	10.8	10.8	13.8	14.4	6.4?	11.6
(MWD/L)/%	34.5	34.4			34	31.1	43.5	43.7	50.4	52.4	55.2?	55.5
远端最大径长 (max APD of DE, MAPDD)	3.8	3.9			5.3	5.3	4.1	4.1	4.4	4.3	2.5	2.5
远端滑车最大宽 (max W at trochea, MWT)	7.1	7.1			9	9.1	6	6	7.8	8	4.5	4.2
(MWT/L)/%	21.8	21.8			21.4	19.8	24.2	24.3	28.5	29	21.4	20.1
肱骨内上髁长 (L of medial epicondyle, LME)	4.6	4.5			5.2	5.1	4 (3.4)	4.1 (3.4)	4.1	4	2.6	2.7
(LME/L)/%	14.2	13.8			12.4	11.1	16.1 (13.7)	16.6 (13.8)	15	14.5	12.4	12.9
肱骨内上髁宽 (W of medial epicondyle, WME)	4.4	4.4			5.1	5	4.2	4.4	5	4.9	2.7	2.5
(WME/L)/%	13.5	13.5			12.1	10.9	16.9	17.8	18.2	17.8	12.9	12
肱骨小头宽（W of capitulum）	3.8	3.8			5	5	4	3.9	3.8	4.1	2.4	2.5
滑车内髁径长（med. condyle APD)	3.8	3.9			5.3	5.2	3	3.1	3.2	3.2	2.3	2.4
滑车外髁径长（lat. condyle APD)	3.8	3.9			5.3	5.1	3.1	3.2	4.2	4.1	2.1	2.1

* 肱骨上部长（UL）：肱骨头顶至三角肌隆起下端距离（upper border of caput-distal end of deltoid tuberosity)。

brachii）附着；大结节的外侧后面也有一明显的凹区，供冈下肌附着，其下方的小隆凸可能为小圆肌附着处。小结节（lesser tuberosity, ltu）位于肱骨头的前内侧，比大结节低小，为肩胛下肌的附着处。臂二头肌沟（bicipital groove, bg）[现多称之为结节间沟（intertubercular sulcus）]，明显，无中纵棱分隔，供臂二头肌长头肌腱经过。

肱骨骨体约为一纵向稍扭曲的、不规则的三角柱形。肱骨体的前面上宽下窄。其上部面向前内方倾，纵向较平，横向凹入，二头肌沟沿此向下一直延伸到三角肌粗隆（deltoid tuberosity, dtu）附近。前面的下部，在三角肌粗隆之下，以微弱的肱骨嵴（crest of humerus, chu）为界，分为前内和前外两部分。大结节嵴（crest of greater tuberosity, cgtu）很发达，由大结节前缘向下延伸，直至三角肌粗隆（dtu）。大结节嵴前面表面粗糙，供胸浅肌和部分胸深肌附着。三角肌粗隆（dtu）很发达，在肱骨大结节嵴的下端、骨体的中上部，呈三角形板状向外突出，供三角肌 [m. deltoideus; Greene（1935）称之为肩峰三角肌，m. acromiodeltoides] 附着。三角肌粗隆在骨体上的位置与山河狸的相近，而比竹鼠的稍高（在竹鼠类中约位于骨体正中部）。三角肌粗隆之下的外方形成宽阔的臂肌沟（musculo-spiral groove, msg），供臂肌附着和通过。臂肌沟外侧以发达的外上髁嵴（lateral epicondylar crest, lepcr）为界。该嵴长约占骨体的下 1/3。嵴的背侧面有两个较明显的凹陷面，供腕桡侧伸肌和指总伸肌等附着。肱骨后面无明显的扭转现象，纵向较平直，仅在靠近肱骨头附近稍向后弯。肱骨骨体后面上部较窄，横向圆凸，往下逐渐变宽，横向稍凹。肱骨骨体前面和后面在内缘的界线较明显。骨体内面上部为不太明显的小结节嵴（crest of lesser tuberosity, cltu），向下直伸，约达骨体上 1/3 处。该嵴前面较粗糙，可能是背阔肌的附着处。该嵴下端的后侧为粗糙的凹面，其前、后均有嵴，可能是圆肌隆起（teres tuberosity, tt），供大圆肌附着。圆肌隆起前下方的嵴线（粗糙面）则可能为喙臂肌（m. coraco-brachialis）的附着处。再往下则为内上髁嵴（medial epicondylar crest, mepcr），为其前的肱肌和其后的肱三头肌的内头的分隔界线。内上髁嵴下端前侧与内上髁上方之间形成一明显的凹面。

远端较近端宽扁，其横径与肱骨近端的横径近于平行，与肱骨头关节面的前后长轴的水平投影线的夹角小于直角。远端关节面前面外侧的外髁（lateral condyle），也被称为肱骨小头（capitulum of humerus, capm），约呈半纺锤形。远端关节面的内髁（medial condyle），也被称为肱骨滑车 [trochlea (tro) of humerus]。该滑车除了外侧一小部分与桡骨近端关节面的内缘关节外，大部分与尺骨的半月切迹（semilunar notch）相关节。滑车稍斜伸、不对称，纵向轴状圆凸，中部横向明显凹入。肱骨滑车在前面的部分较窄，其宽度稍小于肱骨小头的宽度；向下逐渐加宽，转向后方时则占据了整个远端关节面。肱骨小头和滑车之间仅在远端的下后方以弱棱为界，在前面无明显的界线。冠状窝（coronoid fossa, cof）约为圆三角形，其宽度约为远端关节面宽度之半。窝内表面粗糙，但未见有沟、棱的痕迹，只是内侧有一小孔，可能为供神经、血管通过的滋养孔。肘窝（olecranon fossa, olf）较宽而深，宽于冠状窝，但仅稍宽于滑车后部宽。冠状窝和肘窝的深度相近，彼此不穿通。内上髁（medial epicondyle, mep）很发达，向内远伸。其上有三个明显的凹面：上前者最小，可能供旋前圆肌附着；上后者较大，可能供腕桡侧屈肌附着；下者最大，可能供指深屈肌、指浅屈肌和腕尺侧屈肌等附着。内上髁后面与滑车间有一很宽的沟，可能是尺骨神经沟（groove for ular nerve）。内上髁上未见内上髁孔。外上髁（lateral epicondyle, lep）比内上髁小很多，也比滑车的直径小很多。其外侧面上也有三个小凹面：外上者可能供指侧伸肌附着；前内者可能供腕尺侧伸肌附着；下后方者可能供旋后肌附着。

测量见表 14。

桡骨（radius, Ra；图 28, 30, 39；表 15） V 16293 保存有左、右桡骨。右桡骨保存较好，只是其远端骨骺与骨体在原标本中是完全分开保存的，现经修理后粘合在一起。左桡骨只保存了上半部和远端骨骺。

桡骨细长，比肱骨细小得多，也短于肱骨。近、远两端均较宽厚。桡骨近端桡骨头（head of radius）横宽大于前后径。近端与肱骨的关节面为横宽的卵圆形。矢状嵴（sagittal crest, sacr）为较圆缓的、纵向稍凹的弧形纵棱，位于关节面靠近内缘处。在矢状嵴外侧，与肱骨小头的关节面（facet for capitulum of humerus, fcapm）大，占据近端关节面的大部分，表面凹入。矢状嵴内侧与肱骨滑车的关节面（facet for trochlea of humerus, ftro）横向很窄，面向内上方，仅与肱骨滑车外部相关节。矢状嵴前端的小头隆凸（capitular eminence, capem），亦即桡骨的冠状突（coronoid process），自前面微弱可辨。桡骨近端后面与尺骨的关节面（facet for ulna, ful）为横宽的长条形，几乎占据了桡骨头的整个后面，表面纵向平直，横向稍圆凸。桡骨近端的外隆起较内隆起显著。桡骨颈（neck of radius, nra）很明显。

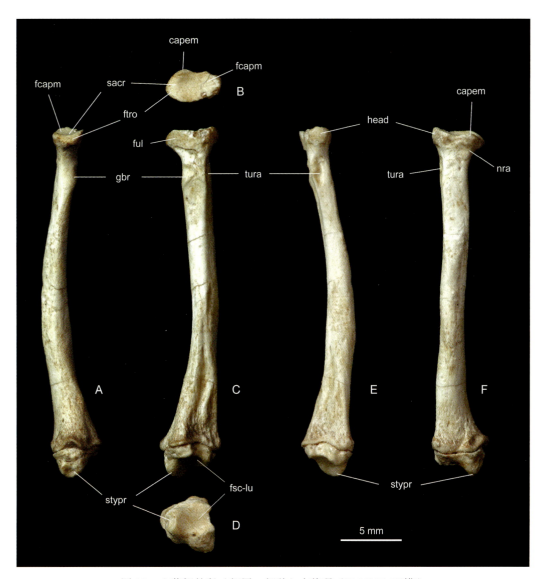

图 39　土著假竹鼠（新属、新种）右桡骨（V 16293, 正模）

Fig. 39　Right radius of *Pseudorhizomys indigenus* gen. et sp. nov. (V 16293, holotype)

A. 内侧面（medial view）, B. 近端（proximal view）, C. 后面（posterior view）, D. 远端（distal view）, E. 外侧面（lateral view）, F. 背面（dorsal view）

缩写（**Abbreviations**）：capem. 小头隆突（capitular eminence）, fcapm. 与肱骨小头的关节面（facet for *capitulum of humerus*）, fsc-lu. 与腕舟 - 月骨关节面（facet for scapho-lunar）, ftro. 与肱骨滑车关节面（facet for trochlea of humerus）, ful. 与尺骨关节面（facet for ulna）, gbr. 臂肌沟 (groove for m. *brachialis*）, head. 桡骨头（head of radius）, nra. 桡骨颈（neck of radius）, sacr. 矢状嵴（sagittal crest）, stypr. 桡骨茎突（styloid process of radius）, tura. 桡骨粗隆（tuberosity of radius）

　　桡骨骨体呈稍向前凸的扁圆柱形，横宽大于前后径。上部较窄，从中部往下逐渐变宽大。骨体前面纵向稍圆凸，横向在上部较平、下部较圆凸。骨体前面中部近内缘处有一较粗糙的凹陷区，可能是旋前圆肌和旋后肌止端附着处。在骨体前面中部近外侧缘有一粗糙隆凸，斜向内下方延伸，可能是拇短伸肌（ *m. extensor pollicis brevis* = 拇展长肌, *m. abductor pollicis longus*）和拇展肌（ *m. abductor pollicis*）等肌肉的附着处。骨体内面上部较窄，往下变宽；纵向上部较直，下部稍向后弯；横向圆凸。在内面上部靠近桡骨颈处有一光滑面，可能是臂肌终端腱经过处（该腱止于尺骨冠状突下方）。笔者称其为桡骨的臂肌沟（groove for *m. brachialis*, gbr）。骨体外面纵向较平直；上部较窄，横向稍凹；往下逐渐变宽，下部并稍转向后外方；表面具较发达的沟、棱，供桡骨和尺骨的骨间韧带（ *lig.*

表 15　临夏盆地副竹鼠类桡骨测量和比较（单位：mm）
Table 15　Measurements and comparison of radii of pararhizomyines of the Linxia Basin (in mm)

测量项（Parameter）	假竹鼠（新属）Pseudorhizomys gen. nov.			三趾马层副竹鼠 Pararhizomys hipparionum
	土著假竹鼠（新种）Ps. indigenus sp. nov.		甘肃假竹鼠（新种）Ps. gansuensis sp. nov.	
	正模（Holotype）			
	V 16293		HMV 1942	V 16286.2
	Ⓛ	Ⓡ	Ⓡ	Ⓛ
全长（total L）		28		
近端最大宽（max W of PE）	4.6	4.6		
近端径长（max APD of PE）	2.8	2.7		
近端关节面宽（PAF W）	4.5	4.3		
近端关节面前后径（PAF APD）	2.8	2.7		
骨体中部最小宽（min W at mid-shaft）	~2	2.2		2.7
骨体中部最小宽处厚（APD at min W at mid-shaft）	2.2	2		2.5
远端最大宽（max W of DE）	4.2+	5	~4.9	4.7
远端前 - 后最大径长（max APD of DE）	3.6+	4.5		4.5

interosseum antibrachii）或骨间膜附着。桡骨粗隆（tuberosity of radius, tura），又称臂二头肌结节（bicipital tuberosity）为一条形凹坑，位于外缘上部桡骨颈下方，供臂二头肌止端附着。骨体的后面纵向凹入，并稍扭转。其上部面向后方，横向凹入，成一较宽的纵沟。在后面上部与尺骨关节面的下方和背肌沟的外侧有一粗糙区可能供指深屈肌桡骨头的附着。后面下部稍向内转，与内面的下部相汇合。

远端比近端粗大，为不规则的四边形。远端与腕舟 - 月骨关节面（facet for scapho-lunar, fsc-lu）宽大，外侧的前 - 后长稍大于内侧者。关节面表面的外部较圆凹，内侧部大部分较平或纵向稍圆凸，面向下外方倾斜。远端内侧的桡骨茎突（styloid process, stypr）向下方伸出，其内侧面稍破损，但其下面有一明显的纵沟，可能供指深屈肌腱通过。外侧面较宽，横向稍凹，表面粗糙，与尺骨远端关节。远端前面正中也有一纵沟，可能供指总伸肌腱通过。

尺骨（ulna, Ul；图 28, 30, 40）　V 16293 的右尺骨保存较好，但肘突上部破损，其远端部分与骨体分离。左尺骨仅保存了上半部，肘突部分也破损。

尺骨与桡骨完全分离。其近端比桡骨者粗大，前后厚，横向窄，呈扁三角柱形，往下逐渐变为扁形骨。肘突（olecranon, ol）只保存了下半部，其向上延伸的情况不清。但从保存的部分看，肘突明显高出钩突（processus anconaeus, pran）很多。肘突前缘很薄，从钩突向上方延伸。肘突内面平直；前外面纵向凹入，横向圆凸。钩突位于肘突前缘外侧，较圆隆。半月切迹（semilunar notch, semno，又称滑车切迹，trochlear notch）深。从前面看，半月切迹上窄下宽；表面横向圆凸，纵向呈半圆弧形凹入，其上部的弯曲度稍大于下部；切迹中纵隆起不明显；内、外关节面不对称。内关节面很大，向下延伸为长卵圆形，其内缘约呈"S"形，上部稍凹入，下部稍圆凸，关节表面面向内方。外关节面比内关节面短小许多，主要面向外下方。尺骨的冠状突（coronoid process, copr）明显。尺骨与桡骨近端关节的桡切迹（radial notch, rano），位于外关节面的下方，冠状突的外侧，约为横向伸长的长条形，其前内端稍低，而后外端稍高；表面纵向平直，横向稍凹，面向前外方。

尺骨骨体直，无扭曲或弯曲现象。上部较大，横切面为窄扁的三角形，三边均明显凹入。三角形的三个角分别形成上部的三条棱：前棱、后棱和内侧棱。上部的三个面中，以外面最宽，表面横向凹入，形成明显的尺骨侧窝（fossa lateralis ulnae, flul）。该窝的深度与 Rhizomys 的相近，但要比 Myospalax 和 Geomys 的浅得多。骨体的前内面上部很窄。尺骨粗隆（tuberosity of ulna, tuul）为一很发达的粗糙凹面，位于骨体前内面的近端，尺骨半月切迹（冠状突）的下方，供臂肌止端附着。骨体的前棱在上部较宽，

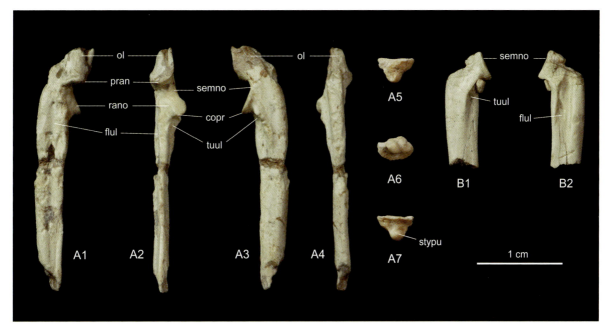

图 40　土著假竹鼠（新属、新种）尺骨（Ｖ 16293，正模）

Fig. 40　Ulna of *Pseudorhizomys indigenus* gen. et sp. nov. (V 16293, holotype)

A. 右尺骨（right ulna）：A1–A4. 右尺骨部分近端和骨体（partial proximal end and shaft of right ulna），A1. 外侧面（lateral view），A2. 前面（anterior view），A3. 内侧面 (medial view)，A4. 后面（posterior view）；A5–A7. 右尺骨远端（distal end of right ulna），A5. 前面（anterior view），A6. 远端面（distal view），A7. 后面（posterior view）。B. 左尺骨部分近端和部分骨体（partial proximal end and partial shaft of left ulna）：B1. 内侧面（medial view），B2. 外侧面（lateral view）

缩写（Abbreviations）：copr. 冠状突（coronoid process），flul. 尺骨侧窝（*fossa lateralis ulnae*），ol. 肘突（olecranon），pran. 钩突（*processus anconaeus*），rano. 桡切迹（radial notch），semno. 半月切迹（semilunar notch），stypu. 尺骨茎突（styloid process of ulna），tuul. 尺骨粗隆（tuberosity of ulna）

往下逐渐变薄锐。骨体的内侧棱上部较发达，与半月切迹的内关节面斜相交，向下后方斜伸，往下逐渐减弱，在骨体中部与后侧棱合并。骨体后内面上部较宽，向下变窄，并逐渐消失。骨体前内面下部变宽，占据了整个内面，使骨体中部往下变成扁形骨。后缘从上到下一直较圆钝。尺骨骨体中下部的横切面呈卵圆形。

　　右尺骨的远端只保存有骨骺部分，而且与骨体完全分离开。从远端看，尺骨远端为扁椭圆形。远端后面为纵向凹入、横向稍凸的曲面。内侧面后部横向稍圆凸，与桡骨远端相关节。前面外部可能稍破损，表面受磨蚀变平，缺失纵沟。尺骨茎突（styloid process of ulna, stypu）明显，远端与楔骨和豌豆骨的关节面约为横向延伸的半轴形面，前后向圆凸。

　　前脚（manus, Ms；图 28，29，41）　V 16293 保存了部分前脚，其中只有右 McII–McV 基本是原位保存，仅其远端稍位移；其余大多数肢骨都或多或少移位，彼此分得较开或彼此重叠。根据原标本，很难准确地判断它们在前脚中的位置。为了了解 V 16293 前脚各骨的形态结构，笔者首先根据它们保存在标本上的位置和相互关系进行初步判断，此后又以现生啮齿类（如竹鼠、鼢鼠、麝鼠和山河狸等）以及 HMV 1942 较完全的前脚作为参考（详见下面 111–119 页的记述），进一步确定了前脚各骨的位置（见图 41）。

　　腕骨（carpal bones, Carp；图 41 A）　V 16293 所保存的腕骨中，目前能确定的只有右侧的大多角骨和钩骨。

　　大多角骨（trapezium, Trm；图 41 A, 42）　背面约呈一不等边四边形。近端与腕舟 - 月骨关节的关节面（fsc-lu）约呈扇形。其背侧很窄，往掌侧变宽大。表面背 - 掌向稍圆凸，横向稍凹。外侧与小多角骨关节的关节面（ftrd）呈上下较窄的扇形，背侧很窄，往掌侧稍变宽。表面背 - 掌向凹入。远端与

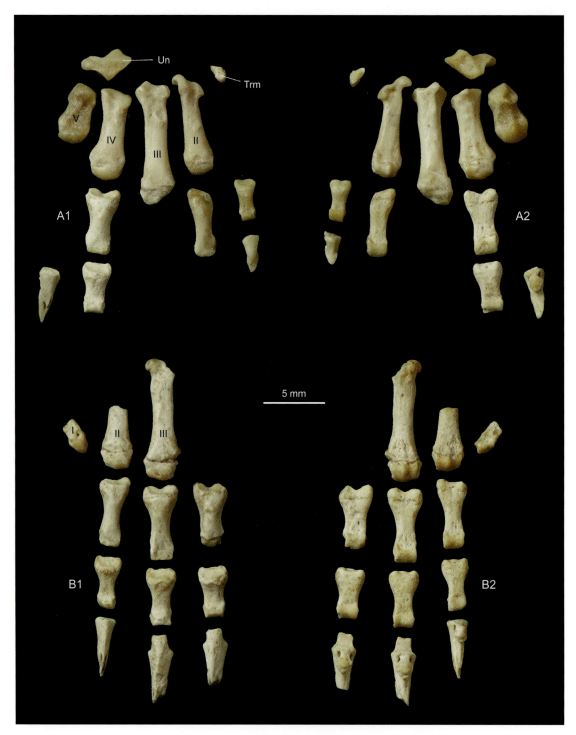

图 41 土著假竹鼠（新属、新种）前脚（V 16293，正模）

Fig. 41　Manus of V 16293（holotype) of *Pseudorhizomys indigenus* gen. et sp. nov.

A. 部分右前脚，包括大多角骨，钩骨，第二至第五掌骨，第一指第一指节骨—爪指骨，第二指第一指节骨，第四指第一和第二指节骨，第五指爪指骨 (partial right manus including trapezium, unciform, McII–McV, FI-Ph1–2, FII-Ph1, FIV-Ph1–2, FV-Ph3)：A1. 背面（dorsal view），A2. 掌面（volar view）；B. 部分左前脚，包括第一掌骨，部分第二和第三掌骨，第二指第一至第三指节骨，第三指第一至第三指节骨和第四指第一至第三指节骨（partial left manus including McI, partial McII and McIII, FII-Ph1–3, FIII-Ph1–3, and FIV-Ph1–3）：B1. 背面（dorsal view），B2 掌面（volar view）

缩写（Abbreviations）： I. 第一掌骨（McI），II. 第二掌骨（McII），III. 第三掌骨（McIII），IV. 第四掌骨（McIV），V. 第五掌骨（McV），Trm. 大多角骨（trapezium），Un. 钩骨（unciform）

第二掌骨相关节的关节面（fmcII）为背掌向伸长的椭圆形。表面也是背-掌向凹入，向外下方倾斜，与外侧的与小多角骨的关节面呈钝角相交，两关节面间有弱棱嵴相隔。内侧有一关节面可能与第一掌相关节 (fmcI)。关节面约为半圆形，表面背-掌向稍凸，上下稍圆凹，面向内下方。该关节面与腕舟-月骨关节面和第二掌骨关节面间都以背-掌向的棱嵴相隔。

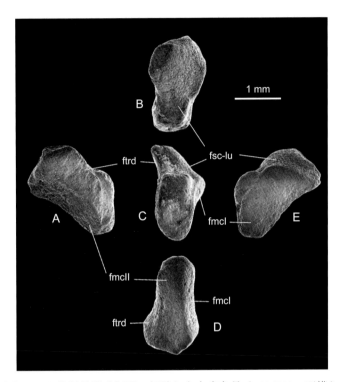

图 42　土著假竹鼠（新属、新种）右大多角骨（V 16293，正模）

Fig. 42　Right trapezium of *Pseudorhizomys indigenus* gen. et sp. nov. (V 16293, holotype)

A. 外侧面（lateral view），B. 近端面（proximal view），C. 背面（dorsal view），D. 远端面（distal view），E. 内侧面（medial view）

缩写（**Abbreviations**）：fmcI. 与第一掌骨关节面（facet for McI），fmcII. 与第二掌骨关节面（facet for McII），fsc-lu. 与腕舟-月骨关节面（facet for scapho-lunar），ftrd. 与小多角骨关节面（facet for trapezoid）

钩骨（unciform, Un；图 41 A, 43）　V 16293 右侧的钩骨保存完整，形状为不规则的四面体，横向很宽大，纵向很短。

钩骨背面形状约为一横宽的顶端向下的倒三角形。这样，钩骨的另外三面由近端面、远端内面和远端外面组成。远端的两面约呈直角相交。

近端面约呈三角形，其中部掌缘有一明显向上的尖突。尖突内侧与腕舟-月骨的关节面（fsc-lu）面向内上方，其表面背-掌向较强烈圆凸，横向稍凹。尖突外侧与楔骨（cuneiform）的关节面（fcu）约为三角形，面向外上方。该关节面比与腕舟-月骨的关节面大，但倾斜度较小、较平缓；关节表面横向明显凹入，背-掌向圆凸，其掌端向下延伸，几乎伸达远端与McIV的关节面。近端两关节面前部间无明显的界线。尖突后面为粗糙面，将腕舟-月骨关节面与楔骨关节面的后部分隔开。

远端内面约为梯形，表面较平，面向内下方倾斜。上有两个关节面。内上方者与头状骨的关节面（fmg），约为方形。外下者与McIII的关节面（fmcIII) 约为外侧宽的梯形。两关节面间的界线不很明显。远端内面与近端面在内侧呈锐角相交，两者仅以短的背-掌向的薄棱为界。

远端外面约为三角形，面向外下方。其上也有两关节面。内侧者为与McIV的关节面（fmcIV），约为宽的、弧形条带，背侧宽于掌侧。表面明显凹入。与其内侧的与McIII关节面间有明显的弧形凹入的棱分开。远端外面外侧者为与 McV 的关节面（fmcV）。该关节面约为卵圆形，背侧稍宽于掌侧。其长轴向背外 - 掌内方向延伸。关节表面沿长轴方向凹入，短轴方向较平。与 McV 的关节面较与 McIV 的关节面更稍向外倾斜。该两关节面间以凹入的弧形纵棱为界。远端外面与近端面在外后侧以锐角相交。两者间以弱的弧形凹棱为界。该凹棱主要为背 - 掌向，并稍向上外 - 下内方向斜伸。

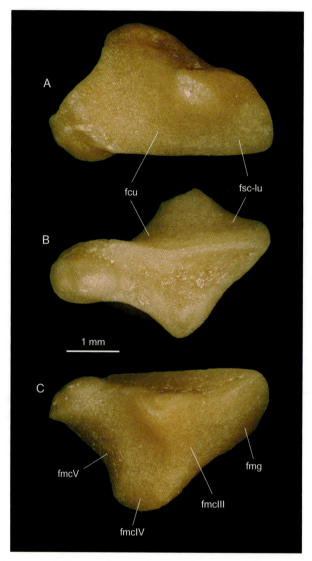

图 43　土著假竹鼠（新属、新种）右钩骨（V 16293，正模）

Fig. 43　Right unciform of *Pseudorhizomys indigenus* gen. et sp. nov. (V 16293, holotype)

A. 近端面（proximal view），B. 背面 (dorsal view)，C. 远端面（distal view）

缩写（**Abbreviations**）：fcu. 与楔骨关节面（facet for cuneiform），fmcIII. 与第三掌骨关节面（facet for McIII），fmcIV. 与第四掌骨关节面（facet for McIV），fmcV. 与第五掌骨关节面（facet for McV），fmg. 与头状骨关节面（facet for magnum），fsc-lu. 与腕舟 - 月骨关节面

掌骨（metacarpals, Mc's）　V 16293 可以确定的有右侧的第二至第五掌骨和左第一至三掌骨。

第一掌骨（McI；图 41 B, 44；表 16）　V 16293 左第一掌骨保存完好。McI 很短，形状不规则。近端有两个关节面，均面向外上方。内上方者与大多角骨关节（ftrm）。关节面约呈圆角三角形，背 - 掌向稍凸，横向稍凹。外下方的关节面与 McII 关节（fmcII）。该关节面较前者稍小，约为半圆形，表面稍凹。两关节面间有弱的背 - 掌向的棱分开。

远端明显窄于近端，其横径不与近端面者平行，而以锐角相交。远端与第一指节骨的关节面（fphI）为背 - 掌向曲度很大的弧面。其背部约为半圆形，背部上缘为不对称的圆弧形：内半部的圆凸度较大，外半部较平缓；表面圆凸，上无纵向隆嵴。远端关节面掌部的矢状嵴很弱，仅在掌端明显凸出，其位置较靠近外侧。矢状嵴两侧的凹槽很不对称：内侧的较发达，往后并逐渐加深；而外侧的凹槽很微弱，不明显。

　　McI 骨体很短。背面靠近内半部有一很深的坑。掌面为宽阔的凹面，表面粗糙。骨体内侧较高。其上部有一个大的结节状的隆起，表面凹入，可能供拇展肌（*m. abductor pollicis*, abp）附着。内侧下部近关节面处有一压迹和周边结节，为侧副韧带附着区。骨体外侧面很低，近、远端关节面仅被一供韧带附着的压迹分开。

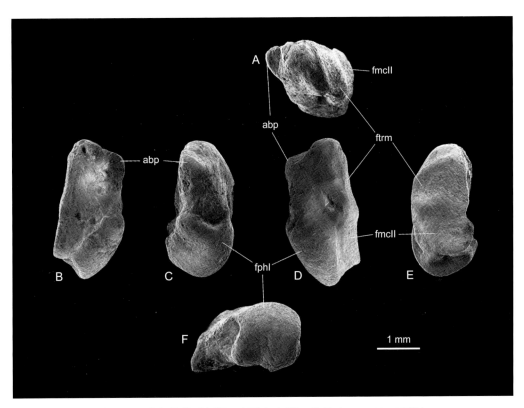

图 44　土著假竹鼠（新属、新种）左第一掌骨（V 16293，正模）

Fig. 44　Left McI of *Pseudorhizomys indigenus* gen. et sp. nov. (V 16293, holotype)

A. 近端面（proximal view），B. 掌面（volar view），C. 内侧面（medial view），D. 背面（dorsal view），E. 外侧面（lateral view），

F. 远端面（distal view）

缩写（**Abbreviations**）：abp. 拇展肌附着处（attachment area for m. *abductor pollicis*），　fmcII. 与 McII 关节面（facet for McII），

fphI. 与第一指节骨关节面（facet for Ph1），ftrm. 与大多角骨关节面（facet for trapezium）

第二掌骨（McII；图 41，45；表 16）　V 16293 保存有左、右 McII。其中右 McII 保存完好，仅远端稍破损；而左 McII 仅保存了下半部。

　　McII 近端较大。近端顶面为不规则梯形，外缘在背 - 掌方向上明显长于内缘，上有三个关节面。最内侧者为与大多角骨的关节面（ftrm），约为窄的椭圆形，表面圆凸，面向内上方。中间者为与小多角骨关节面（ftrd），最大，约呈四边形，其外缘背 - 掌向长于内缘，其背缘有一明显的凹陷；关节面稍向前内方倾斜，表面背 - 掌向稍凸，横向明显凹入。外侧与中央骨关节面（fce）约呈横向窄而背 - 掌向伸长的长条形，窄于中间和内侧的两关节面，而背 - 掌向的深度大于内侧的两关节面，其掌端明显向后突出；关节面表面圆凸，与其内侧的与小多角骨关节面以纵棱分开。近端内侧的与 McI 关节面

（fmcI）约为三角形，表面圆凸，面向内方，与大多角骨的关节面以弱的界线分开。近端外侧在与中央骨关节面的下方有与 McIII 关节面（fmcIII）。该关节面也为背 - 掌向伸长的、稍弯的长条形，前宽后窄。表面背 - 掌向凹入，面向外下方，与中央骨关节面近于垂直。在与 McIII 关节面的下方，为稍凹的粗糙面。近端掌面上缘紧接近端与小多角骨关节面处有一横向伸长的面，可能是在腕部过度弯曲时与小多角骨等腕骨关节的面。其下为较宽的凹面。近端背面中间稍凹，其内、外侧缘有明显的韧带嵴（crest for ligament, cl）供近端掌骨间韧带附着。

骨体较直。上部横向较窄，背 - 掌向较厚。其横切面约为四边形，内侧稍窄于外侧。骨体往下横向逐渐变宽，背 - 掌向在骨体中部稍变薄，往下又变厚。骨体下部横切面约为横宽的卵圆形。骨体背面纵向直，横向稍圆凸。与 *Rhizomys* 和 *Aplodontia* 相似，在背面中部靠内侧缘有明显的结节状隆起，可能是腕桡侧伸肌长头（*m. extensor carpi radialis longus*）附着处，笔者也称其为掌骨粗隆（tuberosity of metacarpal, tmc）。骨体掌面纵向稍凹，横向较平。在掌侧上部内侧缘有一明显的结节状隆起，可能是腕桡侧屈肌（*m. flexor carpi radialis*）和拇收肌（*m. adductor pollicis*, adp）的附着处。掌面下部中央为

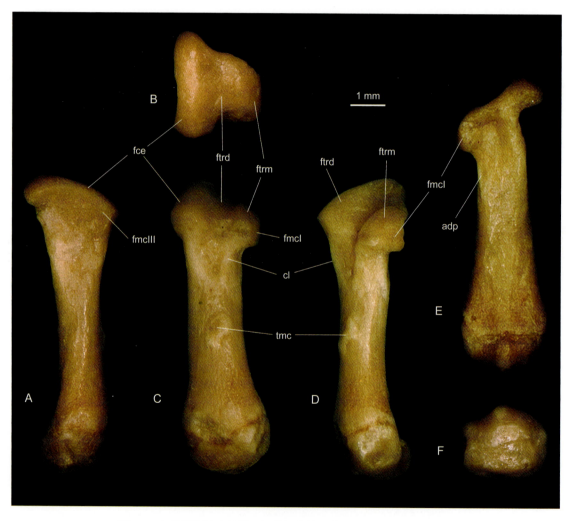

图 45　土著假竹鼠（新属、新种）右第二掌骨（V 16293，正模）

Fig. 45　Right McII of *Pseudorhizomys indigenus* gen. et sp. nov.（V 16293, holotype）

A. 外侧面（lateral view），B. 近端面（proximal view），C. 背面（dorsal view），D. 内侧面（medial view），E. 掌面（volar view），F. 远端面（distal view）

缩写（**Abbreviations**）：adp. 拇收肌附着处（attachment area for m. *adductor pollicis*），cl. 韧带嵴（crest for ligament），fce. 与中央骨关节面（facet for *centrale*），fmcI. 与第一掌骨关节面（facet for McI），fmcIII. 与第三掌骨关节面（facet for McIII），ftrd. 与小多角骨关节面（facet for trapezoid），ftrm. 与大多角骨关节面（facet for trapezium），tmc. 掌骨粗隆（tuberosity of metacarpal）

表16 临夏盆地副竹鼠类掌骨测量和比较（单位：mm）

Table 16 Measurements and comparison of metacarpals of pararhizomyines of the Linxia Basin (in mm)

测量项 (Parameter)	假竹鼠 (新属) Pseudorhizomys gen. nov.												三趾马层副竹鼠 Pa. hipparionum		
	土著假竹鼠 (新种) Ps. indigenus sp. nov.							甘肃假竹鼠 (新种) Ps. gansuensis sp. nov.					V 16306.16	V 16306.18	V 16306.17
	正模 (Holotype)														
	V 16293							HMV 1942							
	McI	McII		McIII		McIV	McV	McI	McII	McIII	McIV	McV	McIII	McIII	McIV
	(L)	(L)	(R)	(L)	(R)	(R)	(R)	(R)	(R)	(R)	(R)	(R)	(R)	(L)	(L)
全长 (total L)	2.1	8.3		9.9	9.9+	7.5	5	1.9+	9+			5.5	10.6		7.6+
近端最大宽 (max W of PE)	2.6	2.5			3	2.7	2.4	~2.5	2.9	2.6	~2.9		3		
近端前后径长 (max APD of PE)	1.5	2.7		2.3+	2.8	2.7	2						3.1		
远端最大宽 (max W of DE)	1.6	2.7	2.5	~3	3+	3.1	2.8		3	3.2	3.2+	2.7	3.3		3.2
远端关节面处宽 (DAF W)	1.6	2	2.1	2.3		2.4	2.1		2.2	2.5	~2.5	2.1	2.7	2.7	2.5
远端前后最大径长 (max APD of DE)	1.6	2	2.1	2.2	2.2+	2.2	2.1		2.3	2.5		~2	2.7	2.7	2.9
骨体中部最小宽 (min W at mid-shaft)		1.7	1.7	1.7	1.8	2	2.2		1.7	1.9	1.9	1.5	1.8		2.1
骨体中部最小宽处厚 (APD at min W at mid-shaft)		1.3	1.3	1.5	1.6	1.4	1.2		1.5	1.6	1.5	1.2	1.8		1.9
掌骨背侧最大长 (max L on dorsal side of Mc)		7.5		9.5	9.6+	7.3	4.5		8.5	10.7	8	4.5	10.4		

稍隆凸纵棱，其两侧有明显凹面。内侧的凹面大于外侧者。骨体的内、外侧缘纵向均较直，彼此不平行，往远方张开。

McII 的远端较骨体宽大，其宽与近端相近，但背-掌向稍薄于近端。远端与第一指节骨的关节面呈横轴状。背部表面横向平直，纵向强烈圆凸，无矢状嵴。关节面掌部有很显著的矢状嵴，嵴两侧的凹槽不对称，内侧的凹槽稍窄，但凹入深，外侧者稍宽，但凹入浅。该矢状嵴和两侧的凹槽均与骨体掌侧下部的纵棱和两侧的凹面相连续。该凹槽与掌骨远端的籽骨相关节。远端内、外侧有明显的压迹和周边结节，供内、外侧韧带（ *lig. collaterale mediale et laterale* ）附着。左、右 McII 远端的骺线均明显存在。

第三掌骨（McIII；图 41，46；表 16） V 16293 保存有左、右 McIII，只是左 McIII 近端和右 McIII 的远端稍有破损。

McIII 比 McII 更长而粗大。近端宽大，顶视约为不规则四边形，外侧缘长于内侧。顶面也有三个关节面。中间为与头状骨关节面（fmg），最大，约为四边形，外侧缘背-掌向长于内侧缘。关节表面背-掌向圆凸，横向明显地凹入，主要面向上，稍向内方；关节面的后外角有一明显转向掌方的三角形区，表面圆凸。外侧与钩骨的关节面（fun）约为小半圆形，背部较宽，掌部变窄。关节面大部分平直，主要面向外上方，稍向掌侧方倾斜，而其掌端的一小部分转向前外方倾斜。该关节面与头状骨关节面的界线为圆凸的弧形。内侧与 McII 的关节面（fmcII）约成三角形（前外角稍破损），主要位于近端内侧的背部，也是背部宽掌部窄。表面稍圆凸，主要面向内上方，与头状骨的关节面间的界线不明显。近端内侧，在与 McII 关节面的后下部，表面粗糙。近端外侧面有与 McIV 的关节面（fmcIV），为背-掌向长而横向窄的长条形。表面背-掌向强烈凹入，垂向稍凹，主要向外下方，稍向掌方倾斜，与钩骨关节面的夹角近于直角。近端外侧在与 McIV 关节面之远处的部分为宽大的凹面。近端背面为表面稍凹的粗糙面。在背面外侧与钩骨关节面的远侧方的韧带嵴（cl）很明显，向下延伸较长。近端掌面稍凹。

骨体直，内、外侧缘对称，上部彼此近于平行，下部彼此往远方张开。上部较窄，横切面为四边形，长宽相近。骨体往下横向逐渐变宽，其背-掌向的厚度在骨体中部稍变薄，再往下部又逐渐变厚。骨

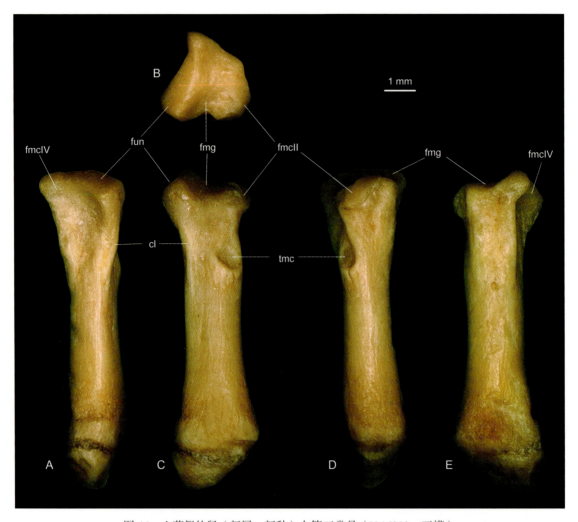

图 46 土著假竹鼠（新属、新种）右第三掌骨（V 16293，正模）

Fig. 46 Right McIII of *Pseudorhizomys indigenus* gen. et sp. nov. (V 16293, holotype)

A. 外侧面（lateral view），B. 近端面（proximal view），C. 背面（dorsal view），D. 内侧面（medial view），E. 掌面（volar view）

缩写（Abbreviations）：cl. 韧带嵴（crest for ligament），fmcII. 与第二掌骨关节面（facet for McII），fmcIV. 与第四掌骨关节面（facet for McIV），fmg. 与头状骨关节面（facet for magnum），fun. 与钩骨关节面（facet for unciform），tmc. 掌骨粗隆（tuberosity of metacarpal）

体下部的横切面为横宽的椭圆形。骨体背面纵向平直，横向稍凸。掌骨粗隆（tuberosity of metacarpal, tmc）也与 *Rhizomys* 和 *Aplodontia* 的相似，很发达，在近端关节面下方约 1 mm 的背侧近内缘处形成四周隆凸中央凹陷的大粗糙区，供腕桡侧伸肌短头附着。这表明该肌肉相当发达。掌面纵向稍凹，其上部内、外侧缘有结节状隆突，表面粗糙，可能分别供腕桡侧屈肌（？）和拇收肌附着。中 - 下部横向圆凸；下端也有中央纵棱和两侧的凹面。

远端较宽大，横向稍宽于近端，背 - 掌向稍薄于近端。其形态结构与 McII 的很相似，只是稍宽大些。左侧 McIII 的远端沿骺线有些移动(其骺软骨尚未完全骨化)，但其形态仍然保存。其形态和 McII 者相近，也是不对称的。

V 16293 的 McIII 的形态结构与 V 16306.16（*Pararhizomys hipparionum*）的很相似，只是其近端与头状骨的关节面横向凹入及与 McIV 关节面下方的凹坑凹入均较浅些，骨体近端背面外侧的韧带嵴较低弱些，掌骨远端也较细小。因 V 16293 为一较年幼的个体（骺线仍明显存在），而 V 16306.16 为成年个体（骺线已消失），这些区别可能与年龄有关，也可能表明 V 16293 的头状骨的远端关节面横向凸度要比 V 16306.16 者相对弱些，活动性稍差。

第四掌骨（McIV；图41，47；表16）　V 16293 仅保存有右 McIV，保存完好。McIV 明显短于 McIII，也比 McII 粗短。近端较宽大。顶视约为三角形，背缘和外缘凹入，而内掌缘稍圆凸。顶面有两个关节面。内侧与 McIII 关节面（fmcIII）为卵圆形，背侧宽而掌侧窄，面向背上方并稍向内方。表面圆凸，以背 - 掌向弯曲度较大。外侧与钩骨关节面（fun）较大，为四边形，外缘背 - 掌向长于内缘，稍向背内方倾斜。表面大致被分成三部分：背部明显凹入；外中部较圆凸；掌部约为三角形，表面微凹。近端外侧与 McV 关节面（fmcV）为背 - 掌向伸长而纵向短的长条形；向外掌方倾斜，与钩骨关节面约以直角相交；其表面背 - 掌向凹入，垂向平直。外面与 McV 关节面远侧有一粗隆，可能供骨间韧带附着。近端背面内、外两侧的韧带嵴（cl）很发达，为粗糙棱形。两韧带嵴之间以一明显的凹陷相隔。近端掌面上缘，紧接与钩骨关节面处有一小的卵圆形面，面向掌侧方，约与近端关节面垂直。它可能是在腕部过度弯曲时与钩骨的关节面。

骨体直。上部相对较窄而厚，横切面约为四边形，横宽大于背 - 掌向厚。骨体往远处迅速变宽，在中部稍变薄，至远侧部又变厚。骨体远部的横切面为横宽的椭圆形，横向明显宽于近端宽。骨体背面纵向直，横向圆凸。掌面纵向稍凹，横向稍凸。其远端部分的纵棱和两侧的凹面延伸较长，约占据骨体的远处 1/3 部分。骨体内侧横向薄而圆凸，近端部分较直，远端部分稍向内方张。外侧缘纵向较内侧缘稍长，纵向凹入，横向也圆凸，但较薄。

远端比近端宽。其形态结构与 McIII 者相似，只是其远端与第一指节骨的关节面两侧是对称的，即掌部的矢状嵴位于正中，其两侧的凹槽大小和凹度都相近。

V 16293 的 McIV 与 *Pa. hipparionum* 的 McIV 的区别也在于其韧带嵴稍低弱和远端较细小些。这也可能与它们的年龄不同有关。

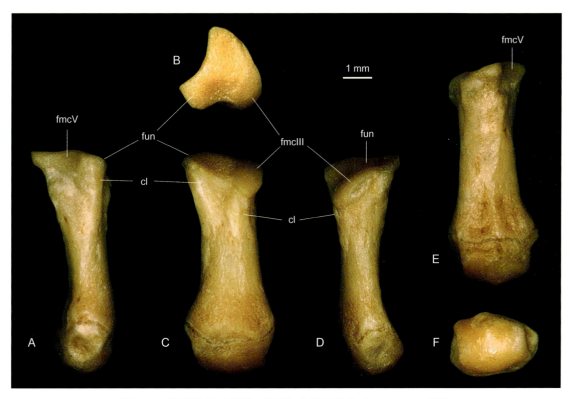

图47　土著假竹鼠（新属、新种）右第四掌骨（V 16293，正模）

Fig. 47　Right McIV of *Pseudorhizomys indigenus* gen. et sp. nov. (V 16293, holotype)

A. 外侧面（lateral view），B. 近端面（proximal view），C. 背面（dorsal view），D. 内侧面（medial view），E. 掌面（volar view），F. 远端面（distal view）

缩写（Abbreviations）：cl. 韧带嵴（crest of ligament），fmcIII. 与 McIII 关节面（facet for McIII），fmcV. 与 McV 关节面（facet for McV），fun. 与钩骨关节面（facet for unciform）

第五掌骨（McV；图 41，48；表 16） V 16293 仅保留了右侧的 McV，保存完好。McV 比 McI 粗大，而短于其他的三枚掌骨（McII，McIII 和 McIV），而且比例上也较后者粗短。McV 的近端顶面有两个关节面。外侧与钩骨关节面（fun）约为卵圆形，几乎占据了整个近端面。其长轴约为背外 - 掌内方向。表面圆凸。近端内侧与 McIV 关节面（fmcIV）很小，约为三角形，表面平，主要面向内方，稍斜向近端方；与近端和钩骨关节面近于垂直，两者间以明显棱为界。近端外侧后部有大而粗糙的压迹窝，周边有结节状隆起，可能供腕尺侧伸肌附着。

骨体直，约为扁圆柱形，上部较窄厚，往下部变宽而薄。骨体背面平直，表面有一条呈内上 - 外下方向斜伸的沟。骨体掌面纵向凹入，中央有一条明显的纵棱，向下内方斜伸，并逐渐加强。纵棱两侧为凹面。内侧的凹面稍大于外侧者。骨体内、外两侧背 - 掌向较薄，分别向内下和外下方延伸，只是外侧缘下端较内侧者分开的更强些。

远端较近端宽大。其形态结构与 McIV 者相似。但其远端关节面内、外不对称。内侧者背 - 掌向长于外侧者，背上缘内侧圆凸的曲度也大于外侧者。关节面掌部中间的纵隆嵴更发达些，与骨体掌侧的中央纵棱相连续。纵隆嵴外侧的凹槽稍大于内侧的凹槽。远端两侧均有供侧副韧带附着的压迹，外侧压迹大于内侧者。

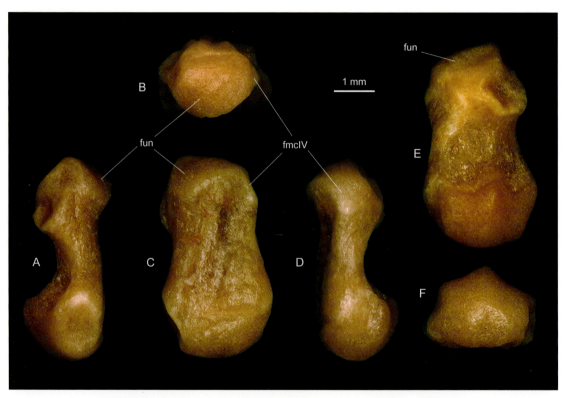

图 48　土著假竹鼠（新属、新种）右第五掌骨（V 16293，正模）

Fig. 48　Right McV of *Pseudorhizomys indigenus* gen. et sp. nov. (V 16293, holotype)

A. 外侧面（lateral view），B. 近端面（proximal view），C. 背面（dorsal view），D. 内侧面（medial view），E. 掌面（volar view），F. 远端面（distal view）

缩写（**Abbreviations**）：fmcIV. 与 McIV 关节面（facet for McIV），fun. 与钩骨关节面（facet for unciform）

指节骨（phalanges of fingers, Ph；图 41） V 16293 左、右前脚都保存有部分指节骨。右前脚保存有六枚指节骨：第一指的两个指节骨，第二指的第一指节骨，第四指的第一和第二指节骨和第五指的第三指节骨。左前脚保存有九枚指节骨，即第二至第四指的三个指节骨。*Pseudorhizomys* 的第一指只有两个指节骨。第二—第五指均有三个指节骨。

第一指（拇指；first finger, FI, Thumb）

第一指第一指节骨（proximal phalanx of first finger, FI-Ph1；图 28, 41, 49；表 17） V 16293 的右侧第一指第一指节骨保存完好。近端较大，横宽稍大于背 - 掌向厚。近端与第一掌骨关节面约为大半圆形，掌侧较宽。表面凹入，不与骨干垂直，而是稍向背上方倾斜。近端掌侧有两个结节，被中央凹沟分开。内侧的结节显著，可能是拇展肌附着处。外侧的结节比内侧者稍高大，可能供拇短屈肌和拇收肌等附着。

骨体直，从侧面看为倒梯形。骨体上部较宽厚，横切面约为正四边形；下部背 - 掌向变薄而横向窄，横切面为横宽的四边形。背面纵向直，稍向下前倾斜。其上部横向圆凸。在背面上部的内半部靠近近端关节面处有明显的结节状隆起，可能是拇短伸肌远端附着处，本书暂称其为伸腱结节（extensor tuberosity, extu）。背面下部横向较直。掌面纵向较直，向背下方斜伸。掌面上部表面横向稍凸，下部较平直。

远端小于近端，但比骨体下部稍宽大些。远端与第二指节关节面约为半圆轴形，其背部稍向背上方延伸，掌部明显向掌上方伸延，掌侧缘明显高于背侧缘。关节表面背 - 掌向圆凸，背部稍窄，横向较平，掌部逐渐变宽，横向稍凹。关节面的背上缘约为向上稍凸的圆弧形，而掌上缘横向较平直。远端两侧有明显的粗糙的压迹，周围有结节，供指侧韧带附着。外侧的压迹较内侧者稍大。

测量见表 17。

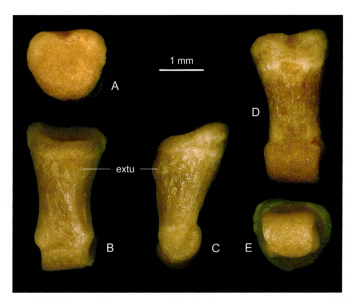

图 49 土著假竹鼠（新属、新种）右第一指第一指节骨（V 16293，正模）
Fig. 49 Right FI-Ph1 of *Pseudorhizomys indigenus* gen. et sp. nov. (V 16293, holotype)
A. 近端面（proximal view），B. 背面（dorsal view），C. 内侧面（medial view），D. 掌面（volar view），E. 远端面（distal view）
缩写（Abbreviation）：extu，伸腱结节（extensor tuberosity）

第一指第二指节骨（爪指骨）（distal phalanx of FI, FI-Ph2；图 29, 41, 50；表 17） V 16293 的前脚的第一指第二指节骨（爪指骨）保存不完整，近端已破损。从保存的部分看，FI-Ph2 约为锥形，近端很宽厚，背 - 掌向的厚度大于横宽。背部和两侧面横向均较圆凸，表面光滑。在背部近端中央仅保留有伸腱突（extensor process, expr）的基部。在掌侧面中部有一很大的结节状的隆凸，在此被称为爪指骨粗隆（tuberosity of ungual phalanx, tunph），可能供指深屈肌附着。这一点与熊猫者很接近，可能说明其挖掘的功能较强。粗隆基部两侧有明显的掌沟（volar sulcus, vos）。内、外侧的掌沟内各见有一掌孔（volar foramen, vof），外侧者较内侧者大。

FI-Ph2 远端为背 - 掌向扁而远端尖薄的铲形，未见中矢切迹（sagittal notch）。其背侧面纵向和横向均圆凸，横向圆凸度较大。掌侧面纵向凹入，横向圆凸。其远端边缘为薄锐的缓圆弧形，上无纵沟

表 17　临夏盆地假竹鼠（新属）指节骨测量和比较（单位：mm）

Table 17　Measurements and comparison of phalanges of *Pseudorhizomys* gen. nov. of the Linxia Basin (in mm)

测量项 (Parameter)		土著假竹鼠（新属、新种）Ps. indigenus gen. et sp. nov. V 16293（正模，holotype）								甘肃假竹鼠（新属、新种）Ps. gansuensis gen. et sp. nov. HMV 1942（正模，holotype）				
		FI	FII		FIII	FIV		FV	ToIII?	FI	FII	FIII	FIV	FV
		®	Ⓛ	®	Ⓛ	Ⓛ	®	®	®?	®	®	®	®	®
全长（total L）	Ph1	3.3	5.6	5.4+	6	5.3+	5.3		8	3.5	5.7	6	5.1	4.1
近端最大宽 (max W of PE)		1.8	2.5		3	2.9	~2.9		2.8	2.1	~2.5	3.3	3.3	3.2
近端前后径长 (max APD of PE)		1.7	2	2.2	2.3	2.7	2.6		2.1	2	2.6	2.8	2.7	2.4
远端最大宽 (max W of DE)		1.4	1.9	1.8+	2		~2.1		1.8	1.4	1.9	2.3	2.3	1.9
远端最大前后径长 (max APD of DE)		1.1	1.2	1.2	1.3		1.5		1.4	1.1	1.2	1.5	1.5	1.3
远端关节面前宽（DAF ant W）		1					1.8		1.6		1.6	1.7	1.8	1.6
远端关节面后宽（DAF post W）		1.2	1.7	1.7	1.8		2		1.5	1.1	1.8	2	2	1.8
骨体中部最小宽 (min W at mid-shaft)		1.2	1.4	1.4	1.5	1.8	1.7		1.2	1	1.3	1.6	1.8	1.6
骨体中部最小宽处厚 (APD at min W at mid-shaft)		1.1	1.4	1.4	1.4	1.6	1.5		1.3	1	1.6	1.5	1.5	1.2
全长（total L）	Ph2 (FII–V)		4.2		4.8	4.1	4.1				4.7	5	4.3	3.6
近端最大宽 (max W of PE)			2.3		2.6	2.5	2.5				2.5	2.7	2.8	2.4
近端前后径长 (max APD of PE)			1.9		2.2	2	1.9				2.4	2.6	2.4	1.9
远端最大宽 (max W of DE)			~1.6		1.9	2	1.9				1.8	2	2.1	1.8
远端最大前后径长 (max APD of DE)			1.3		1.4	1.5	1.4				1.7	1.7	1.6	
远端关节面前宽（DAF ant W）						1.5	1.6				1.2	1.6	1.6	~1
远端关节面后宽（DAF post W）			1.6		1.8	1.9	1.9				1.7	1.9	1.9	1.6
骨体中部最小宽 (min W at mid-shaft)			1.3		1.6	1.7	1.6				1.5	1.6		1.6
骨体中部最小宽处厚 (APD at min W at mid-shaft)			1		1.1	1.3	1.4				1.4	1.4	1.3	1.1
全长 (total L)	Ph2 (FI) or Ph3 (FII–V)	3+					4.2+			3.5+	5.8+	7.5+	7.6+	5.2+
掌侧长 (L on volar side)		~3	4.7				3.9			3.4	4.8+	6.6	6.6	4.5
近端最大宽 (max W of PE)		1.4	1.8		2.2		2.1		1.7	~1.5	1.9	2.2	2.2	1.7+
近端结节处前后径长 (max APD of PE, at tub.)		1.7	2.3		2.6		2.4		1.9	2	3	3.3	3	2.5
近端关节处前后径长 (max APD of PE, at AF)		1.6	1.8		2.4		2.2		1.7	1.5	2.4	2.8	2.7	2.2

或锯齿。该远端边缘被爪尖分为内、外部分。内侧边缘明显长于外侧边缘，其长约为外侧者的两倍。内、外边缘的近端掌侧各有一侧壁孔（lateral foramen, lf）。

第二指（second finger, FII）

第二指第一指节骨（proximal phalanx of McII, FII-Ph1; 图28, 29, 41, 51; 表17）　V 16293 保留有左、右前脚第二指第一指节骨。左 FII-Ph1 保存较好，仅近端背侧和内侧，远端内侧稍破损。而右 FII-Ph1 的近、远端均破损较多。

FII-Ph1 短于 McII，其长约为 McII 长的 2/3，而比 FI-Ph1 粗大得多。FII-Ph1 近端较大，宽大于背 - 掌向厚。顶面稍向前倾（与骨体长轴不垂直）。近端与 McII 关节面的背部表面凹入；掌部则由一中间沟和两侧的结节组成，外侧结节稍大于内侧结节。它们与 McII 远端关节面掌部中央的纵向隆峭和两侧横向凹槽相关节。近端掌侧面中央有一宽的凹槽将两侧的结节分开。

骨体约为半圆柱形，上大下小，横切面近半圆形，上部宽稍大于背 - 掌向的径长；下部逐渐变得更扁，

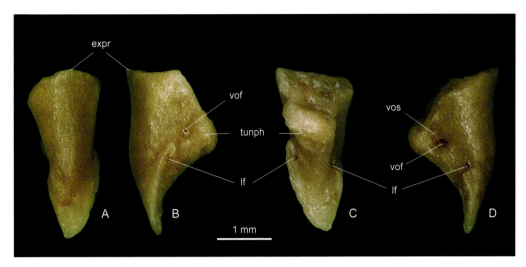

图 50　土著假竹鼠（新属、新种）右第一指爪指骨（V 16293，正模）

Fig. 50　Ungual phalanx of right FI (FI-Ph2) of *Pseudorhizomys indigenus* gen. et sp. nov. (V 16293, holotype)

A. 背面（dorsal view）, B. 内侧面（medial view), C. 掌面（volar view）, D. 外侧面（lateral view）

缩写（**Abbreviations**）：expr. 伸腱突（extensor process）, lf. 侧壁孔（lateral foramen）, tunph. 爪指骨粗隆（tuberosity of ungual phalanx）, vof. 掌孔（volar foramen）, vos. 掌沟（volar sulcus）

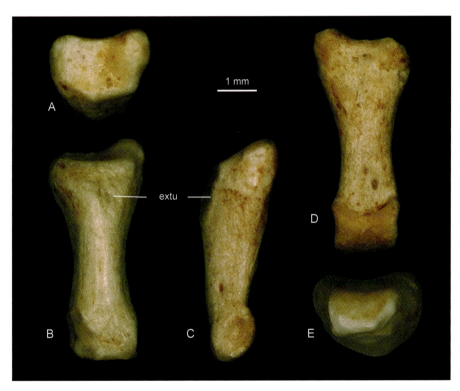

图 51　土著假竹鼠（新属、新种）左第二指第一指骨（V 16293，正模）

Fig. 51　Left FII-Ph1 of *Pseudorhizomys indigenus* gen. et sp. nov. (V 16293, holotype)

A. 近端面（proximal view）, B. 背面（dorsal view）, C. 外侧面（lateral view）, D. 掌面（volar view), E. 远端面（distal view）

缩写（**Abbreviations**）：extu. 伸腱结节（extensor tuberosity）

宽明显大于背 - 掌向径长。骨体背侧纵向较直，横向圆凸，与内、外两侧面平缓过渡，共同形成圆滑的弧面。背侧上部正中有很明显的隆突。依 Greene（1935）、杨安峰和王平等（1985）的记述，褐家鼠的指总伸肌只止于第二至五指的第三指节骨的基部。但有些哺乳动物的第二至五指的三个指节骨都有指总伸肌附着处。很可能 V 16293 的 FII-Ph1 的该隆突也是指总伸肌（*m. extensor digitorum communis*）的附着处，我们也称其为伸腱结节（extensor tuberosity, extu）。该结节较 FI-Ph1 的大，只是表面稍破损。骨体掌侧较平直。掌面下部纵向稍凹，其两侧有纵棱与内、外侧面分开。

　　远端小于近端，但比骨体下部稍宽大些。宽大于掌 - 背径长。远端与第二指节骨关节面与 FI-Ph1 者不同，约为四边形。其背部只有很小的部分转向背侧，而掌部明显向上延伸，使关节面不与骨体纵轴垂直，而是面向掌下方。其背上缘横向较直，而掌上缘为明显向下凹入的弧形。关节表面纵向稍圆凸，被一很浅缓的矢状沟分为内、外两髁。两髁的表面横向稍圆凸。远端两侧有明显的凹陷和边缘的结节状隆起，供侧副韧带附着。

　　第二指第二指节骨（middle phalanx of FII, FII-Ph2；图 28, 29, 41；表 17）　V 16293 仅保存左第二指第二指节骨。该指节骨大部保存完好，仅远端稍破损。FII-Ph2 比 FII-Ph1 短，其近端较粗大。近端关节面稍向前倾，与骨体长轴斜交。近端关节面约为椭圆形，表面凹入，被极弱的矢状棱分为两个凹面。近端前缘有一很厚的横向延伸的枕状隆凸，这可能是指总伸肌肌腱附着处，我们也称其为伸腱结节（extensor tuberosity, extu）。近端掌侧面中央有一很宽的凹槽，可能供指深屈肌腱通过。该沟的两侧为结节状的隆起，可能供指浅屈肌和侧副韧带等附着。

　　FII-Ph2 的骨体约成截角锥形（truncated pyramid），上部较大，往下逐渐变得较细小。骨体背侧和两侧面纵向稍凹，横向平直；掌面较平。

　　远端较近端窄小，但较骨体下部稍宽大，约呈横轴形，明显向背侧突出，超过骨体背面。远端关节面背 - 掌向很圆凸，从侧面看为半圆弧形。其背侧部分稍往背上方延伸，可惜其内侧大部分已破损，其保留的外侧部分表面圆凸。关节面的掌侧部分向上方延伸得较长，明显超过其背侧部，其掌上缘为明显凹入的弧形。掌侧部分与 FII-Ph1 的相似，表面纵向圆凸，并被浅缓的矢状沟分为内、外两髁。两髁表面稍圆凸。远端两侧为明显的凹陷区，供侧副韧带附着。

　　第二指第三指节骨（爪指骨）（distal phalanx of FII, FII-Ph3；图 28, 29, 41；表 17）　V 16293 仅保留左第二指第三指节骨（爪指骨）。该爪指骨大部分保存完好，仅近端稍破损。爪指骨较 FII-Ph2 细长，为尖爪形。近端较厚大，背 - 掌向径长大于横宽。近端与 FII-Ph2 远端的关节面保留的部分为横向的凹槽状，背 - 掌向凹入，横向平直。可惜近端背侧的伸腱突已完全断失。

　　骨体约为三角锥形。背侧有一明显的纵棱伸至爪指骨的远端，将骨体背面分为内、外两侧面。两侧面从侧面看均为远端较尖的窄三角形，表面纵向直，横向圆凸。两侧面的近端部分较平滑，远端部分有明显的纵向沟和棱。骨体掌面供指深屈肌附着的爪指骨粗隆（tuberosity of ungual phalanx, tunph）很发达，其表面上部大部分圆凸，下部稍凹。掌面该粗隆基部两侧有很发达的掌沟（volar sulcus, vos）。内、外侧掌沟的近端有很大的内、外掌孔。爪指骨粗隆的远端有一纵棱伸至爪指骨的尖端。该纵棱的内、外侧有明显的纵向的沟棱伸达尖端。远端上也未见有中矢切迹。FII-Ph3 的远端与 FI-Ph2 者明显不同：远端不呈圆缓薄锐的边缘，而为窄而尖锐的扁锥形；而且远端两侧也未见有侧壁孔的痕迹。

　　第三指（third finger, FIII）

　　第三指第一指节骨（proximal phalanx of FIII, FIII-Ph1；图 28, 29, 41；表 17）　V 16293 仅保存左第三指第一指节骨。FIII-Ph1 保存较好，只是远端背部破损。FIII-Ph1 短于 McIII，其长约为 McIII 长的 3/5，但较 FII-Ph1 粗长些，其近端很宽大。顶面约为半圆形，不与骨体中轴垂直，而是向前倾斜。近端与 McIII 关节面约为横宽的椭圆形，表面凹入。近端掌侧由中央凹槽和两侧的结节状隆凸共同组成近端关节面的掌缘。外隆凸稍大于内隆凸。近端两侧面稍隆凸而粗糙。FIII-Ph1 的其他形态结构均与 FII-Ph1 的很相似，只是伸腱结节更大、更显著，远端关节面的矢状沟凹入较浅。

　　第三指第二指节骨（middle phalanx of FIII, FIII-Ph2；图 41；表 17）　V16293 的左第三指第二指节骨保存较完好，仅远端前外侧稍磨损。FIII-Ph2 比 FIII-Ph1 短，近端稍窄。FIII-Ph2 的形态结构与 FII-Ph2 的很相似，所不同的是 FIII-Ph2 比 FII-Ph2 要长而粗大些；其骨体较宽扁；近端关节面凹入较深，矢状棱更弱而不明显。远端关节面呈半圆轴面，不与骨体中轴垂直，而是向前下方倾斜，其前缘稍向背侧翻转，掌侧上缘横向较平直。远端关节面表面背 - 掌向圆凸，被宽缓的矢状沟分为内、外两髁。内髁稍大于外髁。

第三指第三指节骨（爪指骨）（distal phalanx of FIII, FIII-Ph3；图 29, 41, 52；表 17） V 16293 仅保存了左第三指爪指骨。爪指骨大部分保存完好，只是骨体远端部分断失。FIII-Ph3 比 FIII-Ph2 长而窄，但比 FII-Ph3 粗大而长。FIII-Ph3 近端较窄，背 - 掌向厚。伸腱突（extensor process, expr）很发达，约为瓦片形，往顶端稍变窄。其背、掌视约为顶端较窄、稍凹的梯形。顶端面较粗糙。FIII-Ph3 近端与 FIII-Ph2 的关节面（ffIII-2）为背 - 掌向的凹轴面。它大致可被分成为两部分：掌侧大部分面向上方者约为椭圆形，被中央弱的矢状棱分为两凹面，内侧凹面大于外侧凹面；背部者位于伸腱突的掌侧，较窄，横向平直。FIII-Ph3 的其他形态结构与 FII-Ph3 相似：掌孔和掌沟都很发达，无侧壁孔，而且远端为窄而尖锐的扁锥形。只是 FIII-Ph3 的骨体背侧的纵棱有些向远内侧方向斜伸，因此其背外侧面明显宽于背内侧面，横向也较后者更圆凸些。该两侧面的近端部分也均较平滑，远端部分也有明显的纵向的沟和棱。但其具沟棱的部分要比 FII-Ph3 的更发达、延伸较长些。掌面的纵棱较短，不伸至爪指骨的尖端。

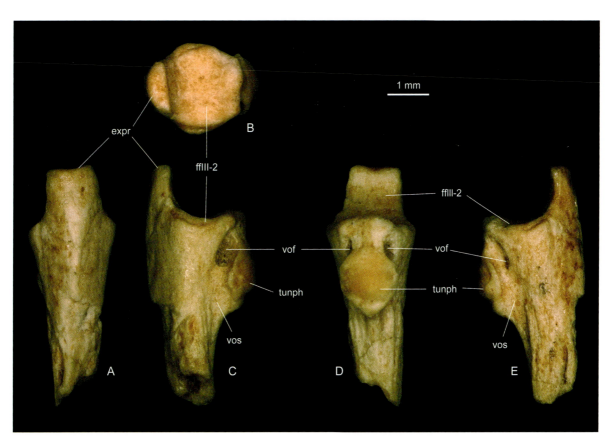

图 52 土著假竹鼠（新属、新种）左第三指爪指骨（V 16293，正模）

Fig. 52 Ungual phalanx of left FIII (FIII-Ph3) of *Pseudorhizomys indigenus* gen. et sp. nov. (V 16293, holotype)

A. 背面（dorsal view），B. 近端面（proximal view），C. 外侧面（lateral view），D. 掌面（volar view），E. 内侧面（medial view）

缩写（Abbreviations）：expr. 伸腱突（extensor process），ffIII-2. 与第三指第二指节骨关节面（facet for FIII-Ph2），tunph. 爪指骨粗隆（tuberosity of ungual phalanx），vof. 掌孔（volar foramen），vos. 掌沟（volar sulcus）

第四指（fourth finger, FIV）

第四指第一指节骨（proximal phalanx of FIV, FIV-Ph1；图 28, 29, 41, 53；表 17） V 16293 保留左、右第四指第一指节骨。其中右 FIV-Ph1 保存完好，而左 FIV-Ph1 远端破损。FIV-Ph1 短于 McIV，其长约为 McIV 的 2/3。FIV-Ph1 比 FI-Ph1 粗大和长很多，较 FII-Ph1 和 FIII-Ph1 稍宽厚，而短于 FIII-Ph1。

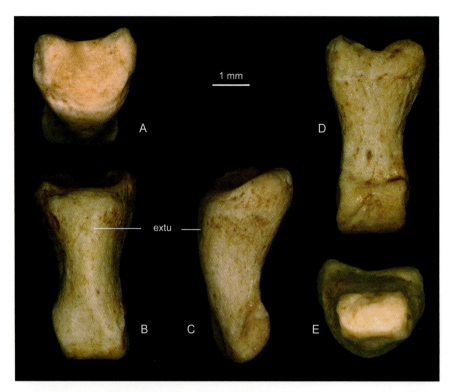

图 53　土著假竹鼠（新属、新种）右第四指第一指节骨（V 16293，正模）

Fig. 53　Right FIV-Ph1 of *Pseudorhizomys indigenus* gen. et sp. nov. (V 16293, holotype)

A. 近端面（proximal view），B. 背面（dorsal view），C. 内侧面（medial view），D. 掌面（volar view），E. 远端面（distal view）

缩写（Abbreviation）： extu. 伸腱结节（extensor tuberosity）

FIV-Ph1 的近端比 FII-Ph1 者更宽大，而与 FIII-PhI 者相近。FIV-Ph1 的其他的形态结构与 FII-Ph1 和 FIII-Ph1 者相似，骨体背侧的伸腱结节（extensor tuberosity, extu）也很明显。与 FII-Ph1 和 FIII-Ph1 不同的是：FIV-Ph1 的近端掌侧供韧带附着的隆凸是内侧的稍大于外侧者。

第四指第二指节骨（middle phalanx of FIV, FIV-Ph2；图 28, 29, 41, 54；表 17）　V16293 左、右第四指第二指节骨均保存完好。FIV-Ph2 短于并稍细于 FIV-Ph1，比 FII-Ph2 稍粗大，比 FIII-Ph2 短。FIV-Ph2 的形态结构与 FII-Ph2 和 FIII-Ph2 者相似。近端关节面也大而凹入，矢状棱也很微弱。近端前缘的伸腱结节（extensor tuberosity, extu）也为厚的横向延伸的枕状隆凸。近端掌侧面中央有可能供指深屈肌腱通过的很宽的凹槽。该槽的两侧有可供指浅屈肌和侧副韧带附着的结节状的隆起。

远端关节面也为半圆的轴状面，主要向掌上方延伸。关节面纵向圆凸，矢状沟也很宽缓，内、外髁表面稍圆凸，内髁也稍大于外髁。远端关节面的背部变窄，并转向背侧，其背缘为稍上凸的圆弧形。掌侧上缘横向较直。远端两侧供侧副韧带附着的凹陷区和周边的结节都很明显。

第四指第三指节骨（爪指骨）（distal phalanx of FIV, FIV-Ph3；图 29, 41；表 17）　V 16293 仅保存左第四指爪指骨。该爪指骨的伸腱突近端和骨体远端都部分断失。FIV-Ph3 保存部分的形态结构与 FIII-Ph3 相同，其近端的宽度和厚度也彼此相近。

第五指（fifth finger, FV）

V 16293 的第五指的第一指节骨和第二指节骨均未保存，仅保存右第五指的第三指节骨。

第五指第三指节骨（爪指骨）（distal phalanx of FV, FV-Ph3；图 41 A；表 17）　右 FV-Ph3 的伸腱突近端部分断失，其余部分保存完好。FV-Ph3 的形态结构与 FIII-Ph3 和 FIV-Ph3 都很相似，只是比上述两爪指骨都要短而瘦小。FV-Ph3 骨体远端部分为宽扁的三角锥形。其背侧纵棱的位置明显偏向内侧，往远方并逐渐与内侧缘相连。其背外侧面横向明显宽于背内侧面。背内侧面的远端部分明显变窄、变

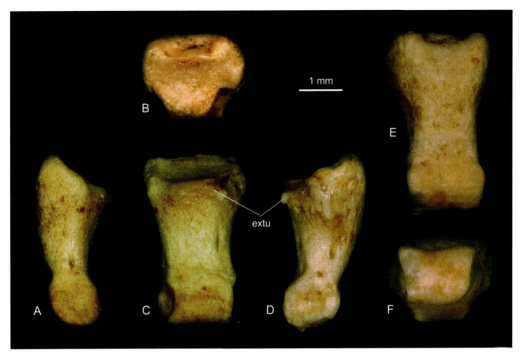

图 54　土著假竹鼠（新属、新种）左第四指第二指节骨（V 16293，正模）

Fig. 54　Left FIV-Ph2 of *Pseudorhizomys indigenus* gen. et sp. nov. (V 16293, holotype)

A. 内侧面（medial view）, B. 近端面（proximal view）, C. 背面（dorsal view）, D. 外侧面（lateral view), E. 掌面（volar view）, F. 远端面（distal view）

缩写（Abbreviation）：extu. 伸腱结节（extensor tuberosity）

尖。其下部与掌面之间以明显的内侧缘为界。而掌面与背外侧面之间的界线在骨体的远端部分也变得很明显，形成薄的外侧缘，与内侧缘在尖端汇聚，使其末端成尖锐的铲形。远端缘的外侧缘长于内侧缘，内、外两侧缘均平滑、无锯齿。尖端无中矢切迹。掌侧指深屈肌附着的爪指骨粗隆（tunph）往远方延伸的纵棱不发育。但具有二侧壁孔：外侧壁孔较明显，位于左侧掌沟的远端；而内侧者较小，位于内侧缘近端掌侧。FV-Ph3 远端背面纵向稍凸，横向平直；掌面纵向稍凹，横向圆凸。纵向的棱和沟仅在内、外侧缘附近发育，较粗糙。

后肢

在 V 16293 的零散的材料中有一枚第一趾（或指）节骨。它比 V 16293 的前脚的所有各指的第一指节骨都细长，其长比前脚中最长的第三指第一指节骨（FIII-Ph1）还要长 1/3。它有可能属于 V 16293 个体，也可能不是，甚至根本不属于副竹鼠类。如果它的确属于 V 16293 这一个体的话，笔者推测它可能是后脚的趾节骨，而不是前脚的指节骨。从该趾节骨内、外侧较对称判断，它可能是中间脚趾（ToII，ToIII 或 ToIV）的第一趾节骨 [primary phalanx of toe，To(II，III 或 IV)-Ph1]。其近端关节面后缘两侧的结节状隆起大小稍有区别。笔者参照前脚的情况判断，它可能是右侧的第三趾（或第二趾）的第一趾节骨，也可能是左侧的第四趾第一趾节骨，前者的可能性更大些。笔者现暂时将其作为右的第三趾（？）第一趾节骨（RToIII?-Ph1）来描述。

RToIII?-Ph1（图 55；表 17）仅近端背侧和远端稍破损。近端宽大，顶视约为浑圆的等边三角形。近端关节面前部已破损，其保留的后部的表面，横向明显圆缓凹入。近端蹠侧中央为宽大的凹沟，两侧有结节状隆起。内侧隆起稍大于外侧隆起。近端背侧伸腱结节（extensor tuberosity, extu）很发达，明显隆凸，表面粗糙，可能是趾长伸肌附着处。近端两侧近蹠缘处有粗糙面，其中外侧者较内侧者稍大而明显。

骨体细长而直。其上部稍宽大，横切面为四边形。中部变细。再往下部又稍加宽，但背 - 蹠向变薄。下部横切面为横宽的椭圆形。骨体背面纵向直，横向圆凸，与内、外两侧面圆滑过渡。两侧面纵向稍凹，

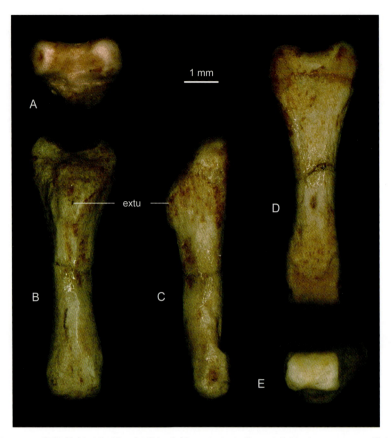

图 55 土著假竹鼠（新属、新种）右第三趾（？）第一趾节骨（V 16293, 正模）

Fig. 55 First finger of right third toe？(RToIII?-Ph1) of *Pseudorhizomys indigenus* gen. et sp. nov. (V 16293, holotype)

A. 近端面（proximal view），B. 背面（dorsal view），C. 内侧面（medial view），D. 蹠面（plantar view），E. 远端面（distal view）

缩写（**Abbreviation**）：extu. 伸腱结节（extensor tuberosity）

横向圆凸。骨体蹠面较平直。

　　远端比近端细小。远端视为横宽的长方形。与第二趾节骨关节面为半圆轴面，其背部一小部分转向背侧，稍向上延伸；而蹠侧向上延伸的部分明显超过背侧者。蹠侧上缘为凹入的弧形。关节面表面背-蹠向圆凸，被宽浅的矢状沟分为内、外两髁。远端两侧有很明显的凹陷区，可能供侧副韧带附着。

2. 比较

　　上述 V 16293 等标本与 *Pararhizomys* 在头骨和牙齿的基本形态上都有很多相似之处：如头骨较低，颧-咬肌结构属鼠型，具窄的吻部和宽的颅部，头骨前部与后部长度相近，前颌骨具侧背嵴，颧弓板限于上颌骨内，眶下孔具腹裂隙；视神经孔很小，咬肌与颊肌神经管很短，卵圆孔与中破裂孔愈合，颈内动脉孔位于听泡内缘中部稍前，头骨与下颌髁关节的关节窝很长，枕面平而微前倾；下颌骨为松鼠型，下颌水平支较粗短，咬肌嵴很发达，角突呈向后伸出的尖角状，i2 后端在下颌骨上升支颊侧形成明显的隆凸；有的臼齿为前-后侧高冠型，冠面结构简单等等。

　　但上述标本与 *Pararhizomys* 仍有明显的区别。它们不同于 *Pararhizomys* 的主要特点是：鼻骨背面纵向较平直，横向在前部圆凸度较小，在后部平直；前颌骨的前端前伸，明显超过鼻骨前端；前颌骨侧背嵴较短，向前近中方向斜伸，与鼻骨-前颌骨缝汇合后消失；前颌骨背侧面为窄的三角形；在背侧的前颌骨-上颌骨缝向外弯，与上颌骨-额骨缝共同形成一圆弧；眶下孔圆形；颧弓前根和颧弓板的位置较靠前；颧骨长，前端伸达泪骨；顶骨为短宽的六边形；硬腭后缘与M3后缘约在同一横线上；中翼窝与翼窝的宽度相近或仅稍宽；内翼突与腹面垂直，不倾斜，左、右内翼突彼此平行地纵向延伸；颏深神经孔与咬肌和颊肌神经孔愈合，而前者的后孔则与后二者的后孔分开。下颌齿隙的后部凹入较浅。下颌骨水平部下缘为圆弧形；颏孔位置较高；下颌切迹较深；臼齿齿冠较低，仅在M1和m3有较明显的前-后侧高冠现象；I2纵向弯曲较小，其前端不向后弯，为前伸型齿，其后端起自上颌骨的中部，M1之前；I2唇侧无明显的纵棱，等等（见表9）。这些区别清楚地表明，上述标本与*Pararhizomys*不同，应为一新属、种，在此称其为土著假竹鼠*Pseudorhizomys indigenus* gen. et sp. nov.。

甘肃假竹鼠（新属、新种）*Pseudorhizomys gansuensis* gen. et sp. nov.

（图56–70, 76；表1, 6–8, 10–19）

正模　HMV 1942，同一个体的含下颌头骨和部分颅后骨骼，产于LX 200502（宋家脑）；柳树组中部下层（详见表1, 10）。

副模　1）V 16297，含下颌头骨，产于LX 200007（古城村）；2）V 16298，较完整的头骨，产于LX 200011（大深沟）；3）V 16299，同一个体的部分头骨和左半下颌支，产于LX 200046（何家庄）；4）V 16300，头骨前部，产于LX 200037（潘杨村）。以上四件标本均产于柳树组中部下层（详见表1和表10）。

归入标本　1）V 16296，含下颌头骨，产于LX 200019（小寨村），柳树组中部上层；2）V 16301，部分头骨具右下颌；3）V 16302，部分头骨；4）V 16303，头骨前部。最后三件标本都产于临夏盆地，但详细地点和层位不明（详见表1和表10）。

产地与层位　甘肃和政（买家集乡宋家脑、新庄乡大深沟、关滩沟乡潘杨村）和广河（阿力麻土乡古城村、买家巷乡何家庄），上中新统柳树组中部下层；广河庄禾集乡小寨村，柳树组中部上层。

鉴别特征　吻部和上颌齿隙长；外层咬肌附着区的前缘约为"S"形，其圆凸的下部向前超过眶下孔的腹裂隙。颞嵴较弱，后部变平缓。听泡前内面的凹面很浅。下颌齿隙凹入的后部相对较短。冠状突前后较长，向后的倾斜度较小，其后缘较直。下颌髁关节面通常为椭圆形。前-后侧高冠仅在M1和m3明显。M1和M2的内凹横向长，在M1插入前边凹和中凹之间。M2–3无前边凹。M3内凹通常向前外方斜伸，与中凹斜交。m1–3无下后边凹，下外凹分叉；m3无下原凹，下外凹的位置后于下中凹。I2唇面横向圆凸，无纵棱。

第二掌骨的掌骨粗隆为粗糙的凹陷，位于骨体背面上部桡侧；第五掌骨近端与钩骨的关节面为圆四边形，表面为鞍形；骨体中部较窄细，约为四边柱形。拇指的爪指骨远端缘具锯齿。

区别特征　与*Ps. indigenus* gen. et sp. nov.的区别在于：听泡前内侧的凹面不明显；冠状突向后的倾斜度较小，其后缘较直；下颌髁关节面为椭圆形；M1–2的内凹较长，在M1插入前边凹和中凹之间；M2和M3以及m3的冠面均仅具两条褶沟（M2和M3均无前边凹和m3无下原凹），M3的内凹与中凹彼此斜交；m1–3的下外凹分叉；m3下外凹的位置稍后于下中凹；第二掌骨的掌骨粗隆为粗糙的凹陷，位于骨体背面上部桡侧；第五掌骨近端与钩骨的关节面为圆四边形，表面为鞍形，骨体中部较窄细，约为四边柱形；拇指的爪指骨远端缘具锯齿。

词源　Gansu，甘肃，省名；该类动物的产出地区。

1. 描述

（1）头骨（图56；表11, 18）

HMV 1942等的头骨的形态结构与*Ps. indigenus* gen. et sp. nov.的很相似。头骨很低，具鼠型颧-咬肌结构。

背面观（图 56 A2, B2） 鼻骨背面纵向较平直，在前部横向稍圆凸，后部横向平直，甚至稍凹。前颌骨的侧背嵴短，微微斜向前内方，其前部与鼻骨 - 前颌骨缝汇合。前颌骨的背面为尖端向前的窄三角形。前颌骨 - 额骨缝为强锯齿状，比较直地斜向前外方延伸，与前颌骨 - 上颌骨缝以钝角相交。前颌骨 - 上颌骨缝在头骨背侧也是向后外方弯的曲线，其后端与上颌骨 - 额骨缝的内端相连，共同形成一圆弧形。泪骨在 V 16296、V 16298 和 V 19302 保存较好，其背面出露得很小，上有明显的泪结节。颧骨长，其前端伸达眼眶前缘，并与泪骨相连。颧弓为圆弧形，相当粗壮。颧弓后端向后延伸不达项嵴，

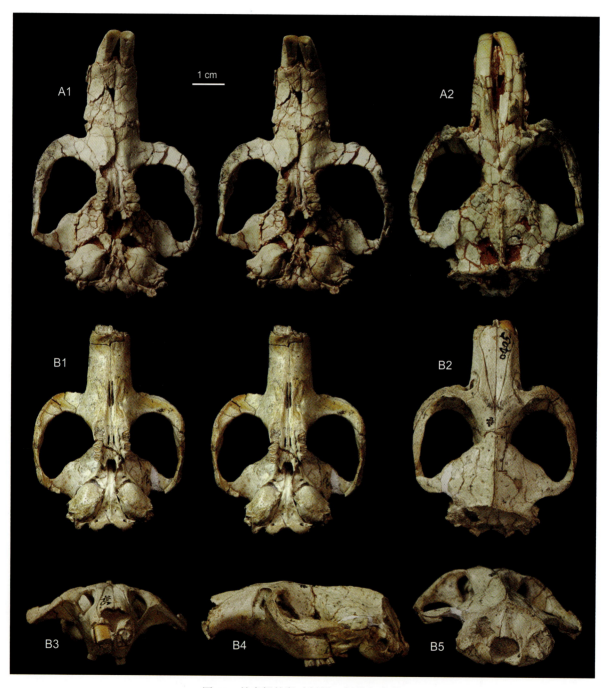

图 56　甘肃假竹鼠（新属、新种）头骨

Fig. 56　Skulls of *Pseudorhizomys gansuensis* gen. et sp. nov.

A. HMV 1942（正模，holotype）：A1. 腹面立体照片（ventral view, stereopair），A2. 背面（dorsal view）；B. V 16298（副模，paratype）：B1. 腹面立体照片（ventral view, stereopair），B2. 背面（dorsal view），B3. 前面（anterior view），B4. 左侧面（left view），B5. 后面（posterior view）

两者间有明显的缺凹分开。颧弓前根的位置靠前，后缘凹入相当深。其弧形后缘的最前点（即眼眶的前缘）的位置明显前于鼻骨和前颌骨后端的位置。眶后收缩显著。额嵴较 *Ps. indigenus* gen. et sp. nov. 者弱。矢状嵴很明显。顶骨在 V 16298 保存完好，约为宽大于长的六边形，也有一前突的前棘。与 *Ps. indigenus* gen. et sp. nov. 相似，颞窝也很宽大。颞嵴短，其后部也变得低缓。

侧面观（图 56 B4）　头骨背缘的轮廓约成中间稍凹的弱弧形：在眶后收缩处稍凹，其前的部分向前下方延伸；其后在矢状嵴处较高，稍圆凸，在近项嵴处稍往下弯。前颌骨前端向前伸，明显超过鼻骨的前端。I2 纵向较少弯曲，其前端向前下方延伸，不向后弯，为前伸型齿。I2 后端起自上颌骨中部，靠近 M1，在 M1 之前形成小隆凸。上颌齿隙很长，比上颊齿列长很多。M1 以前的头骨前部和头骨后部的长度相近（见表 11）。眶下孔限于上颌骨内，也为大的圆形，具明显的腹侧裂隙和骨质隔板。颧弓板表面稍凹，主要向前上方延伸。外层咬肌附着区也限于上颌骨内，前缘也约成 "S" 形，其圆凸的下部向前超过眶下孔的腹侧裂隙。凹坑状的咬肌结节的位置也紧靠门齿孔。眼眶很小。与下颌髁关节的关节窝（glf）很长，向后一直伸达项嵴，将外耳道（eam）挤至其下后方。乳突（msp）明显。项面稍向前倾斜。

眶部诸孔的位置与 *Ps. indigenus* gen. et sp. nov. 者很相似，比较靠前。背腭孔（dpf）很小，或不清楚，约位于 M1 和 M2 之间的上方，后上齿槽孔的后方。它们的咬肌神经孔（mf）和颊肌神经孔（bf）也已愈合，而且与颞深神经孔（dtf）一起位于同一裂隙中。它们的后孔则是分开的：颞深神经管的后孔开口于卵圆孔，而咬肌神经管和颊肌神经管的后孔则开口于翼蝶管的后端，与翼蝶管后孔汇合。其余诸孔的位置，如蝶腭孔、视神经孔、筛孔、前翼裂等，与 *Ps. indigenus* gen. et sp. nov. 者也都一致，在此不再赘述（详见图 56 B4）。

腹面观（图 56 A1, B1）　门齿孔约位于上颌齿隙的中部，其长约为上颌齿隙长的 1/4–2/5。锯齿形的前颌骨 - 上颌骨缝约成向前圆凸的弧形，其内端与门齿孔的中部相交。颧弓板的位置在横向上与门齿孔约在同一水平：其弧形前缘的最前点约与门齿孔前端相对，其弧形后缘的最前点约与门齿孔的后端对齐。腭沟深，在腭沟中只有一对较长的腭后孔（ppf），位于 M1–M2 的内侧。副翼窝浅，界线不明显，后上颌孔（pmf）位于其外后角。后上颌孔之内侧还有小孔。中翼窝（mptf）稍宽于翼窝（ptf）。两内翼突（ipp）不倾斜，而是与腹面垂直、彼此近于平行地纵向延伸。硬腭后缘与 M3 的后缘约在同一横线上。翼蝶管的后孔（ascp）位于翼窝的后外侧，外翼突之后。翼管后孔（pcpf）单一，位于翼窝后内缘，靠近内翼突外侧后部基部，横向稍后于翼蝶管后孔。卵圆孔与中破裂孔汇合。听泡膨胀程度较 *Ps. indigenus* gen. et sp. nov. 者弱，其前内面的凹面较浅，在有的标本（如 V16297 和 V 16298 等）上与 V 16293 的相似，其内侧显示出一条明显的纵沟。听泡内侧从颈内动脉孔（icf）向下后方斜伸的沟大而深，而听泡内侧从窄的颈静脉孔（juf）伸出的沟则不明显。在保存较好的听泡的标本上（如 V 16295– V 16299 和 V 16302 等）在颈静脉孔附近的听泡内后侧壁上均未见有蹬骨动脉孔。

后面观（图 56 B5）　头骨的项面在外形上与 *Ps. indigenus* gen. et sp. nov. 的很相似，也为半圆形，表面也相当平，但具稍圆隆的外枕嵴。枕骨下部较宽，颞骨岩乳部较小。枕骨大孔也为卵圆形，其宽大于高。左、右枕髁的后面彼此分得很开。HMV 1942 的副乳突保存较完好，约呈三面柱形，向外下方伸出。其末端稍膨大，较钝，其腹内面为明显的凹面。乳突较副乳突稍低矮些。

（2）下颌骨（图 57；表 11, 18）

下颌骨松鼠型。水平支下缘在臼齿下方稍圆凸。下颌联合的下缘与水平支下缘约以 40° 相交。下颌颏和下颌颏下缘的斜嵴均明显，其内侧的二腹肌窝为较窄的新月形。下颌齿隙长于下颊齿列的长。下颌齿隙后部凹入也较 *Pararhizomys* 的浅而长。咬肌窝向前伸至 m1 的下方。咬肌嵴很发达，并向外伸张。颏孔 1–2 个，位于 m1 前缘的下方，水平支的中部，与咬肌窝的前端约在同一水平上或稍低。

下颌上升支长，冠状突向后倾斜较弱，前缘与下臼齿齿槽缘的夹角约为 100°，其后缘较平直。冠状突内侧下部的颞肌窝明显。下颌切迹也较深，其深约为下颌高的 2/5。下颌孔大，位于冠状突后缘的下方，与颊齿列的冠面约在同一水平上。下颌髁较冠状突低，向后上方延伸较长，其上部稍向近中方弯曲，

表18 甘肃假竹鼠（新属、新种）头骨和下颌骨测量和比较（单位：mm）
Table 18 Measurements and comparison of skulls and mandibles of *Pseudorhizomys gansuensis* gen. et sp. nov. (in mm)

测量项（Parameter）***	正模 Holotype	副模 Paratypes				归入标本 Referred specimens			
	HMV 1942	V 16297	V 16298	V 16299	V 16300	V 16296	V 16301	V 16302	V 16303
枕鼻长（CNL）						65.9			
颅基长（CBL）			64			67.1			
头骨前部长（LAS）		27.1	32.5		25.6	35	32.5		
头骨后部长（LPS）	35.6		32	30.9		32.9			
腭长（PL）		36.1	42.5			45	41.8+		
上颌齿隙长（LMXD）	33.5, 34	23.2, 23.2	29.2, 29.3		21.2, 21	30.2, 30.2	28.7, 28.9	22.7, 21.6	
腭桥长（LPB）	16.8	13	16.5	15.3		16.4		~13.2	
腭桥前宽（AWPB）	4.2	3	4	3.5	2.8	3.4	~4.2	3*	
腭桥后宽（PWPB）	6.6	4.4	5.2	5.6		~4.3	~6.2	4*	
门齿孔长（LIF）		9.3	8.1		6	7.9		~6.3	
吻部长（RL）		15.9	19.5		14	20	19		20.7
吻部前宽（AWR）	~11.2	11.3	13.3		9.7	13.4	12.3		11.6
吻部后宽（PWR）	12.2	12.6	15.1		11.1	14.8	12.6	10.6	12.8
吻部高（RH）		8.5	11.7		8.3	10.6	10.2		
头骨中部高（HMS）	16.5+	17.3	21.8	17.4	16.2	21.4	19.6	16.2+	
(HMS/LAS)/%		64	68		63	61	60		
鼻骨长（NL）		20.7	23.5		19.8	21.8	22.7		
鼻骨前宽（AWN）		6.8	10		5.3	8.9	7.7		
鼻骨后宽（PWN）		1.6	1.1		1	1	1.1		
眶间宽（IOW）		11.3	14.8		10.2	14.3	13.5	10.4	11.5
眶后收缩处宽（WPOC）	9.2	7.5	7.4	7	7.5	12.8	8.2		
颧弓处宽（ZW）	53.6		49.6			46.6**	48	36.2	
脑颅乳突处宽（WCM）	31.6		28.4			29.5	27.7		
听泡长（LAB）	12, 12.1	11, 10.9	11.5, 11.9	11.9		11.6		10	
听泡宽（WAB）	13.3, 12	10.9, 10.3	12.2, 12	10.7		12.7			
听泡间距（DAB）	5.7	3.2	4.2	4.4				2.6	
项面高（NH）			19.5			16.1+			
枕骨大孔高（HFM）			6.2						
枕骨大孔宽（WFM）	6.5		8.3			7.8			
下颌骨长（MDL）	~49.2	41.2, 40.9							
下颌骨高（MDH）	27.1, ~28.6					~25.8	23.1+		
下颌髁突高（HCPM）	21, 19.4	17.5, 17.8				21.2	19.7+		
下颌切迹深（DMN）	10, 11.1					9.6+			
(DMN/MDH)/%	~36.9, 38.8					37.2+			
下颌齿隙长（LMD）	16, 16.3	11.9, 11.5		13.7		13.7, 13.6	~13.9		
下颌齿隙高（HMD）	10.6	8, 7.9		9.5		10.1, 10.2	9.4		
下颌水平支高（HHR）	13, 11.2	10.3, 10.1		10.4		12, 13.2	~11.4		

* 受压变形（compressed and deformed）。

** 颧弓处宽 = 右半部宽 × 2（zygomatic width = right half width × 2）。

*** 测量见表3（parameters as in Tab. 3）。

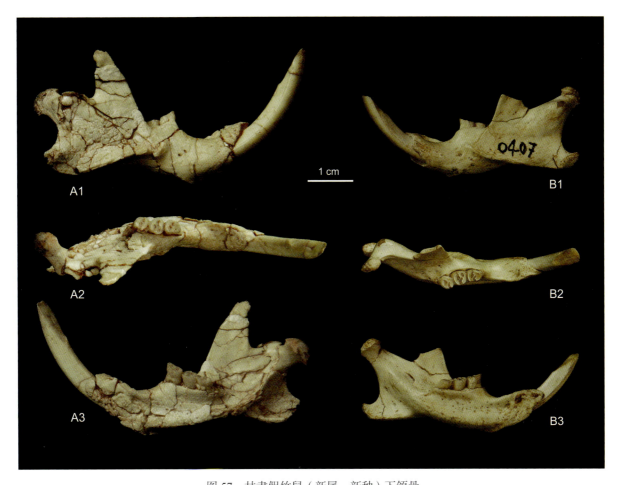

图 57 甘肃假竹鼠（新属、新种）下颌骨

Fig. 57 Mandibles of *Pseudorhizomys gansuensis* gen. et sp. nov.

A. HMV 1942 右下颌（正模）（right hemimandible, holotype）：A1. 颊侧面（buccal view），A2. 冠面（crown view），

A3. 舌侧面（lingual view）；B. V 16297 左下颌骨（left hemimandible）：B1. 颊侧面（buccal view），B2. 冠面（crown

view），B3. 舌侧面（lingual view）

其内侧供翼外肌附着的凹面很浅小。下颌髁顶端关节面为相当宽的椭圆形。下门齿齿槽后端在上升支颊侧，在髁突外下方形成的隆凸很显著。翼内肌窝大而深，其前端伸达冠状突中部的下方。上升支下缘圆弧形。V 16297 的左、右下颌角都保存较好。角突很发育，但比 *Pa. huaxiansis* sp. nov. 者要小。其下缘呈圆弧形隆凸。角突后端与下颌髁后缘约在同一垂线上，角突末端稍向内方弯曲。上升支后缘在髁突与角突之间为平缓凹入的圆弧形，但凹入的程度较 *Pa. huaxiaensis* sp. nov. 者弱。

（3）牙齿（图 58；表 12，19）

齿式 1·0·0·3/1·0·0·3。臼齿的基本形态与 *Ps. indigenus* gen. et sp. nov. 者很相似：为脊形齿，齿冠中等高度，具齿根。臼齿的尺寸从前往后变小。前 - 后侧高冠只是在 M1 和 m3 明显存在。

M1 冠面为四边形，长大于宽，颊侧长于舌侧。齿冠为前侧高冠型，釉质曲线在颊侧有折曲。M1 的冠面也具两条颊侧褶沟（前边凹和中凹）和一条舌侧褶沟（内凹）。三条褶沟在磨蚀较少的标本（如 V 16296、V 16297 和 V 16300）上全向外开口。在 V 16298 中，随着磨蚀加深，内凹变为封闭的盆，但其两颊侧褶沟仍开口，只是前边凹舌端已与颊侧部分分开，形成封闭的釉质坑。在磨蚀更深的标本的 M1 上，三条褶沟已全为封闭的盆。前边凹呈横向延伸，在多数标本上（7/8）仅伸达齿冠的中纵轴处，而在较少磨蚀的 V 16300 则超过中纵轴。中凹是褶沟中最长者，在多数标本上较直，呈横向或稍向后

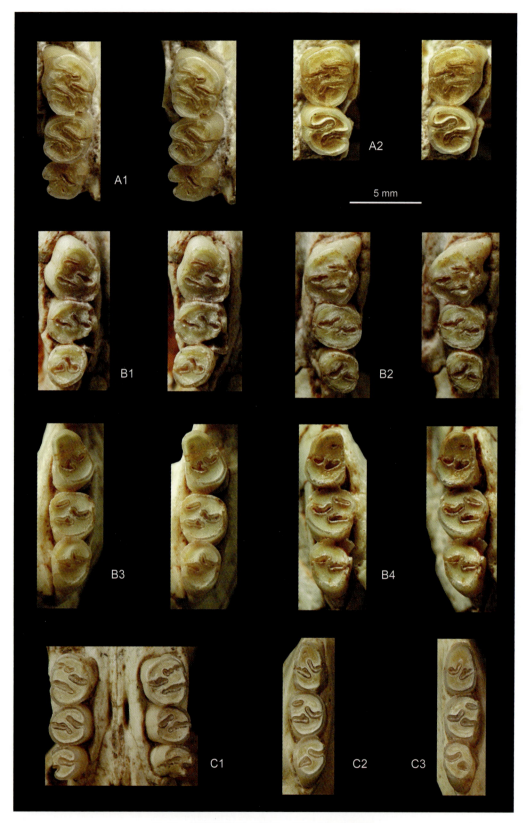

图 58　甘肃假竹鼠（新属、新种）臼齿冠面

Fig. 58　Occlusal view of molars of *Pseudorhizomys gansuensis* gen. et sp. nov.

A. V 16300 上臼齿立体照片（upper molars of V 16300, stereopair）：A1. 右（right）M1–3，A2. 左（left）M1–2；B. HMV 1942（正模, holotype）上、下臼齿立体照片（upper and lower molars, stereopairs）：B1. 右（right）M1–3，B2. 左（left）M1–3，B3. 左（left）m1–3，B4. 右（righ）m1–3；C. V 16297 上、下臼齿（upper and lower molars of V 16297）：C1. 左和右上臼齿（left and right upper molars），C2. 左（left）m1–3，C3. 右（right）m1–3

表19　甘肃假竹鼠（新属、新种）牙齿测量和比较（单位：mm）

Table 19　Measurements and comparison of teeth of *Pseudorhizomys gansuensis* gen. et sp. nov. (in mm)

测量项 (Parameter)*		正模 Holotype		副模 Paratypes								归入标本 Referred specimens							
		HMV 1942		V 16297		V 16298		V 16299		V 16300		V 16296		V 16299		V 16300		V 16303	
		L	R	L	R	L	R	L	R	L	R	L	R	L	R	L	R	L	R
M1–3 L		10	10	8.8		9.5		9	9.1		9.4	9.7	9.9	8.8	8.8	8.4	8.8		
M1	L	3.8	3.9	3.8	3.7	3.7		3.6	3.7	3.7	3.6	4	4	3.6	3.7	3.6	3.6		
	W	3.5	3.7	3.6	3.5	3.4		3.1	3.1	3.5	3.5	3.6	3.5	3.4	3.5	3.2	3.3		
	Hpl									2.4	2.4	2.3	2.3			2.3			
	Hpb1			1.9	1.9	2				2.8	2.8	2.1	2						
	Hpb2			2.1	2.3	1.9				3	3	2.3	2.4						
	(Hpl/Hpb2)/%									80	80	100	95.8						
M2	L	2.9	2.9	2.8	2.7	2.9	2.9	2.8	2.8	2.8	2.9	2.9	2.9	2.6	2.6		2.8		
	W	3.4	3.5	3.2	3.1	3.2	3	3.1	3.1	3.3	3.1	3.2	3.1	3	2.8		3		
	Hpl					2				2.1	2.2	2	1.8						
	Hpb1																		
	Hpb2			1.7		1				2.2	2.1	1.5	1.5						
	(Hpl/Hpb2)/%			182		200				95	105	133	120						
M3	L	2.8	2.8		2.4	2.5	2.6	2.4	2.7		2.5	2.5	2.7	2.4	2.3	2.6	2.4		
	W	2.7	2.7	2.7		2.5	2.5	2.5	2.4		2.7	2.5	2.6	~2.1	2.4	2.8	2.7		
	Hpl			1.2		1.9	1.9					1.7	1.7						
	Hpb2			0.8		1.3		1.3	1.3		1.1	1.2				1.2	1.2		
	(Hpl/Hpb2)/%			150		146						142							
I2	L	4.7	4.2	3.5	3.5	~4.5	~4.5			3.4	3.4	4.4	4.5	4.1	~3.8	3.6	3.7	3.6	3.6
	W	4.5	4.5	3.7	3.7	4.2	4.3			3.5	3.5	4.2	4.2	4.1	4.1	3.9	3.9	3.6	3.6
m1–m3 L		11.4	11.2	10.1	10.5			~9.6				10.7	10.6	10					
m1	L	4	3.9	3.6	3.6			3.5				3.6	3.8	3.6					
	W	2.9	2.9	2.7	2.8			2.7				2.8	2.7	2.9					
	Hpl													1.6					
m2	L	3.5	3.3	3.4	3.3			3				3	3.2	3.2					
	W	3.5	3.5	3.2	~3.2			~2.9				3.3	3.2						
	Hpl													1.8					
m3	L	3.4	3.3	3.1	2.9							3.1	3.3	3					
	W	2.9	3	2.9								2.9	2.7	2.3					
	Hpl	1.8	1.9	1.9	2.1							1.9	1.8						
	Hpb2	1.4		1.5								1.5	1.5						
	(Hpl/Hpb2)/%	129		127								127	120						
i2	L	4.7	4.7	3.6	3.6							5	5.2						
	W	4.3	4.3	3.5	3.6							4	4	~3.9					

　　* 测量项同表4（parameters as in Tab. 4）。

舌侧斜伸（6/8），而在 V 16296 和 V 16300 上，中凹舌端明显向后弯曲。这可能与后二者磨蚀程度较少有关。内凹为横向或稍向前外方斜伸，插入前边凹和中凹之间；其横向长度与前边凹的相近，它不但与中凹，而且与前边凹也部分重叠。原凹明显或不明显。

　　M2 冠面约为卵圆形，宽大于长。前侧高冠现象仅在 V 16300 较明显，而在其他的标本上均不明显。

釉质曲线在颊侧的折曲弱或无。M2 的冠面上均仅见有两条褶沟（中凹和内凹），未见前边凹的痕迹。两褶沟仅在 V 16296–V16298 和 V 16300 标本上开口，而在其余的 M2 中均封闭。中凹或多或少向后弯，横向的长度长于内凹或与内凹相近。内凹在中凹之前向前颊方斜伸，与后者部分重叠。

M3 冠面为卵圆形，后缘稍窄。未见前侧高冠现象。冠面也仅具中凹和内凹，未见前边凹的痕迹。中凹横向很短，稍向前舌方斜伸。中凹的深度要比内凹的深。除了在 HMV 1942 和 V 16301 已封闭外，中凹在其余的 M3 上均向颊方开口。内凹横向较中凹长，向前颊方斜伸，与中凹斜交。内凹通常与后者部分重叠。在现有的标本中，内凹除了在 V 16296 和 V 16298 的 M3，以及 V 16297 的右 M3 开口外，在其他的 M3 上均已封闭。M3 后部也有一些变异，如在 V 16298 的左 M3，V 16299 和 V 16300 的右 M3 的后半部都有一附加的封闭的釉质坑，而在其余的 M3 中未见此坑。另外，V 16298 的左 M3 有一较大的附加的釉质坑，插入内凹和中凹之间，将该两褶沟隔开，彼此不重叠。

I2 的形态与 *Ps. indigenus* gen. et sp. nov. 者很相似。纵向弯曲度较小，其前端不向后弯，而是向前下方延伸，为前伸型齿。I2 横切面为三角形。唇面横向稍圆凸，较圆缓的舌侧角的表面也有明显的纵沟。釉质层主要覆盖在门齿的唇面，仅稍延至内、外侧面。唇侧表面平滑，其上无明显的纵棱等。

m1 的冠面为后缘较宽的卵圆形，长大于宽。后侧高冠在 HMV 1942、V 16299 和 V 16301 上不明显；而在 V 16296、V 16297 颊侧较明显，而且釉质曲线在后二者的颊、舌两侧均明显折曲。冠面具两颊侧褶沟（下原凹和下外凹）和一舌侧褶沟（下中凹）。下原凹在现有的标本上已磨蚀成封闭的坑（6/8），或完全消失（2/8）。下外凹和下中凹仅在 V 16296 的 m1 上开口，而在其他 m1 上均已封闭。下外凹主要向后舌侧斜伸。在多数的 m1 上，下外凹舌部有一明显分叉，仅在老年的 V 16299 和 V 16301 的 m1 上未见明显的分叉。但 V 16299 的 m1 的下外凹舌部的前缘明显向前凸，这可能是分叉的残留部分。由于磨蚀较深，下外凹的分叉会逐渐消失。下中凹的颊部向前弯曲，指向下原凹。其横向的宽度与下外凹的相近。

m2 冠面近圆形，长、宽相近，较 m1 短而宽。未见明显的后侧高冠现象。釉质曲线在 V 16296 和 V 16297 中有折曲，而其他的 m2 上很弱或无。m2 的冠面也具两颊侧褶沟和一舌侧褶沟。该三褶沟仅在 V 16296 的 m2 上是开口的，而在其他的 m2 上均已磨蚀成封闭的盆。三褶沟中下原凹横向最短，主要为横向，仅稍向前舌方斜伸。下中凹是褶沟中最长者，其颊部也向前弯曲，朝向下原凹舌端延伸，但不伸达下原凹。下外凹与 m1 的很相似，其舌部也分叉。但分叉的前分支相对较弱，有时仅形成很微弱的向前的隆凸。只是在磨蚀很深的 V 16301 的 m2 上未见下外凹分叉。

m3 的冠面约呈卵圆形，后缘较窄，长大于宽。后侧高冠明显。釉质曲线无明显折曲。冠面仅见两条褶沟（下中凹和下外凹）。未见下原凹的痕迹。下中凹向前弯曲，横向较下外凹的稍长，也较深。下中凹的舌端仅在 V 16301 的 m3 上封闭，在其余的 m3 上仍开口。下外凹通常为横向。多数 m3 的下外凹的舌部都有分叉现象，只是较 m2 的更弱些，显出下外凹的舌端明显较膨大而前凸。只有在 V 16296 和 V 16301 m3 的下外凹上未见分叉。其中 V 16301 可能是深度磨蚀的结果。下外凹的位置稍后于下中凹，并与下中凹部分重叠。

i2 长而粗壮，纵向弯曲较 *Pa. hipparionum* 者强，与 *Pa. huaxiaensis* sp. nov. 者相近。在齿槽前出露的前部长，并向前上方弯。因此，其前端与 I2 的磨蚀面距齿槽缘的距离较大。i2 的切面也呈三角形，具横向稍圆凸的唇面和圆缓的舌侧角。舌 - 唇向长大于或等于横宽。釉质层主要覆盖在唇面，稍延至两侧面。唇面与外侧面之间无明显的界线，而与内侧面间的界线明显。唇面通常具两条细的纵嵴，而在 V 16296 则只有一条明显的纵嵴。

（4）下门齿微细结构（图 59）

观测的标本为 V 16297 右下颌下门齿的前端。其下门齿釉质层的微细结构的基本形态与 *Pa. hipparionum* 的相似，也属单系型（uniseral）。釉质层的总厚（T）约 140 μm。釉质层也明显分为内、外层。未见无釉柱的最外层（prismless external layer, PLEX）。外层（*portio externa*, PE）很薄，厚约 26 μm，仅约为釉质层总厚的 19%。釉柱间质（interprismatic matrix, IPM）不倾斜，在纵、横切面上均与门齿唇侧表面垂直。内层（*portio interna*, PI）很厚，厚约 114 μm。内、外层的比例（PE/PI）约为 23%。横切面显示：施氏明暗带（Hunter-Schreger bands, HSB）均从釉 - 齿质界面（enamel dentine junction, EDJ）稍向齿的内侧纵嵴的方向倾斜。其倾斜度从舌面往唇面方向稍有变化：近舌侧的倾斜度较大，往唇面倾斜度变小。唇、舌部倾角的差异从内侧纵嵴往门齿的内、外侧方向逐渐变小，甚至变得不明显。

V 16297 的右 i2 唇侧有两条纵嵴。在横切面上可见该二纵嵴处的釉质层均明显变厚（图 59 A–C）。其内侧纵嵴（medial ridge, mr）附近的内部的釉质层的 HSB 的形态结构与 *Pa. hipparionum* 的相似：纵嵴内、外侧的 HSB 斜向唇侧、向纵嵴处汇集。而在内侧纵嵴处，内层的施氏明暗带（HSB）和釉柱间质（IPM）的延伸和倾斜的方向均发生了变化：在近舌部倾角较大，HSB 和 IPM 彼此约呈垂直相交，而往唇侧逐渐转为彼此近于平行向唇侧延伸，并渐渐过渡到外层（PE）。但其外侧纵嵴（lateral ridge, lr）处的釉质层的结构与门齿的其他部分的相似，并无明显变化。

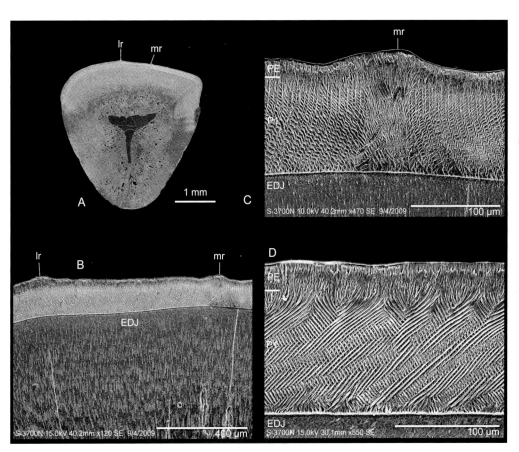

图 59　甘肃假竹鼠（新属、新种）的右下门齿（V 16297）釉质层的微细结构

Fig. 59　Enamel microstructure of right lower incisor (V 16297) of *Pseudorhizomys gansuensis* gen. et sp. nov.

A–C. 横切面（cross section），D. 纵切面（longitudinal section）

缩写（Abbreviation）：EDJ. 釉 - 齿质界面（enamel dentine junction），lr. 外侧纵嵴（lateral ridge），mr. 内侧纵嵴（medial ridge），PE. 外层（*portio externa*），PI. 内层（*portio interna*）

（5）颅后骨骼（图 60–70）

HMV 1942 保存部分颅后骨骼，包括五枚颈椎，还有可能是第一至第四胸椎的破损的椎骨或椎弓、两枚肩胛骨、数枚肋骨和部分右前肢。HMV 1942 的右前肢保存较完好，且多少是原位保存。但其脊椎、肩胛骨和肋骨却在头骨后方杂乱地挤压在一起（见图 60），可能是稍被冲散后再沉积成的。

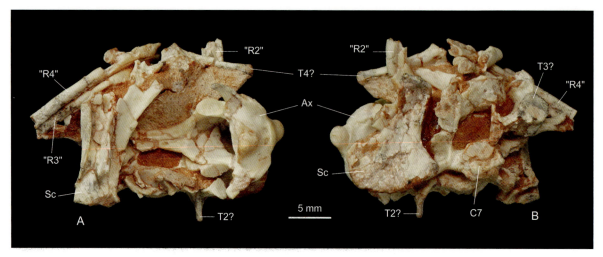

图 60　甘肃假竹鼠（新属、新种）部分颅后骨骼（HMV 1942，正模）

Fig. 60　Partial postcranial bones（HMV 1942, holotype）of *Pseudorhizomys gansuensis* gen. et sp. nov.

A. 上面（upper view），B. 下面（lower view）

缩写（**Abbreviations**）：Ax，枢椎（axis），C7，第七颈椎（7th cervical vertebra），"R2"–"R4"，"第二"至"第四"肋骨（"2nd"–"4th" ribs），Sc，肩胛骨（scapula），T2?–T4?，第二? 至四? 胸椎（2nd? –4th? thoracic vertebrae）

颈椎（cervical vertebrae, C's）

HMV 1942 保存的五枚颈椎包括寰椎、枢椎、第四（或第五）颈椎、第六和第七颈椎。

寰椎（atlas, At；图 61；表 6）　HMV 1942 的寰椎保存较好，但寰椎翼（w）大部分破损。寰椎的外形与 *Ps. indigenus* gen. et sp. nov. 的很相似，也为椭圆的环形，宽大于高，前后很短。背弓（da）也为前后短、横向伸长的条带形。背结节（dt）位于背弓背面正中，也较粗大。但其顶端的位置稍靠后，其后坡较陡而短；前坡较缓而稍长，前面中央供寰枕背侧膜附着的窝较大，而且被纵棱分为两个小坑。侧椎孔和翼孔在背侧的开口也相连，形成侧椎 - 翼孔（lvaf）。但该孔比 V 16293 的要小些，位置也稍后移，约位于背弓背面两侧的中部。此外，HMV 1942 右侧侧椎 - 翼孔（lvaf）的近中侧还有一小孔也与寰椎侧椎管相通。这可能是侧椎孔（lvf）的残余，供枕动脉附属小分支穿过。背弓腹面的形态和内侧椎孔（ilvf）的位置与 *Ps. indigenus* gen. et sp. nov. 者相似，但椎动脉沟（gva）较深。腹弓（va）前后向较短于背弓，上下较厚。近正中处约为稍弯的三角棒形。腹弓背面的齿窝（odf）前后较长而平，横向凹入。腹弓前面背 - 腹向较短，表面较平，构成两前关节凹（craf）间的切迹。腹弓腹面前后较长、中部横向圆凸，两侧部分凹入。该两侧部较 *Ps. indigenus* gen. et sp. nov. 者宽广，凹入也较浅，而且在其两侧端近前缘处有一圆形较深的凹面。在腹弓腹面未见从寰椎翼后部下方向内延伸的棱，近后关节面处也未见小的凹面。腹结节（vt）较 *Ps. indigenus* gen. et sp. nov.（V 16293）者小，其两侧也有一对明显的小凹坑。椎孔与 *Ps. indigenus* gen. et sp. nov. 者相似，但在椎孔的远中侧壁的中部，在椎孔的上、下部之间有小的丘形隆凸，供寰椎横韧带附着。在该丘形隆突的后面和上面有凹入的粗糙面，供寰枢关节内侧韧带附着。

寰椎的前关节凹（craf）与 *Ps. indigenus* gen. et sp. nov. 者相似，大致为肾形。前关节凹上部较宽大，明显凹入，其下部较窄、较平。左、右关节面在上部分得较开，往下彼此靠近。但 HMV 1942 的前关节凹上缘的弧度较 *Ps. indigenus* gen. et sp. nov. 的大，较明显地向前凸，使背弓前缘（从背面看）成为稍后凹的弧形。寰椎的后关节面（caf）也与 *Ps. indigenus* gen. et sp. nov. 者相似，关节面的上部周缘都较薄锐，明显突出。但后关节面的前上角要比 *Ps. indigenus* gen. et sp. nov. 者更向前上方突出些，也更薄锐些。

V 16295 的寰椎翼仅保存了一部分。寰椎翼和寰椎窝的整个形态特征都不清楚。从保存的部分看，寰椎窝也可能较大，但窝内侧连接翼孔和下翼孔的沟较 *Ps. indigenus* gen. et sp. nov. 者窄而深，下

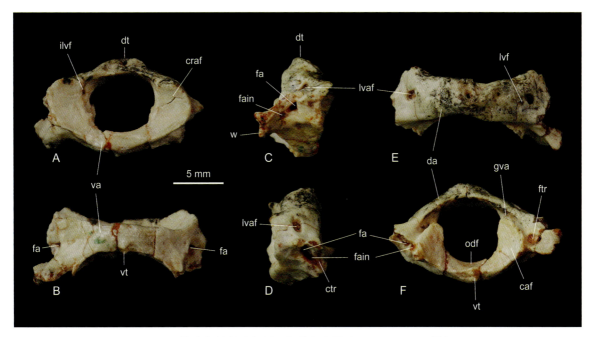

图 61 甘肃假竹鼠（新属、新种）寰椎（HMV 1942，正模）

Fig. 61 Atlas（HMV 1942, holotype）of *Pseudorhizomys gansuensis* gen. et sp. nov.

A. 前面（anterior view），B. 腹面（ventral view），C. 右侧面（right lateral view），D. 左侧面（left lateral view），E. 背

面（dorsal view），F. 后面（posterior view）

缩写（**Abbreviations**）：caf. 后关节面（caudal articular facet），craf. 前关节凹（cranial articular fovea），ctr. 横突管

（*canalis transversarius*），da. 背弓（dorsal arch），dt. 背结节（dorsal tubercle），fa. 翼孔（*for. alare*），fain. 下翼孔（*for.

alare inferior*），ftr. 横突孔（*for. transversarium*），gva. 椎动脉沟（groove for vertebral artery），ilvf. 内侧椎孔（internal

lateral vertebral for.），lvaf. 侧椎 - 翼孔（lateral vertebro-alar for.），lvf. 侧椎孔（lateral vertebral for.），odf. 齿窝 (odontoid

fossa)，va. 腹弓（ventral arch），vt. 腹结节（ventral tubercle），w. 寰椎翼（wing）

翼孔（fain）可能也要小些。翼孔（fa）与下翼孔间的距离与 *Ps. indigenus* gen. et sp. nov. 者相近（约
2 mm）。横突孔（ftr）大，位于后关节面的后外角的外方和椎动脉沟的外端。左侧寰椎翼破损处显露
出连接下翼孔与横突孔的横突管（ctr）。

枢椎（axis，Ax；图 60，62；表 7）   HMV 1942 的枢椎保存较好，仅棘突顶部和腹部稍破损。枢
椎明显长于寰椎。椎体为横宽、上下低的扁圆柱形。齿突（odpr）为横向稍宽，前端钝圆的扁圆锥体。
齿突前端有一对供齿突韧带（ligament of dens）附着的粗糙凹面。齿突背面有一条横沟，供寰椎横韧带
经过。齿突腹面与寰椎齿窝的关节面横向圆凸，纵向稍凹。枢椎的前关节面（aaf）约为上窄下宽的卵
圆形，高大于宽，表面稍圆凸，主要面向前外方。前关节面的外侧缘和腹缘形成圆弧形的薄锐边缘，
向外突出；其内侧缘较平直，下内角与齿突相连；其背缘延伸达椎弓根。椎体背面较平滑；在其后 1/3
处有一对小孔，可能为滋养孔；其前部两侧，近前关节面处各有一小的粗糙区，可能是寰枢关节的内
侧韧带附着的地方。椎体腹面（图 60 A）的外形轮廓与 *Ps. indigenus* gen. et sp. nov. 者相似，也约呈
扁圆形，但比例上更窄长。其前缘也是由齿突的腹缘和两前关节面的腹缘组成的圆弧形棱为界，但该
棱较明显向下突出，而且在齿突两侧与两前关节面间有切迹分开。HMV 1942 的枢椎也无明显的腹棘
（ventral spine），但腹面中央的圆凸隆较 *Ps. indigenus* gen. et sp. nov. 者更明显（可惜后半部已破损）。
此外，在 HMV 1942 的枢椎的腹面中央的前方、紧靠齿突腹缘处有两小坑，被弱的纵棱分开。椎窝（vfs）
为横宽的扁圆形，表面稍凹。椎孔近圆形，高与宽相近。

棘突（sp）很发达，稍向后倾。棘突下部横切面约呈窄的等腰三角形。前缘很薄。后面较宽厚，
中央有竖沟。顶部呈薄板状，但已破损并弯曲变形。

椎弓根较短。后关节突（pzy）位于椎弓根后缘上部。与第三颈椎前关节突关节的关节面（fpzy）

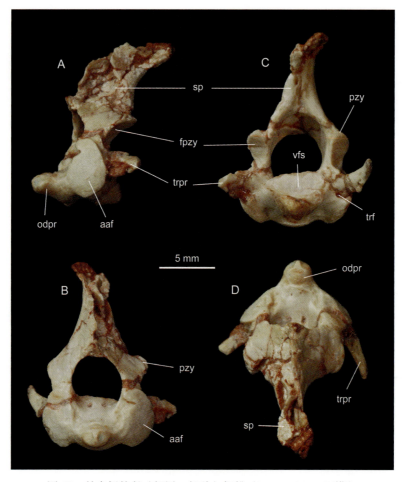

图 62　甘肃假竹鼠（新属、新种）枢椎（HMV 1942，正模）

Fig. 62　Axis (HMV 1942, holotype) of *Pseudorhizomys gansuensis* gen. et sp. nov.

A. 左侧面（left lateral view），B. 前面（anterior view），C. 后面（posterior view），D. 背面（dorsal view）

缩写（**Abbreviations**）：aaf. 前关节面（anterior articular facet），fpzy. 后关节突上关节面（facet on postzygapophysis），odpr. 齿突（odontoid process），pzy. 后关节突（postzygapophysis），sp. 棘突（spinous process），trf. 横突孔（transverse for.），trpr. 横突（transverse process），vfs. 椎窝（vertebral fossa）

面向外下后方。该面约为卵圆形，横轴较短，长轴约沿上下方向延伸，表面大部稍凹。椎弓根后缘，在后关节突与横突之间有一明显的后椎切迹（caudal vertebral notch）。

横突（trpr）约呈窄三角锥形，由椎体和椎弓根结合部侧面向外后方伸出。具背、腹两侧根：背侧根较大，约呈板状；腹侧根为细棒形。横突孔（trf）大，呈横向裂缝状，从横突的背、腹两根中间穿过（可惜在修理时，这一部分被破坏）。

第四（或五）颈椎？（C4/C5?）　HMV 1942 的材料中有一颈椎的右侧椎弓，包括部分椎弓背部和根部，后关节突和部分前关节突。从侧面看，后关节面的外缘与前关节面的外缘近于平行，两关节面间的最短距离约 0.9 mm。后关节面的前、后端分别与前关节面的前、后端大约位于同一水平。它的形态结构与 *Ps. indigenus* gen. et sp. nov. 的 C4 或 C5 的相似。

第六颈椎（C6；图 63 A；表 8）　HMV 1942 的第六颈椎保存了椎体和椎弓的大部分，但椎弓背部破损，左、右横突外端均断失。椎弓和前、后关节突的基本形态与 *Ps. indigenus* gen. et sp. nov. 者很相似。因 HMV 1942 的 C6 是游离的，显示了整个椎体和前、后关节突上的关节面的特征。前关节突上关节面（fprzy）约为椭圆形，前后径大于横径；表面稍凹，主要面向内上方，稍向前方倾斜。后关节突上关节面（fpzy）也为椭圆形，但前后径与横径的差距较小；关节表面较平，仅上部稍凸，主要面向下外方，稍向后方倾斜。从侧面看，两关节面的侧缘彼此近于平行。

椎头和椎窝均为横宽的椭圆形。椎头（cap）表面稍圆凸，中有卵圆形的凹陷。椎窝（vfs）表面稍凹。椎体背面较平，中央部分为稍隆起的纵棱，两侧微凹。椎体腹面纵向较平，横向稍圆凸。C6 的横突（trpr）很发达，具两侧根。横突腹侧根起自椎体侧面，为很宽大的板面。背侧根起自椎弓根，前后明显短于腹侧根。横突外部也分成两支。第六颈椎横突腹板（或称横突下板，lvvc）为前后伸长的板状骨，其外部往下方弯曲；可惜其外端破损，延伸的情况不清楚。第六颈椎横突背支（ldvc）较腹板窄，主要向外方伸出。横突管很短，前、后横突孔（trf）都很大，分别与前、后椎切迹相沟通。

第七颈椎（C7；图 60, 63 B；表 8） HMV 1942 的第七颈椎的棘突、椎体腹面和横突远端部分破损。C7 的椎体约成扁圆截椎体。椎头较椎窝宽大，宽明显大于高，表面稍圆凸，中央有凹陷。椎窝之宽仅稍大于高，表面微凹。椎窝两侧近腹缘处有与第一肋骨头关节的小关节面（facet for head of R1，fhR1）。椎弓背部为短而横宽的条带状，因棘突破损，其形态结构不清楚。椎弓根和前、后关节突，以及前、后关节面的形态结构与 C6 的相似，只是后关节面表面较平。此外，C7 横突的起始点单一，不分背、腹根，中间也无横突孔穿通；横突约为稍扁的圆柱形，向外延伸，其横切面的前后横径稍大于垂向直径。

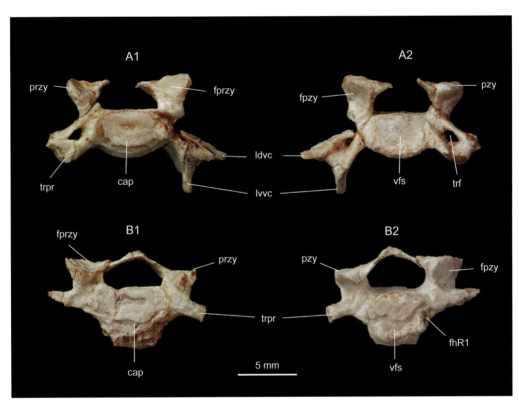

图 63　甘肃假竹鼠（新属、新种）第六和第七颈椎（HMV 1942，正模）

Fig. 63　6th–7th cervical vertebrae (C6–C7; HMV 1942, holotype) of *Pseudorhizomys gansuensis* gen. et sp. nov.

A. 第六颈椎 (6th cervical vertebra, C6)：A1. 前面（anterior view），A2 后面（posterior view）；

B. 第七颈椎 (7th cervical vertebra, C7)：B1. 前面（anterior view），B2. 后面（posterior view）

缩写（**Abbreviations**）：cap. 椎头（caput），fhR1. 与第一肋骨头关节面（facet for head of 1st rib），fprzy. 前关节突上关节面（facet on prezygapophysis），fpzy. 后关节突上关节面（facet on postzygapophysis），ldvc. 第六颈椎横突背支（*lamina dorsalis vertebrae cervicalis* Ⅵ），lvvc. 第六颈椎横突腹板（*lamina ventralis vertebrae cervicalis* Ⅵ），przy. 前关节突（prezygapophysis），pzy. 后关节突（postzygapophysis），trf. 横突孔（transverse for.），trpr. 横突（transverse process），vfs. 椎窝（vertebral fossa）

胸椎（thoracic vertebrae, T's）

HMV 1942 所保存的胸椎都很破碎，只能分辨出有两枚椎体和两枚具棘突的部分椎弓。可惜椎体和椎弓无法完全修理复原，保存的位置也很零乱，很难直接判断它们在脊柱中的位置。幸运的是，*Ps. indigenus* gen. et sp. nov. 的前六枚胸椎均是原位保存。如果其前部的胸椎的变化规律与 *Ps. indigenus* gen. et sp. nov. 的胸椎的变化规律相似的话，根据 V 16293 前部胸椎的椎体的形态和前、后关节突与横突的相互关系的变化规律来判断，HMV 1942 的上述两枚椎体和两枚具棘突的部分椎弓可能是分属 T1 至 T4 四枚胸椎。

第一胸椎（?）（T1?；图 64 A） HMV 1942 保存了第一胸椎（?）的椎体和椎弓根，以及右侧的横突、前、后关节突和左前关节突。椎体约呈扁圆柱形。椎头（cap）表面稍凸，椎窝（vfs）表面稍凹。椎窝两侧有三角形的小面，向外倾斜，是与第二肋骨头关节面（fhR2）。椎体腹面两侧的前部有一凹面，是与第一肋骨头关节面（fhR1）。椎弓根粗壮。前关节突（przy）从椎弓根向前伸出。其上的关节面（fprzy）约呈卵圆形，表面稍凹，面向内上方。后关节突（pzy）上的关节面（fpzy）也呈卵圆形，表面稍凸，面向外下方。横突（trpr）较粗大，从椎弓根侧部横向伸出，位置比前、后关节突都低。

第二胸椎（?）（T2?；图 64 B） HMV 1942 只保存了第二胸椎（?）的椎弓背部和棘突。棘突（sp）约呈窄的扁椎体，其横径窄，前后径长，前、后缘为较窄的嵴；棘突从背弓中央近于垂直地向背方延伸，不向后倾斜。棘突保存部分的高约 5.5 mm。

第三胸椎（?）（T3?；图 64 C） HMV 1942 保存了第三胸椎（?）的椎体和椎弓根。T3（?）的椎体长稍长于 T1（?）的椎体，椎体为扁圆柱形，宽稍大于高。椎体背面较平缓。椎体腹面中央部分纵向直，横向圆凸；两侧部分破损。椎头表面稍圆凸，中央有卵圆形的凹陷区。椎窝（vfs）表面稍凹。其两侧有与第四肋骨头关节面（fhR4）。椎弓根较粗大。前关节突从椎弓根向前伸。前关节面为稍长的卵圆形，表面平直，向上，并稍向外倾。后关节突也从椎弓根向后伸出，可惜其大部分已缺失。横突从椎弓根向外上方伸出，较粗壮，其背侧缘高于前关节面。

第四胸椎（?）（T4?；图 64 D） HMV 1942 只保存了第四胸椎（?）的椎弓右半部分，具有棘突、右横突和右侧的前、后关节突。T4（?）的棘突（sp）很高大（前缘直线高约 11 mm），约为稍向后弯而顶端钝的牛角状。棘突下部的横切面为卵圆形，横径稍大于前后径。前面下部有浅的垂向纵沟。棘突上部的横切面则为横向较窄的卵圆形，前后径大于横径。前关节突从椎弓根向前伸出。前关节面为卵圆形，表面稍凸，面向外上方。后关节突从椎弓根向后伸出。后关节面也约为卵圆形，表面稍凹，面向内下方。粗大的横突从椎弓根向外上方伸出，位置明显高于前关节突，两者间有明显的槽沟分开。而横突背缘稍低于后关节突的背缘。

肋骨（ribs, R's；图 60, 64 D）

HMV 1942 的肋骨保存较好的有三条。其中有一条与第四胸椎椎弓挤压在一起，可能是第四或其附近的肋骨。从肋骨结节和骨体的形态看，该肋骨更像是右侧的第二肋骨。为了叙述简便，笔者暂称其为"第二肋骨"（"R2"）。该肋骨的上部保存得相当完好。肋骨的椎骨端（从肋骨小头到肋骨结节外侧的距离）较长。肋骨颈（costal neck, cn）长而直，将肋骨小头（costal head, ch）和肋骨结节（tubercle of rib, tur）分得较开；靠近肋骨小头附近的表面粗糙，供韧带附着。肋骨小头大于肋骨结节，其顶端保留有一与胸椎椎体的卵圆形关节面。肋骨结节明显隆凸，可惜其顶端的关节面已破损。肋骨体上部很扁平，前面横向稍圆凸，后面较平。肋沟（costal groove）很浅。"R2"的肋骨近端（＝椎骨端）长为 6.9 mm；骨体近端（肋骨结节后）宽 × 厚为 3.2 mm × 1.5 mm。

另外两条被挤压在一起的肋骨可能是右侧第三—四或稍后的肋骨，笔者暂称其为"第三—第四肋骨（"R3–4"）。"第三肋骨"（"R3"）保存有部分肋骨颈和肋骨体的大部分。"R3"的肋骨结节比"R2"的低，其与肋骨体有明显的凹陷区分开（该凹陷区可能是韧带附着处）。肋骨结节顶面为卵圆形，表面稍圆凸。"R3"和"R4"的肋骨体为长而稍弯的扁骨。骨体上部较宽扁，往下部逐渐变得较窄。骨体前面横向稍圆凸。后面在"R3"中较平直，肋沟很浅；而在"R4"中肋沟较明显而长。

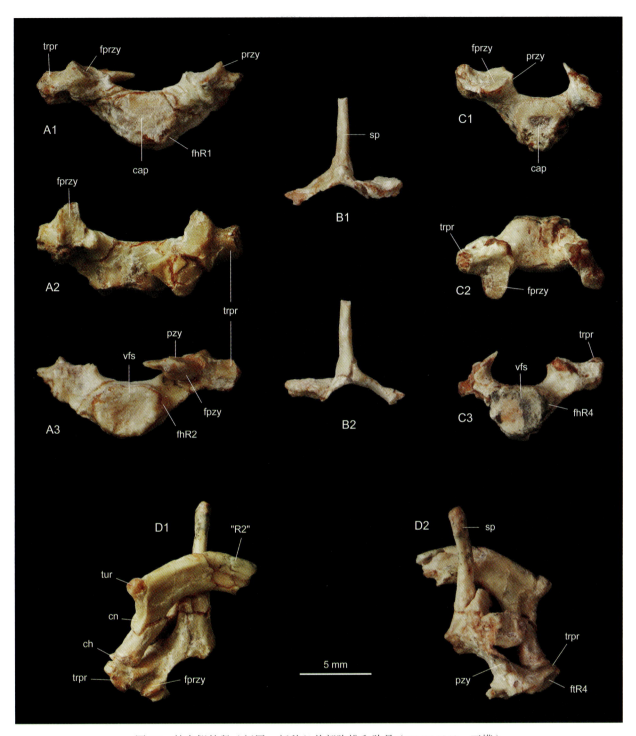

图 64　甘肃假竹鼠（新属、新种）前部胸椎和肋骨（HMV 1942，正模）

Fig. 64　Anterior thoracic vertebrae and rib (HMV 1942, holotype) of *Pseudorhizomys gansuensis* gen. et sp. nov.

A. 第一胸椎（？）[1st thoracic vertebra?（T1？）]：A1. 前面（anterior view），A2. 背面（dorsal view），A3. 后面（posterior view）；

B. 第二胸椎（？）[2nd thoracic vertebra?（T2？）]：B1. 前面（anterior view），B2. 后面（posterior view）；C. 第三胸椎（？）[3rd thoracic vertebra?（T3？）]：C1. 前面（anterior view），C2. 背面（dorsal view），C3. 后面（posterior view）；D. 第四胸椎（？）和"第二肋骨" [4th thoracic vertebra?（T4？）and "2nd rib"（"R2"）]：D1. 前面（anterior view），D2. 后面（posterior view）

缩写（Abbreviations）：cap. 椎头（caput），ch. 肋骨小头（costal head），cn. 肋骨颈（costal neck），fhR1. 与第一肋骨头关节面（facet for head of 1st rib），fhR2. 与第二肋骨头关节面（facet for head of 2nd rib），fhR4. 与第四肋骨头关节面（facet for head of 4th rib），fprzy. 前关节突上关节面（facet on prezygapophysis），fpzy. 后关节突上关节面（facet on postzygapophysis），ftR4. 与第四肋骨结节关节面（facet for tubercle of 4th rib），przy. 前关节突（prezygapophysis），pzy. 后关节突（postzygapophysis），"R2". "第二肋骨"（"2nd rib"），sp. 棘突（spinous process），trpr. 横突（transverse process），tur. 肋骨结节（tubercle of rib），vfs. 椎窝（vertebral fossa）

"R3"和"R4"保存部分直线长分别为 17 mm，18.5 mm。肋骨体近端宽 × 厚分别为 2.5 mm ×
1.4 mm，2.5 mm × 1.3 mm。

前肢（forelimbs）

HMV 1942 保存有部分肩胛骨、右肱骨头，右尺骨和桡骨的远端，以及右前脚。

肩胛骨（scapula，Sc；图 60, 65 A） HMV 1942 只保留了左、右肩胛骨下半部分。其中以右肩胛
骨保存较好。肩臼（glc）约为长卵圆形，其长径大于肩胛颈（nsc）之宽。可惜肩臼的周缘大部分破损，
肩胛结节和喙突也均已缺失。肩胛冈（spsc）只保存了基部。其下端起始点接近肩胛颈，距肩臼外缘约
4.2 mm。肩胛冈下端游离缘高耸，较宽厚。肩峰（acromion）完全缺失。肩胛骨前缘在右肩胛骨上保存
的部分较多。从保存的部分看，肩胛骨前缘向上稍向前弯，其弯曲度与 *Rhizomys* 者相近，而与鼢鼠和
麝鼠的不同。冈上窝（sspf）保存的部分为窄扇形，其下部很窄，宽约为 1.8 mm，向上逐渐变宽。冈
下窝（ispf）也为狭窄的扇形。其下部的宽度与冈上窝者相近，往上也逐渐变宽。后缘横向较厚而浑圆，
其下部受挤压变形。后缘附近有一供冈下肌附着的肌线（lm）。肌线很发达，向下几乎伸达肩胛窝的后
缘。冈下窝在该肌线与肩胛冈之间的部分横向明显凹入，表面光滑；而在该肌线之后的部分较平或稍凹。
在肩胛下窝 (sscf) 近后缘处也有一发达的肌线，往上逐渐与后缘的嵴汇合，往下伸至肩胛颈。该肌线的
下部，在肩胛颈的后面有较粗糙隆起，可能供小圆肌附着。肩胛下窝（sscf）为窄长的三角形，下窄上宽。

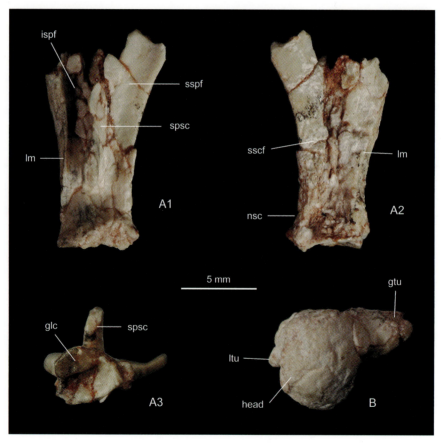

图 65　甘肃假竹鼠（新属、新种）肩胛骨和肱骨（HMV 1942，正模）
Fig. 65　Scapula and humerus (HMV 1942, holotype) of *Pseudorhizomys gansuensis* gen. et sp. nov.

A. 部分右肩胛骨（partial right scapula）：A1. 外侧面（lateral view），A2. 内侧面（costal surface view），A3. 肩臼端面（glenoid
cavity view）；B. 右肱骨近端的近端面（proximal view of proximal part of right humerus）

缩写（**Abbreviations**）：glc. 肩臼（glenoid cavity），gtu. 大结节（greater tuberosity），head. 肱骨头（head of humerus），ispf. 冈
下窝（infraspinous fossa），lm. 肌线（*linea muscularis*），ltu. 小结节（lesser tuberosity），nsc. 肩胛颈（neck of scapula），spsc.
肩胛冈（spine of scapula），sscf. 肩胛下窝（subscapular fossa），sspf. 冈上窝（supraspinous fossa）

左肩胛骨肩胛下窝表面较光滑，未见有明显的中央纵隆的现象。

肱骨（humerus, Hu；图 65 B；表 14） HMV 1942 保存有右肱骨近端的一部分。肱骨头（head）保存较完全，面向后上方，与肩臼关节的面为卵圆的半球面，明显大于肩臼 [肱骨头宽（caput W）5.3 mm，径长（APD）6.9 mm]。肱骨颈在肱骨头的后下方很明显。大结节（gtu）大部分破损，从保存部分看，大结节位于肱骨头的前外侧，向前外方伸，但不明显向上隆凸，其顶端低于肱骨头关节面；与肱骨头以明显的凹面分开。小结节（ltu）也只保留了其基部，位于肱骨头的前内侧，比大结节低小。隆间沟宽大，中无纵棱分隔。

桡骨（radius, Ra；图 66, 67 A；表 15） HMV 1942 保存右桡骨的下半部，远端骺软骨未骨化，骨骺与骨体分离，骨骺破裂为两部分。

桡骨骨体的下部为扁圆柱形，横宽大于前后径，往远端的尺寸增大。骨体下部的外面横向凹入，与尺骨紧连在一起，未修理开，该处可能是桡、尺骨间韧带附着处。桡骨远端较粗大，远端视为不规则的四边形。远端与腕舟 - 月骨的关节面宽大而圆凹。远端最大宽约为 4.8 mm。

尺骨（ulna, Ul；图 66, 67 A） HMV 1942 仅保存了右尺骨的下半部。和桡骨一样，尺骨远端骺软骨也未骨化，骨骺与骨体错位分离。

尺骨与桡骨完全分离。尺骨下半部也为扁圆柱形。远端也较粗大，从前面看，比桡骨的要稍窄一些。尺骨茎突很发达。远端与楔骨关节面较圆凸。远端外侧有明显的肌腱沟，可能供腕尺侧伸肌腱通过。尺骨远端最大宽约为 4 mm。

前脚（manus, Ms；图 66, 67） HMV 1942 保存了较完整的右前脚，其腕骨、掌骨和指骨几乎都是原位保存，只有个别腕骨因受挤压稍移位。

腕骨（carpal bones, Carp；图 66, 67） HMV 1942 的右侧的腕骨挤压在一起，没有完全修理开，只能从整体外观上进行描述。腕骨分为两列。近列有腕舟 - 月骨、楔骨和豌豆骨三件。远列有大多角骨、小多角骨、中央骨、头状骨和钩骨五件。另外在桡侧还有一镰状骨（桡籽骨）。

腕舟 - 月骨 [scapho-lunar, Sc-lu，又名桡中间腕骨（*os carpi radiointermedium*）] 腕舟 - 月骨位于腕部内侧，由腕舟骨（scaphoid）和月骨（lunar）愈合而成，为近列中最大者。HMV 1942 的腕舟 - 月骨保存较好，只是因受压中间产生了一条裂缝。该腕舟 - 月骨横向很宽，纵向很短。近端与桡骨的关节面宽大，为背外 - 掌内向的凸凹面：内侧掌部背 - 掌向凹入；背侧外部背 - 掌向明显圆凸，与背面一起形成圆滑的面，一直延伸到背部的下缘，与远端关节面相接。这表明桡骨与腕舟 - 月骨间的活动相当大。背面和远端面几乎呈直角相交。腕舟 - 月骨远端与远列腕骨（大多角骨、中央骨和头状骨）关节的面形成微凹的曲面，其背缘约为一连续的直线。腕舟 - 月骨外侧分为背、掌两部分：背部与楔骨关节的面为矢状面，表面凹入；而掌部明显向外隆凸。腕舟 - 月骨内侧为一平面，与镰状骨相关节。

楔骨 [cuneiform, Cu，又名三角骨（triquetrum）或尺腕骨（ulnare）] 楔骨在背面出露较小，约呈三角形。近端与尺骨的关节面约呈卵圆形，表面凹入，面向外上方。内侧与腕舟 - 月骨关节的面表面较圆凸。远端与钩骨的关节面内凹外凸。外凸的部分在背面一直延伸达近端的关节面，与后者以锐角相交。楔骨的掌面较平直，只是其外下角似有一凹陷。楔骨的外面出露得很小，为三角形的粗糙面。

豌豆骨 [pisiform, Ps，又名副腕骨（accessory carpal）；图 66 B] 从出露的侧面看，豌豆骨约为背 - 掌向伸长的梯形，背端大于掌端。背端有两个关节面，分别与尺骨和楔骨关节。侧面和掌端均较粗糙。

大多角骨 [trapezium, Trm，又名第 I 腕骨（*carpale* I）] 从背侧看，HMV 1942 的大多角骨有些向背侧位移。该骨为远列腕骨中最小者，其背面约呈四边形。近端与腕舟 - 月骨关节的关节面很小；外侧与小多角骨的关节面较大，表面稍凹。远端与第二掌骨相关节的关节面很小，表面稍圆凹，斜向外下方；和外侧与小多角骨的关节面以钝角相交。内侧有两个关节面：与第一掌骨的关节面较大，占据内侧面的大部分；在内侧近上缘处的关节面很小，可能与镰状骨（falciform）关节。

小多角骨 [trapezoid, Trd，又名第 II 腕骨（*carpale* II）] 小多角骨背面约呈三角形，稍大于大多角骨。其远端仅与第 II 掌骨关节，关节面从背面看为向下圆凸的弧形。小多角骨近端内侧面与大多角骨关节，而近端外侧面和中央骨关节，两关节面均较平直，彼此成锐角相交。

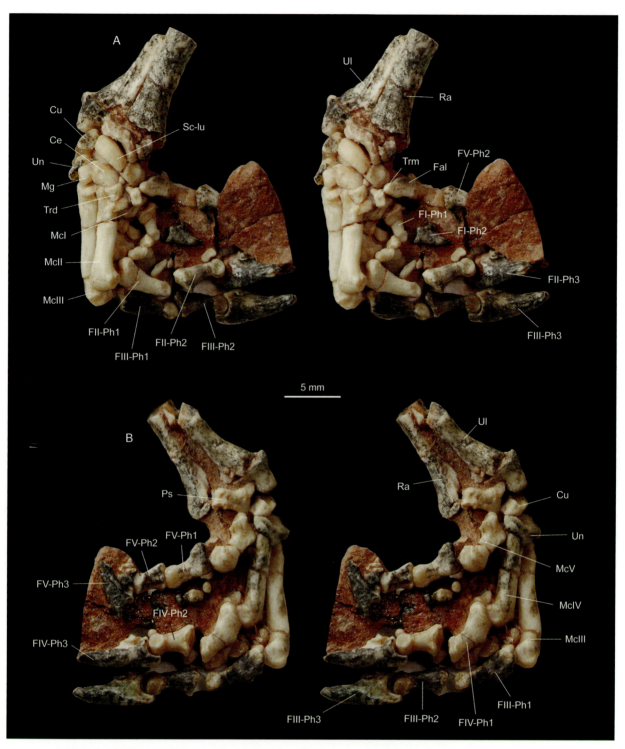

图 66　甘肃假竹鼠（新属、新种）右侧桡骨与尺骨的远端和右前脚（HMV 1942，正模）（立体照片）

Fig. 66　Distal parts of right radius and ulna and right manus (HMV 1942, holotype) of *Pseudorhizomys gansuensis* gen. et sp. nov. (stereopair)

A. 内侧面（medial view），B. 外侧面（lateral view）

缩写（**Abbreviations**）：Ce. 中央骨（*centrale*），Cu. 楔骨（cuneiform），Fal. 镰状骨（falciform），FI-Ph1-2. 第一指第一至第二指节骨（1st–2nd phalanges of 1st finger），FII-Ph1-3. 第二指第一至第三指节骨（1st–3rd phalanges of 2nd finger），FIII-Ph1-3. 第三指第一至第三指节骨（1st–3rd phalanges of 3rd finger），FIV-Ph1-3. 第四指第一至第三指节骨（1st–3rd phalanges of 4th finger），FV-Ph1-3. 第五指第一至第三指节骨（1st–3rd phalanges of 5th finger），McI–McV. 第一至第五掌骨（1st–5th metacarpals），Mg. 头状骨（magnum），Ps. 豌豆骨（pisiform），Ra. 桡骨（radius），Sc-lu. 腕舟-月骨（scapho-lunar），Trd. 小多角骨（trapezoid），Trm. 大多角骨（trapezium），Ul. 尺骨（ulna），Un. 钩骨（unciform）

中央骨（*centrale*, Ce）　中央骨很大，不但明显大于小多角骨，也大于下面要描述的头状骨。背面为横向较宽的三角形。近端与腕舟 - 月骨的关节面大，表面圆凸。远端与小多角骨、第二掌骨和第三掌骨相关节。从背面看，该三关节面似乎在同一平面上，其背缘约成连续的直线。远端关节面与近端关节面在内侧以锐角相交。中央骨外侧与头状骨的关节面表面稍凹，面向外方。

头状骨 [magnum（= *capitate*），Mg，又称巨骨或第 III 腕骨（*carpale* III）]　头状骨从背面看，尺寸比中央骨小，而与小多角骨相近。背面也呈三角形。内侧与中央骨的关节面稍圆凸，外侧与钩骨的关节面较平直。内、外两关节面在近端呈锐角相交。远端只与第三掌骨关节，表面稍圆凸。

钩骨 [unciform（hamate），Un，又名第 IV+V 腕骨（*carpale* IV+V）]　钩骨很大，为远列腕骨中的最大者。形状很不规则，横向很宽，纵向短。近端与楔骨的关节面为内凸外凹的曲面，其背面为横向稍凹的弧形。远端与第三至第五掌骨近端相关节。从背面看，远端的三个关节面的背缘共同形成一浅宽的 V 形。其中央尖突的部分与 McIV 近端的凹入部分相对应。钩骨内侧与头状骨的关节面较平。

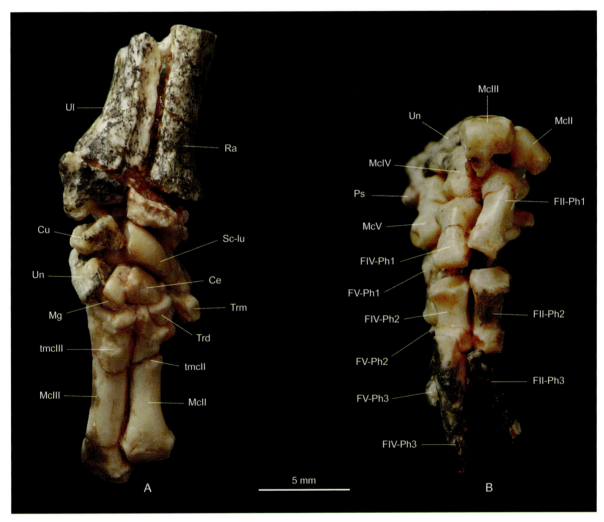

图 67　甘肃假竹鼠（新属、新种）右侧桡骨与尺骨的远端和部分右前脚（HMV 1942，正模）

Fig. 67　Distal parts of right radius and ulna, and partial right manus (HMV 1942, holotype) of

Pseudorhizomys gansuensis gen. et sp. nov.

A. 部分右桡骨、右尺骨和右前脚背面（dorsal view of partial right radius, ulna and manus），B. 部分右侧掌骨远端和部分指节骨背面（第三指的指节骨已被移出）[distal view of partial right metacarpals and dorsal view of partial right phalanges (phalanges of 3rd finger were removed)]

缩写（**Abbreviations**）：tmcII. 第二掌骨的掌骨粗隆（tuberosity of metacarpal II），tmcIII. 第三掌骨的掌骨粗隆（tuberosity of metacarpal III），其余的同图 66（others as in Fig. 66）

镰状骨 [falciform, Fal，又名桡籽骨（*os sesamoideum radii*）] 位于大多角骨之后（图 66 A）。形状为不规则的扁形骨，在腕部内侧向掌侧延伸。就大小而言，是在远列腕骨中仅小于钩骨的一块骨头。其外侧与腕舟 - 月骨和大多角骨关节。

掌骨（metacarpals, Mc） HMV 1942 右侧的五枚掌骨均保存较好。

第一掌骨（McI；图 66 A；表 16） 右第一掌骨被挤压在其他肢骨中，无法修理出来。从露出的部分看，McI 很短，形状不规则。近端顶面斜向外上方，与远端关节面以小角度相交。近端面由两个关节面组成，内者与大多角骨关节，外者与第二掌骨关节。骨体背面和内侧面共同形成圆弧形柱面，纵向稍凹。背面下部有一纵沟。McI 远端与第一指节骨的关节面的背面为圆球形面。

第二掌骨（McII；图 66 A, 67；表 16） 右 McII 保存完好，只是其骨体断裂和近端稍破损。McII 近端较大。近端顶面上有三个关节面。内侧与大多角骨关节的面长条形，背 - 掌向伸长，横向窄，表面稍圆凸，斜向内上方。中间与小多角骨关节的关节面最大，横向凹入，其背缘有一明显的凹陷。外侧与中央骨关节的面也为横向窄而背 - 掌向伸长的长条形，稍窄于内侧与大多角骨的关节面。近端内侧与大多角骨关节的面内缘下方有一面向内下方横向窄而背 - 掌向伸长的弓形面，与 McI 关节。该面下方为凹入的粗糙面。在近端外侧与中央骨关节面的下方有与 McIII 的关节面，面向外下方。

骨体直。骨体上部较窄，横切面约为四边形；向下稍变宽，横切面约为横宽的卵圆形。骨体背面纵向直，横向稍圆凸。掌骨粗隆（tuberosity of metacarpal II, tmcII）较发达，供腕桡侧伸肌长头附着。该掌骨粗隆与 *Ps. indigenus* gen. et sp. nov. 的 McII 的不同，而与后者 McIII 的相似：不呈结节状的隆起，而为四周隆凸的凹坑，且其位置也较高，位于骨体背面上部近内缘处。

McII 的远端比骨体宽大很多，比近端也宽。宽大于背 - 掌径厚。远端与第一指节骨关节的面呈横轴状。其背部约呈半圆形，弧形上缘不对称，外侧的上缘较圆凸，内侧者较平缓，稍向内下方斜伸。关节面横向微凸，纵向强烈圆凸，中央无纵向隆嵴。其掌部有很显著的纵向隆嵴，嵴的两侧各有一宽缓的凹槽。其外侧的凹槽要比内侧者稍窄而深。远端两侧有明显的凹坑，供内、外侧韧带附着。

第三掌骨（McIII；图 66, 67；表 16） 比 McII 长而稍粗大，形态结构与 *Ps. indigenus* gen. et sp. nov. 的 McIII 相似。近端也较大，顶面也有三个关节面。中间与头状骨的关节面最大，横向凹入，背 - 掌向稍圆凸。由于 McIII 的近端相对于 McII 前移，McIII 内侧与 McII 关节的面的前半部显露出来：表面稍圆凸，面向内上方。外侧与钩骨的关节面较窄。近端背面外侧的韧带嵴也较明显，但较少向下延伸。

骨体直。上部的横切面为四边形，向下逐渐变为椭圆柱形。掌骨粗隆（tmcIII）很发达，在背侧上部近内缘处形成大而粗糙的凹陷区，供腕桡侧伸肌短头附着。McIII 的远端与 McII 的相似，只是远端关节面背部的内、外两部分较对称，而掌部的两关节凹槽仍不对称，外侧者仍稍窄于内侧者。

第四掌骨（McIV；图 66 B, 67 B；表 16） 第四掌骨保存完整，只是骨体中部断裂。McIV 比 McIII 短很多，而与 McII 接近等长。McIV 的近端较宽大。近端顶面有两个关节面。外侧与钩骨的关节面占据了近端面的绝大部分。从背面看，该关节面的背缘为下凹的圆弧形。内侧与 McIII 的关节面很窄，面向内上方。近端外侧与 McV 的关节面面向掌下方，表面微凹。近端背侧中央明显凹入。凹陷部分的内、外两侧的韧带嵴也明显存在，但比 *Ps. indigenus* gen. et sp. nov. 和 *Pa. hipparionum* 的 McIV 的都弱。

骨体与 McII 和 McIII 者相似。上部横切面为四边形，往下逐渐变宽大，成卵圆柱形。远端很宽大，比 McII 者稍宽，而与 McIII 者相近。其形态与 McIII 者也相似：与 PhI 的关节面的背部是对称的。与籽骨关节的面的情况不清楚。

第五掌骨（McV；图 66 B, 67 B；表 16） 第五掌骨保存完整，只是骨体中部断裂，其上、下两部分间有些扭转错开：上部稍向外后方扭，而下部稍向内后方扭。McV 很短，比 McIV 短很多。McV 位于 McIV 的外后方，其背面面向背外方。

HMV 1942 的 McV 与 *Ps. indigenus* gen. et sp. nov. 的区别很大。它的近端很粗大。顶端也有两个关节面。外侧与钩骨的关节面也占据了顶面的绝大部分，但形状很不同：约成四边形；表面呈鞍形，横向凹入，背 - 掌向圆凸；稍面向内上方。内侧与 McIV 的关节面较小，表面稍圆凸，面向内上方。近端背面与外侧交界处的结节状隆起很发达，其后也有大而粗糙的凹面，用以供腕尺侧伸肌附着。

骨体很短。其形状也与 *Ps. indigenus* gen. et sp. nov. 者不同：约为四方柱形，比近、远两端细小得多。背面纵向较直，横向圆凸。内面和外面纵向稍凹。远端粗大。其形态结构与 McII–McIV 的相似，但内侧比外侧稍厚。远端关节面的背部内外不对称：其内侧背掌向稍长于外侧者，内侧的前缘也较外侧者更向前凸些。

指节骨（phalanges of fingers, Ph）　HMV 1942 的右前脚的五个指的指节骨都保存完好，且基本上是原位保存，或只稍稍移位。

第一指（拇指）（first finger, FI）

第一指第一指节骨（proximal phalanx of FI, FI-Ph1；图 66 A；表 17 ）　第一指的第一指节骨稍移位。第一指节骨从内侧面看为梯形，近端明显大于远端，比 *Ps. indigenus* gen. et sp. nov. 的 FI-Ph1 稍更粗大（见表 17）。近端背侧为圆凸的半圆形，掌侧横向较背侧宽。近端与第一掌骨的关节面约为半圆形，从侧面看稍向背侧倾斜，卵圆形的凹面占据了其大部分。掌侧有两个明显的隆凸，两隆凸间以中凹相隔。近端内后侧有明显的隆凸，可能是拇展肌附着处。

骨体约为四边形的截锥体，背面窄于掌面，背面横向稍圆凸。背面上部的伸腱结节（extu) 较显著，可能是拇短伸肌远端附着处。背面下部纵向较直。掌面上部较平；下部稍向背下方斜伸，表面纵向较直，横向稍凹。

FI-Ph1 远端明显小于近端，其横宽与骨体相近，但其背 - 掌径则稍长于骨体者。远端与第二指骨的关节面约成半圆轴形，掌侧缘明显高于背侧缘，其横向稍凹，背掌向圆凸。远端两侧有明显的粗糙凹陷区，供指侧韧带附着。

第一指第二指节骨（distal phalanx of FI, FI-Ph2；图 66 A；表 17 ）　第一指的第二指节骨为爪指骨，长与 FI-Ph1 者相近，但较窄。近端很厚，其背 - 掌向厚大于横宽。近端与 FI-Ph1 的关节面为横向稍宽的卵圆形，横向微凸，背 - 掌向凹入。其背侧缘变厚，中央形成明显向上隆突的伸腱突，供拇长伸肌附着。其掌缘厚而圆凸（可能与远端籽骨接触）。

FI-Ph2 骨体背面横向圆凸，从侧面看，背侧纵向上部稍凹，下部稍圆凸。FI-Ph2 的两侧横向圆凸，纵向稍凹；在侧面近掌侧处，有明显的掌沟（volar sulcus）。掌沟向远端分成两支：一支在侧面伸向远方；一支转向掌侧，然后伸向远方。掌沟内有两个掌孔（volar foramen）：一个靠近上部（距沟上端约 0.4 mm）；一个靠近掌沟的分叉处（此孔也可能与 *Ps. indigenus* gen. et sp. nov. 的 FI-Ph2 的侧壁孔相当）。掌侧面中部爪指骨粗隆（tunph）很发达，供指深屈肌附着。该粗隆并向两侧伸出较尖的角突（可能用来增大指深屈肌腱的附着面）。在掌面上部，该隆凸与近端关节面间为一横向直而纵向凹入的凹槽面，其两端与两侧的掌沟相连通。FI-Ph2 远端为背 - 掌向扁的铲形，其背侧面稍圆凸，掌侧面稍凹。FI-Ph2 的远端缘与 *Ps. indigenus* gen. et sp. nov. 者不同：其背侧在远端缘附近有短的纵沟和棱，使远端缘呈锯齿状的圆弧形。远端也未见中矢切迹（sagittal notch）。

第二指（second finger, FII）

第二指第一指节骨（proximal phalanx of FII, FII-Ph1；图 66 A, 67 B；表 17 ）　第二指第一指节骨约为 McII 的 2/3 长。FII-Ph1 比 FI-Ph1 粗大得多 (见表 17)。顶面不与骨体长轴垂直，而是稍向前倾斜。近端面约呈半圆形。其背部为横宽卵圆形的凹面，与 McII 远端关节面的背部相对应；其掌部近中央处有纵向凹槽，凹槽两侧有明显的结节状隆凸，分别与 McII 远端关节面掌部中央的纵向隆嵴和两侧横向凹槽相关节；内隆凸大于外隆凸。近端两侧有稍凹的粗糙面，可能供侧副韧带附着。

骨体约为截圆锥形，上大下小，上部横切面近圆形，宽仅稍大于背 - 掌径长；中部横切面变为扁的卵圆形，宽明显小于背 - 掌径长。骨体背面纵向平直，横向圆凸。在骨体背侧上部约 1/4 处有一很发达的伸腱结节（extu），供指总伸肌附着。骨体侧面横向稍稍圆凸，纵向稍凹。掌面较平。掌面下部两侧有明显的纵棱与两侧面分开。

FII-Ph1 的远端明显小于近端。横宽明显大于背 - 掌径长。远端与第二指骨的关节面不与骨体长轴垂直，而是稍向掌上方倾斜。该关节面纵向稍圆凸，被一很浅的矢状沟分为内、外两髁。内髁稍大于外髁。两髁的表面横向稍圆凸。远端关节面的背缘一小部分稍向背侧弯曲。远端背面平直，与远端关节面呈锐角相交。远端两侧有明显的压迹和边缘的结节状隆起，供侧副韧带附着。

第二指第二指节骨（middle phalanx of FII, FII-Ph2；图 66 A, 67 B；表 17） FII-Ph2 比 FII-Ph1 短，其近端背 - 掌径长虽稍小于后者，但远端背 - 掌径长却大于后者。FII-Ph2 近端较粗大。近端面稍向前下方斜伸，与骨体长轴不垂直，而是斜交。近端与 FII-Ph1 的关节面约呈卵圆形，背 - 掌向稍凹入，矢状棱很宽缓，两侧的关节窝凹入稍深。内侧窝稍大于外侧窝。近端前缘的伸腱结节（extu）很发达，为横向延伸的枕状（pillow-like）隆凸，供指总伸肌肌腱附着。近端掌侧中央有一很宽的凹槽，可能供指深屈肌腱通过。该凹槽的两侧为结节状的隆起，可能供指浅屈肌，掌侧副韧带和纤维鞘（sheath）等附着。内侧隆起明显大于外侧的隆起。近端两侧也有很发达的结节状隆凸供侧副韧带附着。

FII-Ph2 的骨体约成横宽的截圆锥形，上部较大，往下逐渐变得较细小。骨体背侧和两侧面均横向圆凸，纵向稍凹；掌面较平。

远端较骨体下部粗大，约呈横轴形，纵向圆凸，向背方隆凸，明显超过骨体背侧。远端与 FII-Ph3 的关节面从侧面看为半圆弧形。其背侧部分稍往背上方延伸，横向变窄、较平。关节面的背缘为向背上方凸的圆弧形，其内部的圆凸度稍大于外部者。该关节面往掌侧逐渐变宽，表面横向明显凹入，其掌侧部分向上方延伸得较长，明显超过背者。远端背面，关节面之上的部分较平，稍向背上方倾斜。远端两侧为明显的凹陷区，供侧副韧带附着。

第二指第三指节骨（distal phalanx of FII, FII-Ph3；图 66 A, 67 B；表 17） 第二指第三指节骨（爪指骨）保存较好，只是其远端稍破损。爪指骨为尖爪形，比 FI-Ph2 者粗大得多，而比 FII-Ph2 长而窄。近端较厚大，背 - 掌径长明显大于横宽。近端与 FII-Ph2 的关节面约为卵圆形。关节表面为背 - 掌向强烈凹入的半圆形的凹面，横向稍凸。其两侧缘有很狭窄的沟状压迹，可能供侧副韧带附着。近端背侧的伸腱突很发达，比 FI-Ph2 的发达得多。其形状与 FI-Ph2 的不同，而与 *Ps. indigenus* 的 FIII-Ph3 和 FIV-Ph3 的很相似：约为顶端稍窄的瓦片形，其背、掌视约为梯形，顶端稍窄、稍凹，顶端面较粗糙。

骨体背侧有一明显而稍圆凸的纵棱。该纵棱往远端变弱至无。纵棱两侧的背外面和背内面均约呈较窄而远端较尖的三角形，表面稍圆凸。该两侧面的近端部分较平滑，远端部分有明显的纵向沟、棱。掌沟（vos）很发育，位于两侧面的掌侧。内、外两掌沟在近端相连，在掌侧面近端形成稍浅的横沟。掌孔（vof）很大。骨体掌面供指深屈肌附着的爪指骨粗隆（tunph）很发达，但与 FI-Ph2 的不同，它可分成近、远两部分。近端部圆隆，远端部凹入，从侧面看呈凸凹形。爪指骨粗隆远端缘为圆弧形，较薄锐，构成掌沟的掌侧缘。这不但增加了指深屈肌腱的附着区域，而且也加长了力臂，似乎表明 FII-Ph3 的指深屈肌很发达。骨体掌侧远端部分有一纵棱从该粗隆伸达远端。FII-Ph3 的远端与 FI-Ph2 的不同：不呈背 - 掌向扁的铲形，而是较厚，为向远端变尖而横向稍窄的钝锥形。表面有很发达的纵向的沟、棱伸达远端。远端也未见有中矢切迹。

第三指（third finger, FIII）

第三指第一指节骨（proximal phalanx of FIII, FIII-Ph1；图 66, 68；表 17） 第三指第一指节骨保存较好，只是骨体下部断裂。FIII-Ph1 与 FII-Ph1 形态结构很相似，只是更粗长些。近端与 McIII 的关节面的背侧的凹面比 FII-Ph1 的更大而深，掌侧的两结节状的隆凸也更粗大，但内隆凸仍大于外隆凸。近端两侧的侧副韧带附着处不是凹面，而是呈结节状的隆起。内、外侧的隆起对称，大小相近。骨体背侧上部的伸腱结节（extu）比 FII-Ph1 的稍大。骨体和远端的形态结构与 FII-Ph1 的相似，只是远端关节面的纵向圆凸程度要比 FII-Ph1 的大，从侧面看约成半圆弧形。远端关节面上方的背侧面不像 FII-Ph1 那样平，而是较明显地凹入。远端两侧供侧韧带附着的凹面较大。

第三指第二指节骨（middle phalanx of FIII, FIII-Ph2；图 66, 69；表 17） 第三指第二指节骨保存完好，仅骨体下部稍断裂。FIII-Ph2 比 FIII-Ph1 短而细，但比 FII-Ph2 稍长而粗大。FIII-Ph2 的形态结构与 FII-Ph2 的很相似，近端前缘的伸腱结节（extu）很发育，为横向延伸的枕状隆凸。所不同的是 FIII-Ph2 的近端关节面的矢状棱更弱。矢状棱两侧的关节窝大小相近。近端掌侧中央的凹槽更大而深。

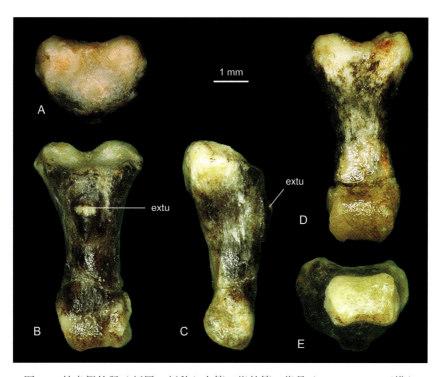

图 68　甘肃假竹鼠（新属、新种）右第三指的第一指骨（HMV 1942，正模）

Fig. 68　Right FIII-Ph1 (HMV 1942, holotype) of *Pseudorhizomys gansuensis* gen. et sp. nov.

A. 近端面（proximal view），B. 背面（dorsal view），C. 外侧面（lateral view），D. 掌面（volar view），E. 远端面（distal view）

缩写（**Abbreviation**）：extu. 伸腱结节（extensor tuberosity）

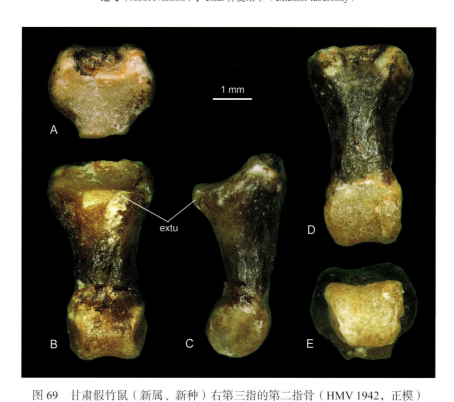

图 69　甘肃假竹鼠（新属、新种）右第三指的第二指骨（HMV 1942，正模）

Fig. 69　Second phalanx of right 3rd finger (FIII-Ph2) of HMV 1942 (holotype) of *Pseudorhizomys gansuensis* gen. et sp. nov.

A. 近端面（proximal view），B. 背面（dorsal view），C. 内侧面（medial view），D. 掌面（volar view），E. 远端面（distal view）

缩写（**Abbreviation**）：extu. 伸腱结节（extensor tuberosity）

凹槽内、外两侧供指浅屈肌附着的结节状隆起的大小也相近。远端与 FIII-PH3 的关节面的背侧部分也稍变窄，但其前缘较少向前圆凸，而较平直。

第三指第三指节骨（distal phalanx of FIII, FIII-Ph3；图 66, 70；表 17）　第三指的第三指节骨（爪指骨）保存完好，只是近端关节面的外侧缘稍破损。FIII-Ph3 较 FIII-Ph2 长很多，具有横向较窄、背 - 掌向较厚的近端。FIII-Ph3 的基本形态与 FII-Ph3 相似，但更粗大而长。FIII-Ph3 近端背侧的伸腱突很发达（可惜其外部已破损）。近端与 FIII-Ph2 的关节面（ffIII-2）约为窄五边形，关节表面为背 - 掌向强烈凹入的半圆形的曲面。FIII-Ph3 的背侧纵棱长，几乎伸达远端，并稍向内弯。纵棱两侧的背外面和背内面均为窄长三角形，但往远端收缩变窄得较缓慢。背外面的远端部分比背内面的稍宽，横向较圆凸。但远端纵向的沟棱在背内面要比背外面更发达些。掌沟（vos）分布于 FIII-Ph3 两侧上部，很发达，均呈稍向掌侧凹入的弧形，两掌沟下端在爪指骨粗隆下方彼此相连。内侧掌沟稍深于外侧者。两侧掌沟内各有一掌孔（vof），内掌孔明显大于外侧者。未见侧壁孔。掌侧的爪指骨粗隆（tunph）很发达。远端缘未见有中矢切迹。

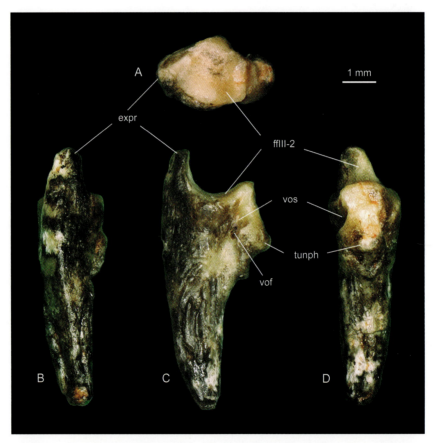

图 70　甘肃假竹鼠（新属、新种）右第三指的爪指骨（HMV 1942，正模）

Fig. 70　Ungual phalanx of right 3rd finger (FIII-Ph3) of HMV 1942 (holotype) of *Pseudorhizomys gansuensis* gen. et sp. nov.

A. 近端面（proximal view），B. 背面（dorsal view），C. 内侧面（medial view），D. 掌面（volar view）

缩写（Abbreviations）：expr. 伸腱突（extensor process），ffIII-2. 与第三指的第二指节骨关节面（facet for 2nd phalanx of 3rd finger, FIII-Ph2），tunph. 爪指骨粗隆（tuberosity of ungual phalanx），vof. 掌孔（volar foramen），vos. 掌沟（volar sulcus）

第四指（fourth finger, FIV）

第四指第一指节骨（proximal phalanx of FIV, FIV-Ph1；图 66 B, 67 B；表 17）　第四指第一指节骨保存完好，仅远端关节面表面破损。FIV-Ph1 约为 McIV 的 2/3 长。FIV-Ph1 的形态结构与 FIII-Ph1 的相

似，粗细也相近，但较短；而比 FII-Ph1 的较粗而短。近端与 McIV 的关节面的凹面较 FIII-Ph1 的稍深。近端两侧的侧副韧带附着的隆起较大而显著，外侧隆起要比内侧者更发达，更向外突出。骨体背侧的伸腱结节（extu）也很发达。骨体背面内、外侧面稍不对称，背内侧面较背外侧面横向倾斜度较大，较陡。骨体掌面下部两侧的侧棱很发达。

第四指第二指节骨（middle phalanx of FIV, FIV-Ph2；图 66 B, 67 B；表 17） 第四指第二指节骨保存完好。FIV-Ph2 短于并稍细于 FIV-Ph1，也比 FIII-Ph2 和 FII-Ph2 都短。其粗细与 FIII-Ph2 的相近，而较 FII-Ph2 的稍宽些。FIV-Ph2 的形态结构与 FIII-Ph2 和 FII-Ph2 的相似。但近端关节面中的矢状棱较微弱，近端两侧供侧副韧带和指浅屈肌等附着的隆凸更发达。特别不同的是：FIV-Ph2 骨体中部较宽扁。远端与 FIV-Ph3 的关节面的背缘不像 FIII-Ph2 的那样平直，而是像 FII-Ph2 的那样呈向背上方凸的圆弧形。但它与 FII-Ph2 不同的是，其外部的圆凸度稍大于内部者。

第四指第三指节骨（distal phalanx of FIV, FIV-PhIII；图 66 B, 67 B；表 17） 第四指第三指节骨（爪指骨）的伸腱突破损，远端稍变形。FIV-Ph3 很长大，比 FIV-Ph2 长很多，其长几乎是 FIV-Ph2 的两倍长，而与 FIII-Ph3 相近。FIV-Ph3 的形态结构与 FIII-Ph3 和 FII-Ph3 的很相似。只是背面上部背外侧面比背内侧面横向要更宽些，凸曲度稍大些。FIV-Ph3 往远端逐渐变得较扁而尖，而且其方向稍有扭转（也可能是受压稍变形的结果？）：其背面横向稍圆凸，面向背外方；而掌侧面较平直，面向掌内方。

第五指（fifth finger, FV）

第五指第一指节骨（proximal phalanx of FV, FV-Ph1；图 66 B, 67 B；表 17） 第五指第一指节骨稍短于 McV。FV-Ph1 的形态结构与 FIV-Ph1 的很相似，但较 FIV-1 短，稍细。FV-Ph1 内侧稍长于外侧。从侧面看，近端关节面和远端关节面均与骨体的纵轴不垂直。近端两侧供侧副韧带附着的隆起不如 FIV-Ph1 的大而向外突出。但外侧的隆起下方有明显的粗糙凹面。骨体较扁平，其宽度明显大于背 - 掌径长。骨体背面的伸腱结节（extu）仍很发达。骨体背面内、外侧面明显不对称：背内侧面相对横向较窄，较陡，主要面向内方；而背外侧面较宽，倾斜度较小，主要面向背外方。骨体掌面横向较平，纵向稍凹。远端背 - 掌向较扁，并稍向外后方向旋转。远端与第二指节骨的关节面较向后倾斜，与背侧面的夹角较小。关节面上的矢状沟明显，内髁横向稍宽于外髁。

第五指第二指节骨（middle phalanx of FV, FV-Ph2；图 66 B, 67 B；表 17） 第五指第二指节骨比 FV-Ph1 短而细。FV-Ph2 的形态结构与 FIV-Ph2 的很相似。所不同的是 FV-Ph2 的尺寸要比 FIV-Ph2 的更短而细。背侧供指总伸肌附着的伸腱结节（extu）隆凸的横棱较发育，但被中部的凹入分成两个结节。远端两侧供侧副韧带附着的凹陷区也是外侧者比内侧者更发达些。

第五指第三指节骨（distal phalanx of FV, FV-Ph3；图 66 B, 67 B；表 17） 第五指第三指节骨仅伸腱突顶端稍破损。FV-Ph3 显然比 FV-Ph2 长。FV-Ph3 的形态结构与 FIV-Ph3 的很相似，其背侧内、外侧面也不对称，也是外侧面较少倾斜，并宽于内侧面；也具有稍斜、扁而稍尖的远端。只是 FV-Ph3 的尺寸比 FIV-Ph3 更小。在前足的五个爪指骨中，FV-Ph3 不但小于 FIII-Ph3 和 FIV-Ph3，也比 FII-Ph3 更短小，但明显大于 FI-Ph2。

2. 比较

HMV 1942 等标本与 *Ps. indigenus* gen. et sp. nov. 在头骨和牙齿的基本形态上都很相似，而与 *Pararhizomys* 的明显不同：如鼻骨背面纵向平直，在前部横向的圆凸度较小；前颌骨的前端明显超过鼻骨前端；前颌骨侧背嵴较短而斜伸，其前部与鼻骨 - 前颌骨缝汇合；前颌骨背侧面为窄的三角形；前颌骨 - 上颌骨缝在背侧为向外弯的曲线，并与上颌骨 - 额骨缝共同形成一圆弧；眶下孔为大的圆形；外层咬肌附着区域的前缘为 "S" 形；颧弓前根的位置较靠前；颧骨较长，伸达眼眶前缘，通常与泪骨相连；顶骨为短宽的六边形；颞深神经孔与咬肌神经孔和颊肌神经孔愈合，但前者的后孔与后两者的后孔彼此分开；下颌骨水平部下缘呈圆弧形；颏孔位置较高；下颌切迹较深；臼齿齿冠较低，仅 M1 和 m3 具明显的前 - 后侧高冠；I2 纵向弯曲较小，其前端不向后弯，属前伸型齿，其后端起自上颌骨的中部，

M1 之前；I2 唇侧无明显的纵棱；等等。显然，上述标本应归入 *Pseudorhizomys* 新属。

但上述标本也具有一些不同于 *Ps. indigenus* gen. et sp. nov. 的一些特征，如：其听泡膨胀程度较小，其前内侧的凹面较浅；下颌髁关节面为较宽的椭圆形；在牙齿冠面形态上区别也很明显（详见区别特征部分）。此外，其第二掌骨粗隆和第五掌骨的形状不同，拇指的爪指骨远端缘具锯齿等。显然，上述标本代表不同于 *Ps. indigenus* gen. et sp. nov. 的另一新种，在此称其为甘肃假竹鼠（*Pseudorhizomys gansuensis* gen. et sp. nov.）。

需要指出的是：V 16296 具有一些不同于 *Ps. gansuensis* gen. et sp. nov. 的正模（HMV 1942）和副模的特征。如它的下颌髁上的关节面的前缘较尖凸和 m3 的下外凹舌部不分叉。另外，它的齿冠较高等。但它的其他特征都与 *Ps. gansuensis* gen. et sp. nov. 的正模和副模一致，故仍将其归入此种。有意思的是，V 16296 产出的层位为柳树组上部（杨家山动物群），属晚中新世灞河期的晚期。而该种的其他标本均产自产有大深沟动物群的柳树组的中部，属晚中新世灞河期的中期。这样，V 16296 这些不同特征也可能代表向不同方向演变的趋势。

在 *Pseudorhizomys gansuensis* gen. et sp. nov. 已知的 9 件标本（见表 10）中有 4 件标本（HMV 1942，V 16296，V 16298 和 V 16301）比其余的 5 件标本 (V 16297，V 16299，V 16300，V 16302 和 V 16303) 在尺寸上稍大，其吻部和门齿也更粗壮。一种可能的解释是，前 4 件标本代表雄性个体，而后 5 件则代表雌性个体。

平齿假竹鼠（新属、新种）*Pseudorhizomys planus* gen. et sp. nov.
（图 71–72，76；表 1，10–12）

正模 V 16304，头骨前部具左、右 I2 和臼齿。

产地与层位 广河县庄禾集乡，详细地点和层位不明。

鉴别特征 吻部和上颌齿隙短。前颌骨三角形的背面较短小。外层咬肌附着区的前缘约为"S"形，其下部的圆形弧向前凸，超过眶下孔的腹侧裂隙。颊齿齿冠较高，M1 前侧高冠较显著。M1–3 内凹横向短，不与前边凹重叠。M2–3 的前边凹为封闭的盆。M3 的中凹和内凹相对，并相贯通。I2 窄小，唇面横向平直，具一很微弱的纵棱。

区别特征 与 *Ps. indigenus* gen. et sp. nov. 和 *Ps. gansuensis* gen. et sp. nov. 的区别在于吻部和上颌齿隙较短；前颌骨三角形的背面短；I2 窄小，唇面横向平直，具一很微弱的纵棱。此外，它与 *Ps. gansuensis* gen. et sp. nov. 的区别还在于其 M1–3 内凹横向短，M1 的内凹不与前边凹重叠，M2–3 具前边凹和 M3 的中凹和内凹横向相对。

词源 Planus，拉丁文，平的，意寓 I2 唇面横向平直。

1. 描述

头骨（图 71；表 11） V 16304 的吻部短而窄，其两侧面稍向前靠近。前颌骨前端向前延伸超过鼻骨的前端。鼻骨向后变窄，后端已破损。但从已保存的部分判断，鼻骨 - 额骨缝可能与前颌骨 - 额骨缝的内端连接。鼻骨背面纵向近于平直，其前部横向圆凸，但隆凸的程度较小。前颌骨的侧背嵴与 *Ps. indigenus* gen. et sp. nov. 的相似，也很短，斜向前内方延伸，其前部与鼻骨 - 前颌骨缝合并。前颌骨的背侧面约呈窄的三角形。前颌骨 - 额骨缝较直，斜向前外方延伸。前颌骨 - 上颌骨缝和上颌骨 - 额骨缝相连，形成外弯的弧形。眶下孔为圆形，具腹侧裂隙和隔板。上颌齿隙的长度比上颊齿列长。门齿孔位于上颌齿隙的中后部，其长约为齿隙长的 1/4。前颌骨 - 上颌骨缝在门齿孔的前 1/3 处插入门齿孔。门齿孔的侧缘形成的嵴向后延伸，一直伸达 M1 齿槽的前缘。M1 之前未见有明显的隆凸。外层咬肌附着区的前缘为"S"形，上部凹入；下部向前圆凸，超过眶下孔腹侧裂隙。颧弓板的后缘位于门齿孔稍后。腭沟的深度在 M1 和 M2 的内侧加深。该沟中的腭后孔大而单一，位于 M1 和 M2 间的内侧。左、右上

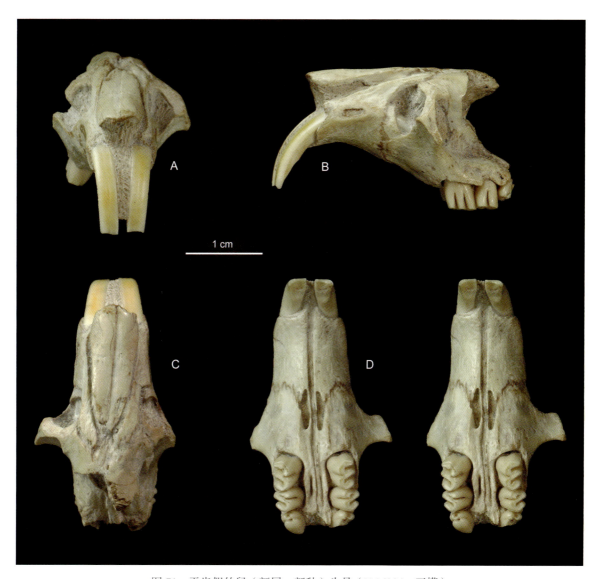

图 71　平齿假竹鼠（新属、新种）头骨（V 16304，正模）

Fig. 71　Skull of *Pseudorhizomys planus* gen. et sp. nov. (V 16304, holotype)

A. 前面（anterior view），B. 左侧面（left lateral view），C. 背面（dorsal view），D. 腹面（立体照片）（ventral view, stereopair）

颊齿列彼此近于平行。

　　牙齿（图 72，表 12）　V 16304 的齿式为 1·0·0·3/。臼齿次高冠，具齿根，大小由 M1 往 M3 递减。M1 冠面为长椭圆形，长大于宽，颊侧长于舌侧。M1 齿冠为前侧高冠型。冠面有三条很发达的褶沟（前边凹、中凹和内凹），均向外开口；原凹也存在，但很浅弱。颊侧的两褶沟（前边凹和中凹）稍向后舌方延伸，比内凹较长而深。中凹长于前边凹，稍向后弯。横向较短的内凹向前尖方向延伸，不伸达前边凹，也不与前边凹重叠。M2 的冠面约近方形，具三条褶沟。内凹向前颊侧方向延伸，与封闭的前边凹的舌端相遇。中凹向后舌方延伸，横向长度大于内凹，而深度与后者相近。M3 为卵圆形，后缘稍窄，冠面也具三条褶沟。前边凹也成封闭的盆状。内凹和中凹横向很短，都为横向延伸，彼此相对。中凹和内凹之间有一釉质盆相连。三者共同形成一横沟，将 M3 分成前、后两部分。内凹的深度浅于中凹。

　　I2 弯曲度较少，其前部向前腹面延伸，前端不向后弯，属前伸型齿。I2 的横切面为三角形，唇面平直。釉质层主要覆盖在唇面，仅稍稍延至两侧面。I2 唇面横向相当平，具有一非常弱的唇侧纵棱。

图 72　平齿假竹鼠（新属、新种）上臼齿冠面（Ⅴ 16304，正模）

Fig. 72　Occlusal view of upper molars of *Pseudorhizomys planus* gen. et sp. nov. (V 16304, holotype)

A. 右上臼齿（right M1–3），B. 左上臼齿（left M1–3）

2. 比较

上面的描述表明，Ⅴ 16304 的一些特征与 *Pseudorhizomys* gen. nov. 很相似，而与 *Pararhizomys* 不同：如前颌骨前端向前延伸超过鼻骨的前端，前颌骨具有短而斜伸的侧背嵴和三角形的背面，鼻骨纵向较平直，和 I2 的弯曲度相对较小，其前端不向后弯，属前伸型齿等。显然，Ⅴ 16304 不属于 *Pararhizomys*，而应被归入 *Pseudorhizomys* 新属。*Pseudorhizomys* 新属现已知包括两个新种：*Ps. indigenus* 和 *Ps. gansuensis*。与该二种比较，Ⅴ 16304 与它们有一些不同：如它的尺寸较小；吻部较短而窄（见表 11），前颌骨三角形的背面较短小；I2 较细小，唇面横向较平和具弱的纵棱等。此外，Ⅴ 16304 与 *Ps. gansuensis* gen. et sp. nov. 的区别还在于：M1–3 的内凹较短，M2–3 具前边凹，M3 的内凹和中凹横向延伸，彼此相对等。Ⅴ 16304 很可能代表不同于上两个种的另一新种。笔者称其为平齿假竹鼠（*Pseudorhizomys planus* gen. et sp. nov.）。

原始假竹鼠（新属、新种）*Pseudorhizomys pristinus* gen. et sp. nov.

（图 73, 74, 76；表 1, 10–12）

正模　Ⅴ 16305，含下颌头骨，产自和政县关滩沟乡潘杨阴洼南约 500 m（LX 201001），柳树组中部下层；晚中新世灞河期中期[①]（见表 1, 10）。

鉴别特征　个体较小而原始的一类 *Pseudorhizomys*。吻部和上颌齿隙较长。外层咬肌附着区的前缘为前凸的圆弧形，达眶下孔腹侧裂隙的外缘。咬肌结节靠近眶下孔的腹侧裂隙。颞嵴较长而明显，较窄锐。下颌齿隙后部的凹入较长而浅。冠状突前后较短，较向后倾，其上部明显向后弯曲，后缘明显凹入。

[①] 标本为地表采集，其层位存疑。

下颌髁上的关节面为肾形。臼齿齿冠较低，颊侧釉质曲线均无明显折曲。仅 m3 有较明显的后侧高冠，而 M1 无明显的前侧高冠。M2–3 具前边凹，M2 前边凹较少退化，仍向颊侧开口。M1–3 内凹插入前边凹和中凹之间。m1–3 具下后边凹。m3 具下原凹。I2 唇面横向圆凸，无明显的纵棱。

区别特征　此种与 *Pseudorhizomys* gen. nov. 的三个新建种（*Ps. indigenus*，*Ps. gansuensis* 和 *Ps. planus*）的区别在于：外层咬肌附着区的前缘为前凸的圆弧形。咬肌结节靠近眶下孔的腹侧裂隙。齿冠较低，M1 无明显的前侧高冠现象；臼齿釉质曲线均无明显折曲。

此外，它与 *Ps. indigenus* gen. et sp. nov. 和 *Ps. gansuensis* gen. et sp. nov. 的区别在于：个体较小；颞嵴较发育，较窄锐；冠状突前后较短，其上部明显向后弯曲，后缘明显凹入；下颌髁的关节面为肾形；m1–3 具下后边凹。

与 *Ps. indigenus* gen. et sp. nov. 和 *Ps. planus* gen. et sp. nov. 的区别还在于 M1–3 的内凹较长，插入前边凹和中凹之间；M2 前边凹较少退化，仍开口。

与 *Ps. gansuensis* gen. et sp. nov. 的区别还在于：冠状突较向后斜。M2–3 具前边凹，m1–3 的下外凹不分叉，和 m3 具下原凹。

与 *Ps. planus* gen. et sp. nov. 的区别还在于：吻部和上颌齿隙较长，前颌骨三角形的背面较窄长，和 I2 唇面横向较凸。

与 *Ps.? hehoensis* 的区别在于：齿冠较高，m3 具后侧高冠现象，i2 唇侧横向较平缓。

词源　Pristinus，拉丁文，早的，原始的。

1. 描述

（1）头骨（图 73 A；表 11）

V 16305 的头骨侧向稍压扁，其左半部分相对前移，吻部的后部相对变窄、变高，门齿孔被挤压成一条缝，两颊齿列之间的腭部也变得很狭窄，牙齿有些错位。

自背面看，吻部长窄，其两侧缘近于平行。鼻骨大部分破损，但其基本轮廓仍得以保存，约呈前宽后窄的楔形。其后端很尖，与前颌骨的后端约在同一横线上。前颌骨 - 额骨缝细锯齿状，较直地斜向前外方。前颌骨 - 上颌骨缝与上颌骨 - 额骨缝相连，共同形成一外弯的圆弧形。前颌骨侧背嵴相对较长，稍向前内方斜伸，其前端与鼻骨 - 前颌骨缝的前部相遇。前颌骨的背面为很窄长的三角形。颧弓细，圆弧形。右侧颧弓因受压，其前部较平缓，后部较向外圆凸。颧骨长，向前伸至眼眶前缘，其前端伸达眶下孔上缘的外后方。泪骨保存不好，特征不清。眶后收缩明显。额嵴不明显。矢状嵴显著。颞嵴很明显，为较长而窄锐的嵴，伸达项嵴。

自侧面看，前颌骨前伸，其前端明显超过鼻骨的前端。前颌骨侧面由 I2 齿槽形成的纵棱很明显。该棱与前颌骨侧背嵴间形成三角形的凹面，与眶下孔相通。眶下孔为大的圆形，限于上颌骨内，具明显的腹侧裂隙。眶下孔中的隔板发育较弱。I2 后端起自上颌骨的中部，位于 M1 之前；其前端向下延伸，而不向后弯曲，属前伸型齿。颧弓板主要向前和上方延伸，也只限于上颌骨内，表面明显凹入。外层咬肌附着区前缘为前凸的圆弧形，沿眶下孔腹侧裂隙的外嵴延伸。咬肌结节较大而明显，为卵圆形粗糙面，位于眶下孔腹侧裂隙的下后方。上颊齿列的位置往后移。上颌齿隙明显长于上颊齿列。M1 以前的头骨前部与头骨后部（包括 M1）的长度相近。与下颌髁关节的关节窝很长，向后伸达项嵴，将外耳道挤至其下后方。臼后孔位于关节窝后部内侧，外耳道的上方。乳突很小。茎乳孔位于外耳道和乳突之间。

两颊齿列间的腭部的结构大部分已破损，而且两齿列的相互位置也有错动。硬腭后缘的位置大约与 M3 的中部相当。中翼窝较窄，其宽度约与翼窝相近。两内翼突不倾斜。

颞深神经孔与咬肌神经孔和颊肌神经孔融合为一条裂缝。翼蝶管后孔（ascp）小，位于翼窝外后角，外翼突后端基部内侧，在融合的颞深神经、咬肌神经和颊肌神经后孔的内侧。翼管后孔（pcpf）位于翼窝内侧，靠近内翼突外侧后部的基部。卵圆孔也与中破裂孔合而为一。

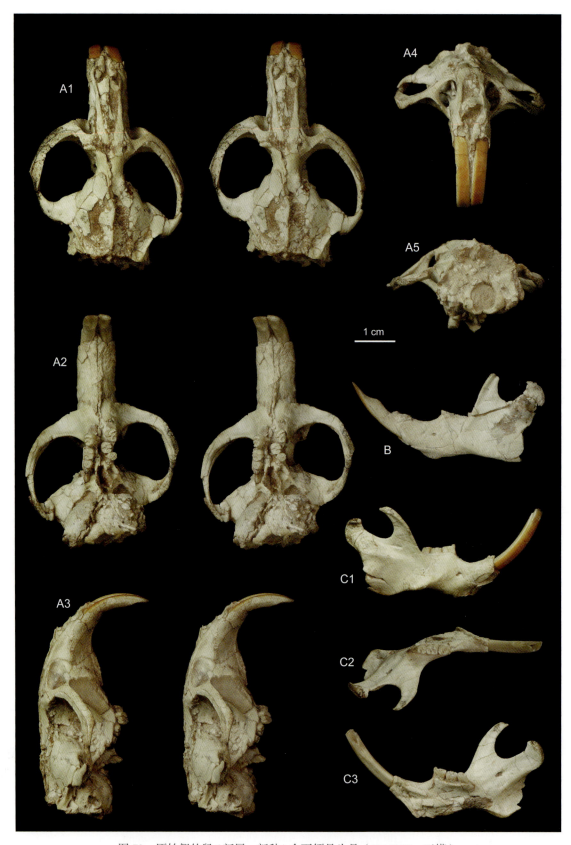

图 73 原始假竹鼠（新属、新种）含下颌骨头骨（V 16305，正模）

Fig. 73 Skull with mandible of *Pseudorhizomys pristinus* gen. et sp. nov. (V 16305, holotype)

A. 头骨（skull）：A1. 背面（立体照片）（dorsal view, stereopair），A2. 腹面（立体照片）（ventral view, stereopair），A3. 右侧面（立体照片）（right lateral view, stereopair），A4. 前面（anterior view），A5. 后面（posterior view）；B. 左下颌骨颊侧面（buccal view of left hemimandible）；C. 右下颌骨（right hemimandible）：C1. 颊侧面（buccal view），C2. 冠面（crown view），C3. 舌侧面（lingual view）

基枕骨保存较好，前窄后宽，但未见明显的纵中棱。基蝶骨破碎，它与基枕骨的界线不清楚。枕髁腹面很宽短，稍破损。左、右两枕髁的腹部彼此很靠近，仅以一纵沟分开。舌下神经孔大。右侧听泡保存较好，外形为卵圆形，中等膨胀，其前面内侧无明显的凹面。听泡表面也无明显的棱嵴。听泡的长轴和棘突均向前内方向延伸。欧氏管孔开孔于听泡棘突前方近中侧。颈静脉孔明显。因听泡与基枕骨连接处较破碎，颈内动脉孔的情况不清楚。

项面约呈半圆形，稍向前倾斜。从侧面观，枕髁位于项嵴的后方。项面大部已破损，仅下部保存较好。枕骨大孔约呈圆形，高稍大于宽。两枕髁在后面彼此分得较开。副乳突很小。

（2）下颌骨（图 73 B，C；表 11）

下颌骨的水平支较粗短。下颌齿隙的上缘也约呈"S"形，其隆凸的前部很短，而凹入的后部明显长于前部，但凹入较浅，具较缓的后坡。咬肌窝向前伸达 m1 和 m2 之间的下方。咬肌嵴很发达，并向外张，但其后部稍变平缓。颏孔位于 m1 的前下方，水平支的中部，与咬肌窝的前端约在同一横线上。冠状突前后较短，较向后倾，其前缘与下臼齿齿槽缘的夹角约为 120°。其上部明显向后弯曲，其后缘凹入较深。冠状突内侧下面的颞肌窝明显。下颌切迹较深，其深约为下颌骨高的 2/5。髁突主要向后上方延伸。其顶端稍向舌侧弯曲。下颌颈较细长。下颌髁的关节面约为肾形曲面，其后部较窄。髁突内侧下方稍凹，表面粗糙，可能为翼外肌附着处。i2 齿槽后端在下颌骨上升支的颊侧形成明显的隆凸。下颌孔大，位于冠状突后缘下方，与下臼齿列冠面约在同一水平上。左、右下颌角虽已缺失，但上升支下缘的大部分仍保存。从保存的部分看，翼内肌窝大而深，其前端伸达冠状突中部下方。上升支下缘在冠状突后缘的下方形成一较明显的拐角，先向下凸，再转向后上方延伸。这样，角突的位置可能比上升支下缘更高，使上升支后缘在下颌髁和下颌角突间的距离变短。

（3）牙齿（图 74；表 12）

V 16305 的齿式：1·0·0·3/1·0·0·3。臼齿的基本形态结构与 *Ps. indigenus* gen. et sp. nov. 很相似，也为脊形齿，齿冠高度中等，具齿根。臼齿从前向后变小。但臼齿的齿冠较低，仅在 m3 有较明显的后侧高冠现象。

M1 冠面约呈四边形，前缘宽于后缘，外缘长于内缘。无明显的前侧高冠现象。前缘有弱的原凹的痕迹。冠面主要具两颊侧褶沟（前边凹和中凹）和一舌侧褶沟（内凹）。三条褶沟均伸达齿冠中部，并向外开口。前边凹和中凹的横向的长度彼此相近，其舌部均向后弯，只是中凹后弯的曲度更大些。内凹横向较两颊侧褶沟稍短，横向延伸，其颊端插在前边凹与中凹之间，与两颊侧褶沟部分重叠。

M2 的冠面的形态结构与 M1 的很相似，也为四边形，但后缘仅稍窄于前缘，其颊、舌侧缘的长度相近。此外，其尺寸稍小。M2 冠面也具三条褶沟，但其两颊侧褶沟的舌部向后的弯曲度稍小些。

M3 的冠面为卵圆形，宽稍大于长，后缘窄于前缘。冠面也具三条褶沟，但它们均已封闭成釉质盆。其中前边凹最小。在左 M3 上，中凹稍大于内凹；而在右 M3 上，则相反，是内凹大于中凹。内凹插在前边凹和中凹之间。

I2 的形态结构与 *Ps. indigenus* gen. et sp. nov. 很相似，其前部向下方延伸，而不向后弯，为前伸型齿。I2 横切面也为舌侧角较平缓的三角形。釉质层也主要覆盖在唇面，仅稍延伸到两侧。其唇面横向圆凸，无纵棱，只是其圆凸度较 *Ps. indigenus* gen. et sp. nov. 和 *Ps. gansuensis* gen. et sp. nov. 者稍弱。

m1 的冠面为长卵圆形，前缘横向窄于后缘。在左 m1 的冠面上见有四条褶沟：两颊侧褶沟（下原凹和下外凹）和两舌侧褶沟（下中凹和下后边凹）。其中下中凹为横向延伸，仅其颊端稍显向下原凹方向弯，其舌侧仍向外开口。下外凹已磨蚀成封闭的盆，其舌部未见分叉现象。下原凹和下后边凹被磨蚀成残余的盆。在右 m1 的冠面仅见两条明显的褶沟：下中凹和下外凹；该二褶沟均已被磨蚀成封闭的盆。下后边凹仅保留有微弱的痕迹，而未见下原凹的痕迹。从磨蚀程度看，右 m1 显然要比左 m1 磨

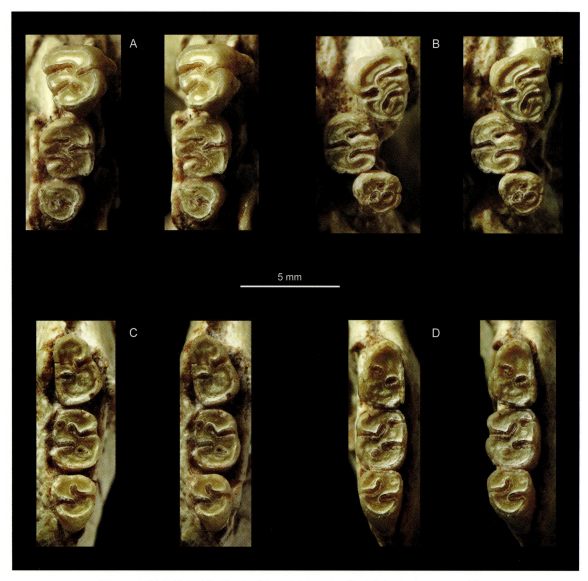

图 74　原始假竹鼠（新属、新种）臼齿冠面（立体照片）（V 16305，正模）

Fig. 74　Occlusal view of molars of *Pseudorhizomys pristinus* gen. et sp. nov. (V 16305, holotype, stereopair)

A. 右（right）M1–3, B. 左（left）M1–3, C. 左（left）m1–3, D. 右（right）m1–3

蚀得更深。很可能它的下原凹原来是存在的，只是已被完全磨蚀掉了。

m2 冠面为椭圆形，长大于宽。冠面具四条褶沟。其中除了下后边凹为封闭的釉质盆外，其他三条褶沟均向外开口。下中凹横向最长，其颊部向前弯向下原凹，但不达后者。下外凹为横向延伸，短于下中凹，长于下原凹。下外凹也未见有分叉现象。下后边凹最小。

m3 冠面约为卵圆形，长稍大于宽，后缘较窄。右 m3 的冠面结构与 m2 的相似，仍可见四个褶沟。其中下中凹和下外凹仍向外开口。下中凹横向长于下外凹，其颊部转向前弯。但它不指向下原凹，而是指向正前方。下外凹稍向后舌侧斜伸。下原凹和下后边凹均被磨蚀成封闭的釉质坑。其中下原凹大于下后边凹。左 m3 的下中凹和下外凹的形态结构与右 m3 的相似，但冠面未见下原凹和下后边凹的痕迹，可能是被深度磨蚀而消失了。

i2 纵向的弯曲度较 I2 的稍弱些，主要向前上方延伸。其横切面约为等边的三角形，唇面稍向外方斜。釉质层主要覆盖在唇面，稍向两侧面延伸。唇面与内侧的界线明显，与外侧面圆缓过渡。唇侧表面有两条纵嵴。

需要指出的是：右下颌骨上的 i2 可能并不属于 V 16305 这一个体。因为该门齿与下颌骨连接处是断开的，连接处并不完全对茬，可能是人为地被接在一起的；因为 V 16305 的右下颌骨前端受压变窄，明显窄于右 i2。但左、右 i2 的形态结构很相似，只是右 i2 稍窄小一些（见表 12）。这一区别可能只是种内的个体变异，也可能是性别差异，即右门齿属于雌性个体，而左 i2 属雄性个体。

2. 比较

上述 V 16305 的头骨、下颌骨和牙齿的形态特征，如前颌骨前端前伸超过鼻骨前端；前颌骨侧背嵴斜伸与鼻骨 - 前颌骨缝前部相连，前颌骨背侧为窄的三角形；前颌骨 - 上颌骨缝与上颌骨 - 额骨缝相连成圆弧形；眶下孔为大的圆形；颧骨很长，伸达眼眶前缘；I2 纵向较少弯曲，属前伸型齿，后部起自上颌中部；下颌齿隙后部凹入浅，后坡平缓；下颌切迹较深等，都与 *Pseudorhizomys* 新属的相似而不同于 *Pararhizomys*，应归入 *Pseudorhizomys* 新属。

V 16305 与新属 *Pseudorhizomys* 已知三个新种（*Ps. indigenus*，*Ps. gansuensis* 和 *Ps. planus*）有明显的区别：外层咬肌附着区域的前缘为圆弧形，表层咬肌附着的疤痕靠近眶下孔的腹侧裂隙，臼齿齿冠较低。

V 16305 还具有一些不同于 *Ps. indigenus* gen. et sp. nov. 和 *Ps. gansuensis* gen. et sp. nov. 的特征：个体小而纤细，颞嵴较发达，为明显的锐嵴；下颌骨冠状突前后较短，具较明显凹入的后缘，下颌髁关节面约为肾形，m1–3 具其下后边凹等。V 16305 与 *Ps. indigenus* gen. et sp. nov. 和 *Ps. planus* gen. et sp. nov. 不同还在于其 M1–3 的内凹较长和 M2 具较少退化的前边凹；与 *Ps. gansuensis* gen. et sp. nov. 的区别还有：其冠状突较向后倾，M2–3 具前边凹，m1–3 的下外凹不分叉和 m3 具下原凹等。此外，它与 *Ps. planus* gen. et sp. nov. 的区别还在于其吻部和上齿隙较长，前颌骨的背侧面较窄长，I2 唇面横向较凸等。以上区别表明，V 16305 应为 *Pseudorhizomys* 新属的一个新种，被称为原始假竹鼠（*Pseudorhizomys pristinus* gen. et sp. nov. ）。

3. 讨论

郑绍华（1980）描述了产自西藏比如县布隆盆地晚中新世灞河期中期布隆组的一段左（实为右）下颌骨（IVPP V 5183），将其命名为黑河低冠竹鼠（*Brachyrhizomys hehoensis*），归入竹鼠科。后来，Jacobs 等（1985）和弗林（Flynn, 2009）认为该标本可能是副竹鼠属（*Pararhizomys*）的成员。

笔者观察了这件标本，发现该标本实为右下颌，并赞同 Jacobs 等和弗林的上述观点，即 V 5183 的确很像副竹鼠类，而与竹鼠科明显不同。V 5183 的下颌骨咬肌窝的咬肌嵴发达；下臼齿稍向颊侧弯，舌侧齿冠稍高于颊侧齿冠；其冠面结构较简单，缺下中脊等特征都和副竹鼠类相同，而不同于竹鼠类。将 V 5183 与 *Pararhizomys* 和 *Pseudorhizomys* gen. nov. 两属进行比较后发现，V 5183 在下颌骨的形态、颏孔的位置和牙齿的结构形态上更像后者，特别是 *Ps. pristinus* gen. et sp. nov.，而与 *Pararhizomys* 的区别要大些。这主要表现在下颌骨下缘向下圆凸，臼齿的齿冠都较低，前 - 后侧高冠现象仅在 m3 中存在，下臼齿均具四条褶沟（舌侧的下中凹和下后边凹及颊侧的下原凹和下外凹），m2–3 的下中凹较长，并向前弯曲；i2 唇侧有两条纵嵴等。V 5183 应改归入 *Pseudorhizomys* 属。

然而，V 5183 与 *Ps. pristinus* gen. et sp. nov. 也有明显区别：其下颌骨水平部和臼齿齿冠都更低，未见明显的后侧高冠现象，i2 唇侧面横向较圆凸等。这些区别表明，V 5183 可能代表一个不同于 *Ps. pristinus* 的另一个种。由于 V 5183 仅是一段下颌骨，尚不知其头骨形态是否也与 *Pseudorhizomys* gen. nov. 者相同，目前笔者只能暂时有疑问地将它归入 *Pseudorhizomys* gen. nov.，保留郑绍华的原始种名，称其为黑河假竹鼠（属存疑）（*Pseudorhizomys? hehoensis*）。

三、副竹鼠类的系统演化和分类

（一）副竹鼠类属级及种级之间的系统关系

1. 副竹鼠与假竹鼠的相互关系

对 *Pararhizomys* 和 *Pseudorhizomys* 的区别特征（见表 9）进行比较和分析后，笔者发现，*Pseudorhizomys* 的诸多特征，如鼻骨背面纵向较平直，前颌骨前端前伸超过鼻骨前端，前颌骨背面呈窄三角形，前颌骨侧背嵴较短弱而斜伸，颧骨较长，前端向前伸达眼眶前缘，硬腭后缘的位置较靠前，I2 较少弯曲，为前伸型齿，其后端起点位置较靠后而唇面无纵嵴，臼齿齿冠较低，保留有较多的褶沟等（详见表 9: 1–4, 7–9, 12, 21–24, 27 等），都是比 *Pararhizomys* 更原始的特征。其中 I2 的形状在副竹鼠类的演化过程中可能起着重要的作用。头骨的许多形态结构（见表 9: 1–4, 7, 8 和 21 等），很可能都是随着门齿的变粗壮和弯曲而变化的。总体上讲，*Pseudorhizomys* 是比 *Pararhizomys* 更原始的一个属。然而，这并不意味着 *Pseudorhizomys* 是 *Pararhizomys* 的直接祖先，因为 *Pseudorhizomys* 还具有一些自有特化特征，如顶骨较宽短，颊肌神经孔、咬肌神经孔和颞深神经孔三孔愈合为一等（见表 9: 10, 15）。此外，现有的资料表明，*Pararhizomys* 最早出现的时间（例如 *Pa. longensis*）可能比 *Pseudorhizomys* 所有已知种都还要早些。很可能它们是在中新世中期或更早从共同的祖先起源后，就向不同的方向演化了。

有趣的是，到目前为止 *Pararhizomys* 和 *Pseudorhizomys* 已知各种均分别产于不同的地点，至今还没有发现这两个属产于同一地点的情况。至于这是采集的偏差，还是由于不同生态类型而产生的，还需要进一步研究。但从它们在临夏盆地产出的层位看，似乎这两个属在柳树组从下到上的层位中都存在。这至少表明，两者至少有一段时期是同时在临夏盆地存在的。

2. 副竹鼠属内各种间的相互关系

副竹鼠属目前已知包括四个种。此外还有两个未定种，*Pararhizomys* sp. I 和 *Pararhizomys* sp. II。两个未定种由于材料太少，特征不够清楚，其确切的分类位置及与已知种的关系无法判断。现仅对该属的四个已命名种的相互关系作一初步的探讨。对于像副竹鼠这样的小属，笔者以为采用"最老式（但完全有效）的组合规则"（Wiley et al., 1991: 15），亦即"亨尼希论证法（Hennig argumentation）"（同上: 45–47）即可有效地找出属内各种间的系统关系。该方法中最关键的是先要找出各种共有、但以不同状态存在的形态特征。其特征极性（character polarity）的确定以鼠形超科中某些原始种类（如 *Pappocricetodon* 等）的特征状态为主要依据。下面是这些主要特征：

1. 前颌骨侧面的纵沟：（0）较浅；（1）较深。
2. 颧弓板的位置：（0）弧形后缘的最前点在 M1 之前；（1）与 M1 中部相对。
3. 下颌骨齿隙与下颊齿列的长度：（0）长度相近；（1）前者长于后者。
4. 臼齿齿冠高度：（0）齿冠较低，仅部分臼齿（M1–M2 和 m3）为前 - 后侧高冠；（1）臼齿均为前 - 后侧高冠齿，但无舌侧高冠现象；（2）齿冠高，均为前 - 后侧高冠和舌侧高冠齿。
5. M3 的内凹和中凹：（0）彼此斜向延伸；（1）彼此相对。
6. m1 下中凹：（0）较短，较少弯曲；（1）较长，较强烈弯曲，其颊部明显向前延伸。

基于上述特征演化趋势，笔者对副竹鼠属的 4 个种的相互关系简要分析如下：

在 *Pararhizomys* 属中，出现得最早的是 *Pa. longensis*。虽然它仍具有一些较原始的特征，如下颌骨齿隙与下颊齿列的长度彼此相近（图 75: 3-0），齿冠较低，仅部分臼齿为较低弱的前 - 后侧高冠齿（4-0），M3 的内凹和中凹彼此斜向延伸（5-0）和 m1 的下中凹较少弯曲（6-0）等；但它已具有某些较特化的特征，如前颌骨侧面具有较深的纵向凹槽（1-1）和颧弓板位置较靠后（2-1）等。这表明，*Pa. longensis* 虽较原始，却已开始特化。稍后，相继出现的是 *Pa. hipparionum* 和 *Pa. qinensis*。两者共有近裔特征是：臼齿齿冠较 *Pa. longensis* 者要高，均为前 - 后侧高冠齿，其舌侧和颊侧齿冠高度和褶沟的深度相近（4-1）。但 *Pa. hipparionum* 具有自近裔特征：m1 的下中凹明显弯曲（6-1），这明显不同于 *Pa. longensis* 和 *Pa. huaxiaensis* 者（可惜，*Pa. qinensis* 未保留下牙，此特征不知）。这表明，*Pa. hipparionum* 和 *Pa. qinensis* 可能组成一对姐妹组。

Pararhizomys huaxiaensis 则是该属中第四个出现的种。它的臼齿齿冠更高，不仅具有较高的前 - 后侧高冠型齿，而且还是舌侧高冠型齿，其舌侧褶沟明显深于颊侧褶沟（4-2）。另外，它的下颌齿隙较长，明显长于下齿列者（3-1），M3 的内凹和中凹彼此相对，不重叠（5-1）等。这些特征显然要比上述三个种都更进步。然而，该种又具有与 *Pa. longensis* 相似，而比 *Pa. hipparionum* 原始的特征，即 m1 具较短且较少弯曲的下中凹（6-0）。

这样，在 *Pararhizomys* 属中，很可能只有 *Pa. qinensis* 和 *Pa. hipparionum* 代表一条演化支系，组成姐妹群；*Pa. huaxiaensis* 则代表了不同于 *Pa. hipparionum*-*Pa. qinensis* 姊妹组的另一个演化支系；而 *Pa.*

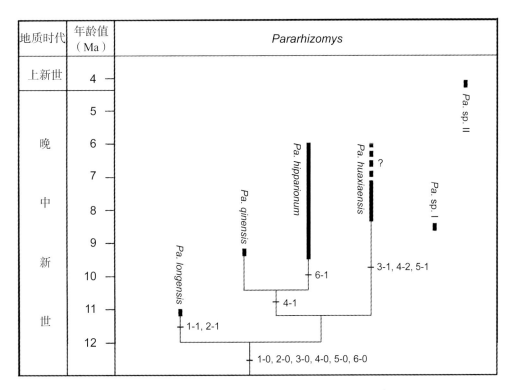

图 75　副竹鼠属各种的地质时代分布和系统关系 *

Fig. 75　Geological occurence and phylogenetic relationships of *Pararhizomys* species[*]

细线（thin line）：推测的系统关系（inferred relationship）；垂直黑粗线（vertical thick black line）：时代分布（geological time duration）；?，庆阳的标本的归属还需证实（systematic position of the Qinyang specimen pending further study）；图中特征 1–6 见上文（for characters 1–6, *vide supra*）

* *Pa. longensis* 和两个未定种的时代基于化石点在剖面中的位置，*Pa. qinensis* 者基于古地磁资料（Zhang et al., 2013），而 *Pa. hipparionum* 和 *Pa. huaxiaensis* 者则基于动物群（geological ages of *Pa. longensis* and *Pararhizomys* spp. I, II based on stratigraphic position of their localities, *Pa. qinensis* based on paleomagnetic dating of Zhang et al., 2013, *Pa. hipparionum* and *Pa. huaxiaensis* based on local faunas）

longensis 则代表早已分出的另一演化支系（见图 75）。用简化的文氏法（Venn diagram）表示为：(*Pa. longensis* (*Pa. huaxiaensis* (*Pa. qinensis* + *Pa. hipparionum*)))。

3. 假竹鼠属内各种间的相互关系

假竹鼠新属目前包括四个新种（*Ps. indigenus*，*Ps. gansuensis*，*Ps. planus* 和 *Ps. pristinus*）和一有疑问归入的种（*Ps.? hehoensis*）。假竹鼠属主要区别特征的演化趋势如下：

1. 颧弓板上外层咬肌附着区前缘形状：（0）前凸圆弧形；（1）上凹下凸的"S"形。
2. 颞嵴：（0）较长而明显，其后部较窄锐；（1）较短，其后部较平缓。
3. 下颌骨齿隙后部凹入部分：（0）相对较长，凹入较浅；（1）相对较短，凹入较深。
4. I2 唇面：（0）横向较宽而圆凸，表面光滑；（1）较窄而平，表面具弱的纵嵴。
5. 臼齿齿冠高度：（0）无前 - 后侧高冠现象；（1）仅 m3 为后侧高冠齿；（2）M1 和 m3 为前 - 后侧高冠齿。
6. M1–M2 内凹：（0）长，横向插入到前边凹和中凹之间；（1）较短，仅在 M1 插入到前边凹和中凹之间；（2）更短，在 M1 和 M2 均不与前边凹和中凹重叠。
7. M2–M3 前边凹：（0）在 M2–M3 很发育，仅在 M3 退化为封闭的盆；（1）均有，但在 M2–M3 均退化为封闭的盆；（2）在 M2–M3 完全退化消失。
8. M3 内凹和中凹：（0）内凹长，斜向延伸，与中凹部分重叠；（1）内凹短，与中凹相对。
9. m1–m3 下后边凹：（0）有；（1）无。
10. m1–m3 下外凹：（0）不分叉；（1）分叉。
11. m3 下原凹：（0）有；（1）退化或消失。
12. m3 下外凹与下中凹的位置：（0）下外凹位于下中凹之后；（1）下外凹与下中凹的舌部大致相对。

基于上述特征演化趋势的认识，笔者对假竹鼠属目前已知种的相互关系分析如下：*Ps. gansuensis*，*Ps. pristinus* 和 *Ps.? hehoensis* 三种出现的时代较早，均为晚中新世灞河期的中期。如果 *Pseudorhizomys* 属中臼齿确有由低冠向较高冠和冠面结构由复杂向简单方向演变的趋势，而且 *Ps.? hehoensis* 确应归入 *Pseudorhizomys* 属的话，*Ps.? hehoensis* 应是该属目前已知最原始的种，因为它的臼齿齿冠最低，尚未出现前 - 后侧高冠现象（5-0）；而且臼齿冠面的褶沟较多，m1–m3 的下后边凹和 m3 的下原凹均仍存在（9-0, 11-0）。*Ps. pristinus* 较 *Ps.? hehoensis* 稍进步些，其齿冠稍高，m3 已具后侧高冠现象 (5-1)。然而 *Ps. pristinus* 仍具有一些比 *Pseudorhizomys* 其余三种（*Ps. indigenus*、*Ps. planus* 和 *Ps. gansuensis*）更原始的特征，如：颧弓板上外层咬肌的前缘为前凸的弧形（1-0），颞嵴较发达，具较窄锐的后部（2-0），下颌骨齿隙后部较长而凹入较浅（3-0），M1–M2 内凹较长，插入前边凹和中凹之间（6-0），M2–M3 均具前边凹，而且在 M2 中不退化（7-0）。而 *Ps. indigenus*、*Ps. planus* 和 *Ps. gansuensis* 三种则具有一些较 *Ps. pristinus* 进步的共近裔特征：如外层咬肌附着区的前缘由圆弧形变为"S"形（1-1）、颞嵴变低缓（2-1）、下颌骨齿隙后部变短，凹入变深（3-1），臼齿齿冠进一步增高，不但 m3 为后侧高冠，而且 M1 也出现前侧高冠现象（5-2），m1–m3 的下后边凹已退化、消失 (9-1) 等。这些都显示了该三种的确比 *Ps.? hehoensis* 和 *Ps. pristinus* 进步。而在该三种中 *Ps. indigenus* 与 *Ps. planus* 共有如下近裔特征：M2–M3 的前边凹均已退化成封闭的坑（7-1），M1–M3 的内凹，横向变短，在 M1–M2 中不达前边凹内端水平（6-2），在 M3 中和中凹相对（8-1）等。很可能，*Ps. indigenus* 和 *Ps. planus* 组成一姐妹群。但 *Ps. indigenus* 和 *Ps. planus* 又各自具有不同的特征：*Ps. indigenus* 的 m3 的下外凹的位置较前移（12-1），比 *Ps.? hehoensis*、*Ps. pristinus* 和 *Ps. gansuensis* 的进步（可惜 *Ps. planus* 未保存下牙）；而 *Ps. planus* 的 I2 较小，唇面横向较平（4-1）等特点不但与 *Ps. indigenus* 不同，也与 *Ps. pristinus* 和 *Ps. gansuensis* 均不同。有意思的是，尽管 *Ps. gansuensis* 出现的时代较 *Ps. indigenus* 早，而且保留了若干较 *Ps. indigenus* 和 *Ps. planus* 更原始的特征，如 M1–2 的内凹横向较长，在 M1 中插入到前边凹和中凹之间（6-1），在 M3 中与中凹部分重叠（8-0），以及具有一些与 *Ps. indigenus* 和 *Ps. planus* 共有的近裔特征（1-1, 2-1, 3-1, 5-2, 9-1）；但 *Ps. gansuensis* 具有一些较其他四种都更进步的特征，如它的 M2–M3 的前边凹和 m3 的下原凹均完全消失（7-2, 11-1），而且其 m1–m3 的下外凹还出现分叉（10-1）等。这表明，*Ps. gansuensis* gen. et sp. nov. 代表了早已分出的另一特化分支。

综上所述，*Pseudorhizomys* 属除 *Ps.? hehoensis* 外的四个种中，*Ps. pristinus* 代表较原始的早已分出的一支，其余的三个种可能又分为两支：*Ps. indigenus* 和 *Ps. planus* 为形成姐妹群的一支；而 *Ps. gansuensis* 则为另一支。*Ps. gansuensis* 产出的时代要早于 *Ps. indigenus*，为晚中新世中期，中灞河期。这样，这两支可能在晚中新世早期（早灞河期）或中新世中期甚至更早就分开了。遗憾的是，*Ps. pristinus* 的产出确切地点和层位还不能肯定，而 *Ps. planus* 的产出地点和层位也都不清楚，加之 *Ps.? hehoensis* 和 *Ps. planus* 已知的材料还太少，这为全面了解它们的形态特征、分类位置和相互关系增加了一定的难度。图 76 只是根据现有资料对 *Pseudorhizomys* 各种之间系统演化关系进行了粗略推测。用简化的文氏法（Venn diagram）表示为：(*Ps.? hehoensis* (*Ps. pristinus* ((*Ps. indigenus* + *Ps. planus*) *Ps. gansuensis*)))。

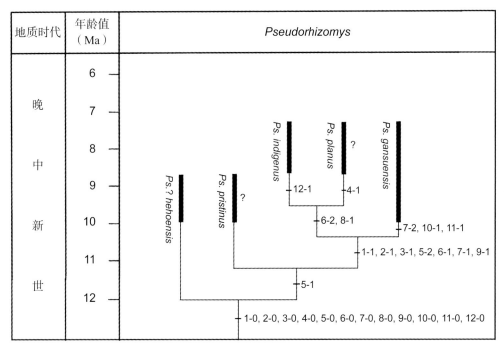

图 76　假竹鼠属各种的地质时代分布和系统关系

Fig. 76　Geological occurence and phylogenetic relationships of *Pseudorhizomys* species

（二）副竹鼠类与鼠超科内相近科之间的系统关系

1. 目前对副竹鼠类分类地位的各种意见

副竹鼠类的演化和系统分类位置一直令人困惑不解。在创建副竹鼠属（*Pararhizomys*）时，德日进和杨钟健（Teilhard de Chardin et Young, 1931: 12）认为该属与盲鼹鼠亚科（Spalacinae）的竹鼠（*Rhizomys*）有较近的系统关系，但是他们仍只将该属归入鼠超科（Muroidea），而没有确定其科和亚科的地位。后来，辛普森（Simpson, 1945）、杨钟健和刘东生（Young et Liu, 1950）以及 Kowalski（1968）都把副竹鼠属归入到竹鼠科（Rhizomyidae）中。稍后，Flynn（1982, 1990）提出，副竹鼠属应是盲鼹鼠科（Spalacidae）的成员，将它从竹鼠科中排除。McKenna 和 Bell（1977）在其哺乳动物分类的专著中则将副竹鼠属归入到包含量非常大的鼠科（Muridae）的竹鼠亚科（Rhizomyinae）中。张兆群等（2005）指出，副竹鼠属的分类位置存在两种可能性：既可能是竹鼠科的成员，也可能代表鼠科中独立的一支；并暂时将其归入鼠超科，但科不确定。这一观点得到李强（Li, 2010）的支持。2009年，Flynn又一次提出，副竹鼠属可能还是竹鼠科的成员。那么，副竹鼠类的系统分类位置究竟应在何处？与哪一科级单元有更近的系统关系呢？

2. 副竹鼠类及鼠超科内相近科的分支分析

（1）分类单元的选取

本书讨论的核心问题是副竹鼠类在鼠超科内科一级的系统分类位置，所以我们将内类群分类单元设定为属级。

根据头骨的颧 - 咬肌结构和齿式，副竹鼠类无疑应归入鼠超科（Muroidea）。关于鼠超科内的科级分类，到目前为止仍然分歧很大（见 Ellerman, 1940; Simpson, 1945; Wood, 1955; Schaub, 1958; McKenna et Bell, 1997; Michaux et al., 2001; Jansa et Weksler, 2004; Norris et al., 2004; Steppan et al., 2004; Flynn, 2009; Jansa et al., 2009; Blanga-Kanfi et al., 2009; Lin et al., 2014; de Bruijn et al., 2015）。《中国古脊椎动物志》负责小哺乳动物部分的编委们讨论后认为：为了方便起见，鼠超科（Muroidea）可暂定包括仓鼠科（Cricetidae）、鼠科（Muridae）、田鼠科（Arvicolidae）、沙鼠科（Gerbillidae）、鼢鼠科（Myospalacidae）、盲鼹鼠科（Spalacidae）、拟速掘鼠科（Tachyoryctoididae）、竹鼠科（Rhizomyidae）和刺山鼠科 Platacanthomyidae 9 个科。本卷也采纳上述意见。从解剖特征上看，在这些科中，以拟速掘鼠科、盲鼹鼠科和竹鼠科与副竹鼠类有更多的相近特征。它们除了头骨具鼠型颧 - 咬肌结构、向后伸向项嵴的很长的关节窝、松鼠型下颌骨、无 P4/p4 和门齿具单系的显微结构等鼠超科所共有的特征外，还具有如下的共近裔特征：下门齿后端在下颌骨颊侧形成隆凸，颊齿为简单的脊形齿，M1（m1）的前边尖（下前边尖）不明显，融合到横向延伸的前边脊（下前边脊）中，m1 下原尖与下后尖具有通过下前边脊的前连接等。这些都表明，在 Muroidea 中副竹鼠类与该三科有最近的亲缘关系。

这样，内类群的分类单元，我们只选择了上述三科中较重要和 / 或保存较完整的几个属。拟速掘鼠科目前仅知晚渐新世和早中新世的三个属（*Tachyoryctoides*、*Ayakozomys* 和 *Eumysodon*）。其中只有 *Tachyoryctoides* 属有少数几个不完整的头骨化石可供对比，所以选取了这个属作为该科的代表。盲鼹鼠科的已知化石材料中几乎没有头骨可供对比。笔者仅采用了较原始的 *Heramys* 属作为该科的化石类代表。在盲鼹鼠科中，尽管目前已知最早的化石是晚渐新世的 *Vetuspalax*，但它仅保存有单个的牙齿，而且牙齿也已相当特化（de Bruijn et al., 2013, 2015）。另外，*Debruijnia* 出现的时代也稍早于 *Heramys*，臼齿的特征也稍更原始些，但 *Heramys* 目前已知保存的材料要稍好于 *Debruijnia*（如有较好的下颌骨等），故而选定了 *Heramys*。盲鼹鼠科晚期化石种类之间的关系和演化趋势都比较清楚，但在牙齿的结构形态上的变化比较大（Ünay, 1999; Sarica et Sen, 2003）。基于上述情况，我们只选取了现生属，即 *Spalax*。在竹鼠科中，除选取了一个保存较好的具有代表性的化石属 *Miorhizomys* 外，则选取了现生属 *Rhizomys* 和 *Tachyoryctes*。

关于外类群，始新世的祖仓鼠（*Pappocricetodon*）应是最合适的代表。这是鼠超科中已知最早、也是最原始的种类。但在该属已知的材料中，仅牙齿和下颌保存较好，头骨仅保存有残破的吻部，所能提供的头骨特征极性的信息不多。为此，笔者还选取了更原始的副鼠（*Paramys*）作为首选外类群。副鼠材料保存得很完整，且已有很详尽的记述（Wood, 1962），其在啮齿类中所处的较基干的地位也是当前啮齿类专家所普遍承认的。许多祖仓鼠所不能提供的特征极性状态都可以副鼠作为参考。虽然亚洲已知最早、最原始的啮齿类是钟健鼠（*Cocomys*）和外鼠（*Exmus*），而且它们都保存有较好的含下颌骨的头骨（Li et al., 1989; Wible et al., 2005），但它们的头骨和牙齿冠面形态都较 *Paramys* 更原始，与鼠形超科之间的差别悬殊，并有可能代表更原始的、不同于鼠形类的另一分支的祖先类型，故笔者未选其作为本研究类群的外类群。

这样，在本书的数据库中就只包括以下 10 个分类单元：*Paramys*, *Pappocricetodon*, *Heramys*, *Spalax*, *Miorhizomys*, *Rhizomys*, *Tachyoryctes*, *Tachyoryctoides*, *Pararhizomys* 及 *Pseudorhizomys*。

（2）特征选取

为了更深入地了解副竹鼠类与上述其他 3 个科之间的系统发育关系，笔者选取了尽可能多的头骨及下颌（0–31）和牙齿（32–53）的特征，共 54 个。其中 22 个为 2 态特征（binary），32 个为多态特征（multistate）。在多态特征中有 4 个牙齿的特征（45，M3 内凹；48，m2 下原凹；50，下臼齿下前边凹；53，m1–m2 下中脊）显示出反向演化的现象，即特征由弱或无（编码 0）先变为较发育（编码 1），又变为弱或呈残疾状（编码 2）。上述特征我们没有按照有 - 无的二态性状编码，而是按照我们对化石观察所得的演化顺序（由无→发育→退化为残疾）进行三态编码。

在特征的选取和分析中，*Paramys* 的主要参考资料是 Wood（1962）的专著和 Wahlert（1974, 1985b）的文章；*Pappocricetodon* 的资料主要依据的是童永生（1992, 1997）、王伴月和道森（Wang et Dawson, 1994）、道森和童永生（Dawson et Tong, 1998）的文章；Spalacidae 和 Rhizomyidae 的资料来自 Ellerman（1940, 1941）、Flynn（1982, 1990, 2009）、Carleton 和 Musser（1985）、Hofmeijer 和 de Bruijn（1985）、郑绍华（1993）、Ünay（1999）、Sarica 和 Sen（2003）、Wesselman 等（2009）的文章；而 Tachyoryctoididae 的资料则取自于 Bohlin（1937, 1946）、Kowalski（1974）、李传夔和邱铸鼎（1980）、Bendukidze 等（2009）、王伴月和邱占祥（Wang et Qiu, 2012）及 Daxner-Höck 等（2015）的文章。还有一些资料是笔者根据有关的化石和现生的标本的观察而得。*Tachyoryctoides* 的头骨和下颌骨的一些特征是笔者观察中国科学院古脊椎动物与古人类研究所内蒙古考察队最近新采的标本所得。

本书选取的用以构建系统发育树的特征及其极性判断如下（按照 WinClada 的预设要求，特征计数从 0 开始）：

0. 吻部长度：（0）长约等于宽；（1）长明显大于宽。
1. 头骨中 - 后部高度：（0）低，M2 处颅高（不包括 M2 冠高）明显小于上颌齿隙长；（1）较高，M2 处颅高约等于上颌齿隙长；（2）高，M2 处颅高明显大于上颌齿隙长。
2. 鼻骨：（0）两侧缘（除最后端外）大体互相平行，后端与前端接近等宽或稍窄；（1）后端显著窄于前端，两侧缘向后趋中靠近，但后端不尖；（2）两侧缘显著向后趋中，后端变尖。在最原始的啮型类中，鼻骨的后端大多是比较宽的，如 *Rhombomylus* 和 *Matutinia*（Meng et al., 2003）。在 Wood（1962）记述的大部分副鼠类的材料中也是如此。在上述内类群中后端变尖显然是近裔性状。
3. 鼻骨后端的位置：（0）明显后于前颌骨后端；（1）与前颌骨后端约在同一横线上；（2）明显前于前颌骨后端。*Paramys* 的鼻骨后端向后延伸较长，明显超过前颌骨的后端。这应是近祖性状。
4. 前颌骨侧背嵴：（0）无；（1）较短，仅后部有；（2）在整个吻部背侧面形成长的侧背嵴。
5. 门齿孔长度及位置：（0）门齿孔长约等于上颌齿隙长的 1/2，距门齿近，距第一颊齿远；（1）门齿孔短，其长约为上颌齿隙长的 1/4–1/3，约位于上颌齿隙的中部；（2）门齿孔极小，不及上颌齿隙的 1/5，约位于上颌齿隙的中部。*Paramys* 和 *Rhombomylus* 的门齿孔都较大，距门齿近，而距第一颊齿远。这应该是近祖性状。
6. 眶下孔：（0）小而圆，无下裂隙，位于吻部后下角、眼眶之前下方；（1）宽大圆形，无下裂隙，位于眼眶之前、颧弓板上方；（2）宽大的圆形或卵圆形，位于吻部侧面后部的上半部、颧弓板上前内方，有下裂隙；（3）自前上方看为横宽的三角形，无下裂隙，位于吻部侧面后部之上端、颧弓板的上方。现有的化石材料表明，眶下孔的下裂隙在最原始的啮型类（*Paramys* 和 *Pappocricetodon*）中是没有的；在竹鼠中，则可能是从有（*Miorhizomys*）再到消失（*Rhizomys*）。
7. 颧弓板：（0）未形成；（1）有，较小，宽大于长，面向下方；（2）较宽大，宽大于长，面向前下方；（3）很宽大，长宽接近，面向前下外方。

8. 外层咬肌前上端附着区：（0）限于上颌骨内；（1）从上颌骨向前延伸到前颌骨。

9. 颧弓后端向后延伸的程度：（0）不达项嵴，两者之间以明显的凹缺相分；（1）颧弓后端和项嵴直接相连。

10. 颧骨：（0）长，前端伸达眼眶前缘，接近或接触泪骨；（1）很短，其前端远离眼眶前缘和泪骨。在头骨保存较好的副鼠类中（Wood, 1962），颧骨总是很长。*Cocomy* 和 *Exmus* 的颧骨也都很长，其前端向前超过眼眶前缘。这应该是近祖性状。*Rhombomylus* 的颧骨已很特化，具有多个突起，但仍然很长大（Meng et al., 2003）。

11. 硬腭后缘：（0）在 M3 之前；（1）约与 M3 或其后缘在同一水平；（2）在 M3 之后。*Paramys*，*Rhombomylus*，*Cocomys* 和 *Exmus* 的硬腭后缘全在 M3 之前，这应该是近祖性状。

12. 腭面上的腭沟：（0）在腭后孔附近不加深；（1）在腭后孔附近变为很深的窄槽。

13. 副翼窝：（0）无；（1）有。

14. 脑颅前外壁：（0）光滑圆凸；（1）在额骨 - 鳞骨缝附近形成约近垂向的棱嵴。

15. 翼窝：（0）远小于中翼窝，其前缘远远后于硬腭后缘；（1）与中翼窝大小相近或稍小，其前缘仅稍后于硬腭后缘；（2）大而深，底面深凹；（3）大而深，无底，通向脑颅。*Paramys* 的翼窝很小，其外翼突发育很弱，所以其外界限不很清楚。由于头骨中部很长，中翼窝也很长，所以翼窝远在中翼窝前端之后（见 Wood, 1962: 4–5；Wahlert, 1974: Fig. 4）。*Rhombomylus*, *Cocomys* 和 *Exmus* 的外翼突均已相当发育，翼窝小，但已很明显，其前缘也是远离中翼窝的前端。小而后位的翼窝显然是近祖性状。

16. 颊肌神经孔、咬肌神经孔和颞深神经孔：（0）咬肌神经孔与颞深神经孔愈合，而与颊肌神经孔分离，与其后孔相距较远（亦即神经管较长）；（1）愈合的咬肌神经孔和颞深神经孔仍与颊肌神经孔分离，但距其后孔的距离短（亦即神经管很短）；（2）三神经管前开孔深陷在同一凹陷中，离后开孔很近（神经管很短）。

17. 卵圆孔：（0）为单一孔，位于翼蝶骨侧壁；（1）与中破裂孔愈合。

18. 外耳道：（0）孔状；（1）锥形管状，耳孔围以不规则突起；（2）短管状；（3）长管状。

19. 外耳孔的位置：（0）位于关节窝后下方；（1）位于伸长的关节窝后端下方；（2）位于关节窝后方或后上方，紧靠项嵴。

20. 关节窝（= 下颌关节窝）：（0）长宽相近，或长稍大于宽，其后端远离项面；（1）长大于宽，其后端以一短的凹缺或外耳道与项嵴相隔；（2）长远大于宽，其后端接近或伸达项面。

21. 项嵴和项面：（0）项面微凸或平，项嵴和项面总体接近垂直；（1）项面平，项嵴与项面稍斜向前上方伸展（项面稍面向上方）；（2）项嵴稍斜向前下方延伸，而项面显著斜向前上方延伸（面向后上方），项面向后凸出。

22. 两枕髁腹部之间的距离：（0）彼此分得很开；（1）彼此很靠近，仅以很窄的纵沟相隔。

23. 副乳突：（0）很小，（1）较宽大。

24. 下颌骨髁突：（0）向后上方延伸；（1）主要向上伸。

25. 下颌角突后缘：（0）位于下颌髁后缘的下方或之后；（1）位于下颌髁后缘之前。

26. 下颌角突的形状：（0）尖角状，主要向后伸；（1）圆叶状，角突末端常有一伸向后上方的小尖突。Wood（1962）记述的 *Paramys* 的下颌中其角突远端都没有保存，但是在属于 *Paramyinae* 的 *Leptotomus costilloi* 中可以看到，其下颌角呈尖角形，主要向后伸（Wood, 1962: 79, Fig. 24）。*Cocomys*，*Rhombomylus* 和 *Exmus* 等亚洲一些早期啮齿动物的下颌角也都是向后伸展的尖角形。所以向后伸的尖角状的下颌角应为原始状态。

27. 下门齿后端：（0）在下颌骨上不形成明显的隆突；（1）在上升支颊侧形成较小的隆突，隆突明显低于下颌髁，与下颌髁之间也不以深沟相隔；（2）隆突粗大，其顶端稍低于下颌髁顶端，与后者有沟相隔；（3）隆突与下颌髁同高或更高，两者间以深沟隔开。

28. 下颌骨咬肌嵴：（0）弱嵴形或线形；（1）明显向外伸张。

29. 颞肌在下颌骨上的附着区（颞肌窝）的位置：（0）附着面位于冠状突上部，向后伸达髁突内侧；（1）附着面增大，向下延伸到冠状突内侧面的基部；（2）附着面更大，向前下方延伸到冠状突前缘内侧面基部，并形成明显疤痕或凹槽。

30. 翼内肌窝：（0）大而深；（1）窄小而浅。

31. 下颌齿隙长：（0）短于下颊齿列长；（1）等于或长于下颊齿列长。

32. I2 唇侧：（0）无明显的纵棱；（1）具纵棱。

33. i2 釉质层在外侧延伸的宽度（唇舌向）：（0）很宽，明显宽于很窄的内侧釉质层条带；（1）和内侧釉质层条带一样，都很窄。

34. 上前臼齿：（0）具 2 枚（P3 和 P4），（1）仅 1 枚（P4），（2）无。

35. 臼齿齿冠：（0）低冠；（1）冠高中等，常具前 - 后侧高冠现象；（2）冠高中等，具单面高冠现象（上臼齿舌侧冠高于颊侧者，而下臼齿颊侧冠高于舌侧者），（3）高冠齿。

36. M1–M3 齿型与中脊：（0）丘型齿，无中脊；（1）丘 - 脊型齿，具短小中脊；（2）脊型齿，中脊很发育；（3）脊型齿，中脊退化变短小以至消失。

37. M1 前边尖：（0）无；（1）小丘形，具前齿带；（2）前边尖与前边脊相连成横脊形，无前齿带。

38. M2 前边凹：（0）有；（1）无。

39. M3 前边凹：（0）有；（1）无。

40. M1 和 M2 后脊：（0）与原尖后臂相连；（1）与次尖或次尖附近的前、后臂相连；（2）与后边脊相连或愈合，使后边凹退化或消失（后边凹的位置见 Wang et Qiu, 2012: 110, Fig. 1）。

41. M3 后脊：（0）与原尖相连；（1）与次尖前臂相连；（2）与后边脊相连或愈合，后边凹退化或消失。

42. M1 和 M2 中凹：（0）前后宽≥横长；（1）前后宽 < 横长；（2）前后宽≤横长的 1/2。

43. M3 中凹：（0）前后宽≥横长；（1）前后宽 < 横长；（2）前后宽≤横长的 1/2。

44. M3 内脊：（0）有；（1）无。

45. M3 内凹：（0）浅小；（1）较深而宽；（2）退化至无。从表面上看，这好像是向不同方向演化的结果。实际上在牙齿由低冠向高冠演化的过程中，这种现象经常发生，确是循序演化的结果（特征 48，50，53 与此相同）。

46. 第一个下颊齿（有 p4 时为 p4，无 p4 时为 m1）的下前边尖：（0）无；（1）丘形；（2）横脊形（= 下前边脊）。

47. 第一个下颊齿（含义同上）下后尖与下原尖：（0）两尖不连接；（1）两尖直接相连（通过下后脊相连）；（2）两尖前连接，即下后尖与下原尖通过下前边脊相连；（3）两尖双连接，即除通过下前边脊相连外，两尖还直接相连。

48. m2 下原凹：（0）无；（1）明显存在；（2）退化至残迹或消失（见特征 45）。

49. 下臼齿下外凹：（0）前后向较长，横向较短，不斜伸；（1）前后向较长，横向较短，但明显向后舌方斜伸；（2）前后变短，形成横向伸长的褶沟。

50. 下臼齿下前边凹：（0）无；（1）明显发育；（2）退化成残迹状或消失（见特征 45）。

51. 下臼齿下中凹：（0）前后向较宽；（1）前后向变窄，横向长明显大于前后宽，形成横向或向前颊侧延伸的褶沟。

52. 下臼齿下后边凹：（0）在 m1-m3 均较发达；（1）在 m1-m2 发达，在 m3 退化变小或无；（2）在 m1-m3 退化变小或无。

53. m1 和 m2 下中脊：（0）未形成；（1）明显存在；（2）退化变小或为残迹（见特征 45）。

根据以上选定的分类单元和特征及其状态构成的数据库如表 20 所示。

表 20 数据矩阵（分类单元及特征）
Table 20　Data matrix（tabulation of taxa and characters）

Taxa \ Characters	0	1	2	3	4	5	6	7	8	9	10	11	12	13	14	15	16	17	18	19	20	21	22	23	24	25	26	27	28	29	30	31	32	33	34	35	36	37	38	39	40	41	42	43	44	45	46	47	48	49	50	51	52	53	54	
Paramys	0	0	0	0	0	0	0	0	0	0	0	0	0	0	0	0	0	0	0	0	0	0	0	0	0	0	0	0	0	0	0	0	0	0	0	0	0	0	0	0	0	0	0	0	0	0	0	0	0	0	0	0	0	0	54	100%
Pappocricetodon	?	?	?	?	?	0	0	1	0	?	?	0	0	0	?	?	?	?	?	?	1	0	?	0	?	?	0	?	0	0	0	?	0	0	1	0	0	0	0	0	0	0	0	0	0	1	1	1	0	1	0	0	0	1	30	56%
Heramys	?	?	?	?	?	0	1	1	1	?	1	1	0	0	?	1	?	?	?	?	1	0	?	0	?	?	1	0	0	0	1	?	0	0	2	0	2	0	0	0	1	1	1	0	0	1	2	1	0	1	0	0	0	2	23	43%
Spalax	1	1	1	0	2	2	2	2	2	0	1	1	0	1	3	1	1	2	1	2	1	0	2	0	0	2	0	2	0	2	0	0	0	1	3	2	3	2	0	0	2	2	2	0	0	2	2	2	2	2	2	2	2	2	54	100%
Miorhizomys	0	1	1	0	1	2	2	3	3	0	1	1	0	1	2	1	?	?	1	1	1	2	1	0	0	2	0	2	0	2	0	0	0	1	3	2	3	2	0	2	2	1	2	1	0	2	2	2	2	2	2	1	0	1	46	85%
Rhizomys	0	2	1	1	0	1	2	3	3	1	1	0	0	0	2	1	2	2	3	2	1	0	2	0	0	2	0	2	0	2	0	0	0	1	2	2	2	2	0	2	2	1	2	2	1	2	3	2	2	2	2	1	0	1	54	100%
Tachyoryctes	1	2	1	2	0	1	2	2	0	1	1	1	0	0	2	1	2	2	2	2	1	1	0	0	0	2	0	2	0	2	0	0	0	1	2	3	2	2	0	2	2	1	2	1	1	2	3	3	2	2	2	1	0	1	50	93%
Tachyoryctoides	1	2	1	2	1	1	2	3	1	0	1	1	0	0	2	1	2	2	2	2	1	2	2	0	0	2	0	2	0	2	0	0	0	1	3	2	3	2	0	2	2	2	2	2	0	2	2	2	1	2	2	2	2	2	50	93%
Pseudorhizomys	1	0	2	1	2	1	2	3	0	0	1	1	0	1	2	1	2	1	1	2	2	1	2	0	0	2	0	2	0	2	0	1	0	1	2	2	2	2	0	2	2	2	2	2	2	2	2	2	2	2	2	2	1	2	54	100%
Pararhizomys	1	0	2	1	2	1	2	3	0	0	1	1	0	1	2	1	2	2	2	2	2	1	2	0	0	2	0	2	0	2	0	1	0	1	3	2	3	2	0	2	2	2	2	2	2	2	2	2	2	2	2	2	2	2	54	100%

注：? = 缺失数据（missing data）；倒数第 2 列：编码特征数（除去缺失项）[penultimate column: number of scored states (excluding ?)]；最后 1 列：编码完整性 [last column: completeness（= number of scored states/54）]。

（3）系统发育树构建

由于本系统发育分析所涉及的只有 10 个分类单元和 54 个特征，属于小数据分析范畴，没有必要采用复杂的搜索程序，我们采用了表达手段比较简便实用的 WinClada 软件（Nixon, K. C. 1999-2002. WinClada ver. 1.00.08）。WinClada 中的 Heuristics 搜索法虽然仍然属于经验式范畴，但增加了多重树二分重接方法（Multiple TBR+TBR [multi*max*]），应该很容易搜索出最简约树。

在预设有序 [ordered = additive（累加）] 特征状态下，搜索后获 1 个最简约树（图 77 A）。该树长（L，或称步数，steps）为 119，一致指数（consistency index, CI）为 78，保留指数（retention index, RI）为 74。获得的一个次最简约树参数为：L=121, CI=77 和 RI=72。它和最简约树的主要区别是将 *Miorhizomys* 作为 (*Spalax* (*Tachyoryctoides* (*Pseudorhizomys*+*Pararhizomys*))) 的姊妹群将它们聚合为一支，其唯一的共近裔形状为 22-1（头骨枕髁互相靠近）。

将特征全部改为无序 [unorderd = inadditive（非累加）] 状态后也只搜索出 1 个最简约树，参数为：L=114, CI=82, RI=73，树型和分类单元的分支地位也和上述有序状态下所获得的最简约树完全一样。获得的一个次最简约树的参数为：L=116, CI=81, RI=70。它的树型则和上述的次最简约树完全一样。

在分别对头骨和下颌的特征加权为 2 后进行搜索，得出多种树型，其中最简约的仍然和上述不加权的最简约树一致，只是其树长增加很多（可达 188），CI 和 RI 稍有变化（CI 和 RI 分别为 78 和 71）。

在同样的优化条件下，我们也用 TNT 软件进行了尝试。采用隐含枚举搜索法（implicit enumeration）后在两个最简约树中有一个与 WinClada 中所得出的结果完全相同。

最终我们决定选取 WinClada 在累加特征状态下所获得的最简约树作为最佳

方案。

在 WinClada 中采取累加和非累加两种不同对特征状态处理的情况下所得到的最简约树，虽然树型（topology）相同，但其构树的参数（步长、CI 和 RI）不同，特征的分布和解释也不尽相同。在非累加条件下，构树参数总体较好（见上），而在累加条件下其参数只有 RI 稍好一点（RI=74）。但在特征分布和解释上，非累加条件下所产生的结果却远比不上累加条件下产生的结果。其一是构树过程中起作用的特征数较少，其二是在非累加条件下会出现很多不连续状态的特征（例如 0→2，1→3，而非 0→1，1→2 等），因为这样的特征在非累加条件下仍然作为一个步长计算，在简约原则中与连续状态具有同等被选择的机会。在我们的例子中，在节点 1→2 中出现 6 个 0→2，在节点 2→3 中出现 8 个这样的特征。这样的特征在极性比较肯定的哺乳类化石中不应该大量出现。

在特征分布的优化处理上，WinClada 提供了 3 种模式：无歧义优化（unambiguous optimization，软件预设模式；图 77 A）、速转换优化（ACCTRAN，图 77 C）和迟转换优化（DELTRAN，图 77 B）。无歧义优化所提供的特征，虽然在算法上最准确，但排除了许多可能反映真实演化过程的特征，使得可供分析和检验的特征的数量减少，特别是在分支图的基部。速转换优化，由于其方法本身的原因，更有利于产生反向演化，而使更多的特征向分支树的基部集中。例如，在我们的例子中，速转换优化下的节点 0→1 就有 26 个特征，在迟转换优化下则只有 13 个特征；同样在节点 1→2 中它们分别是 15 个和 10 个。相反，在节点 2→3 中情况正好相反，它们分别是 9 和 20。在我们的分支分析中，重点不是在基部，而是在中部。所以速转换优化模式不利于我们的分析。此外，速转换模式有利于产生反向演化，而不利于平行演化。在啮齿类中虽然反向和平行演化都经常发生，但平行演化无异更为经常发生。迟转换优化模式在上述两点上更适合于本分支分析的实际情况和要求。所以我们最后采用的是在累加条件下和迟转换优化下所产生的最简约树（见图 77 B）。

WinClada 对同功／同塑（homology/homoplasy）有两种表现方法：一是根据特征本身，即只要产生多余的步数即为同塑（any extra steps make it homopasious）；另一方法则是根据特征状态，即只有不连续的状态以同塑表示（only discontinuous states mapped as homoplasy）。后者是软件预设项，也是本书所采用的。

（4）特征分析

WinClada 在"特征鉴别（character diagnoser）"命令下为迟转换优化树提供的特征信息（见 77B）可综述如下 [每一个特征列举的顺序为：首先为特征序号（约定以粗体显示，如 **6**；若带有 * 号表示在无歧义优化树中也有者，如 ***46**；若为斜体代表平行演化，如 *33*；若有下划线代表反向演化，如 **41**），次为编码（转换以箭头表示，如 0→1，或 1→2），然后为 CI 和 RI（如 75, 66 等），uninf = uninformative, autap = autapomorphy]。

节点的特征分布：
Node 0→1: **6**, 0→1, 100, 100; **7**, 0→1, 75, 66; **34**, 0→1, 100, 100; **36**, 0→1, 100, 100; **37**, 0→1, 100, 100; **40**, 0→1, 66, 66; **41**, 0→1, 66, 50; **45**, 0→1, 66, 0; **46**, 0→1, 100, 100; **47**, 0→1, 75, 66; **48**, 0→1, 66, 66; **50**, 0→1, 66, 66; **53**, 0→1, 100, 100;

Node 1→2: **29**, 0→1, 66, 0; ***34**, 1→2, 100, 100; ***35**, 0→1, 75, 66; ***36**, 1→2, 100, 100; ***37**, 1→2, 100, 100; ***41**, 1→2, 66, 50; ***43**, 0→1, 100, 100; ***46**, 1→2, 100, 100; ***47**, 1→2, 75, 66; ***49**, 0→1, 66, 75;

Node 2→3: **1**, 0→1, 66, 66; **2**, 0→1, 100, 100; **3**, 0→1, uninf; **5**, 0→1, uninf; **6**, 1→2, 100, 100; **7**, 1→2, 75, 66; **10**, 0→1, 33, 0; **11**, 0→1, 66, 0; **12**, 0→1, 50, 0; **15**, 0→2, 75, 50; **17**, 0→1, uninf; **18**, 0→2, 60, 33; **19**, 0→1, 100, 100; **20**, 0→1, 100, 100; **27**, 0→1, 75, 75; ***40**, 1→2, 66, 66; ***42**, 0→1, 100, 100; ***49**, 1→2, 100, 100; ***50**, 1→2, 66, 66; ***51**, 0→1, 100, 100;

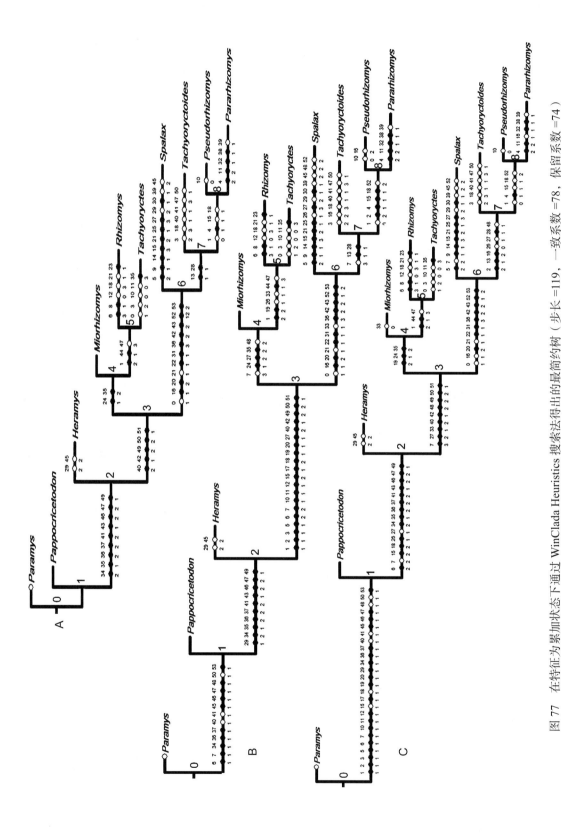

图 77 在特征为累加状态下通过 WinClada Heuristics 搜索法得出的最简约树（步长 =119，一致系数 =78，保留系数 =74）

Fig. 77 Most parsimony trees obtained by using WinClada Heuristics（L=119, CI=78, RI=75）

A. 无歧义优化分支图（MPT under unambiguous optimization），B. 迟转换优化分支图（MPT under DELTRAN optimization），C. 速转换优化分支图（MPT under ACCTRAN optimization）

实心圆代表近裔特征（共近裔或自近裔）；空心圆代表同塑特征（不连续特征状态）[black circles indicate apomorphies (synapomorphies or autapomorphies)]; 空心圆代表同塑特征（不连续特征状态）[white circles indicate homoplasies

(discontinuoud states mapped as homoplasies)]

Node 3 → 4: *7*, 2 → 3, 75, 66; ***24**, 0 → 1, 100, 100; **27**, 1 → 2, 75, 75; ***35**, 1 → 2, 100, 100; ***48**, 1 → 2, 66, 66;

Node 4 → 5: ***1**, 1 → 2, 66, 66; **19**, 1 → 2, 100, 100; **26**, 0 → 1, 50, 50; **33**, 0 → 1, 50, 66; ***44**, 0 → 1, 100, 100; ***47**, 2 → 3, 75, 66;

Node 3 → 6: ***0**, 0 → 1, 50, 50; ***16**, 0 → 1, 66, 50; ***20**, 1 → 2, 100, 100; ***21**, 0 → 1, 66, 50; ***22**, 0 → 1, 100, 100; ***31**, 0 → 1, 100, 100; **33**, 0 → 1, 50, 66; ***36**, 2 → 3, 100, 100; ***42**, 1 → 2, 100, 100; ***43**, 1 → 2, 100, 100; ***52**, 0 → 1, 66, 80; ***53**, 1 → 2, 100, 100;

Node 6 → 7: *7*, 2 → 3, 75, 66; ***13**, 0 → 1, 100, 100; ***28**, 0 → 1, 100, 100;

Node 7 → 8: ***1**, 1 → 0, 66, 66; **2**, 1 → 2, 100, 100; ***4**, 0 → 1, 100, 100; ***15**, 2 → 1, 75, 50; ***18**, 2 → 1, 60, 33; **52**, 1 → 2, 66, 80.

分类单元：

Heramys: ***29**, 1 → 2, 66, 0; ***45**, 1 → 2, 66, 0;

Rhizomys: ***6**, 2 → 3, 100, 100; ***8**, 0 → 1, uninf (autap); ***12**, 1 → 0, 50, 0; ***18**, 2 → 3, 60, 33; ***21**, 0 → 1, 66, 50; ***23**, 0 → 1, uninf (autap);

Tachyoryctes: ***0**, 0 → 1, 50, 50; ***3**, 1 → 2, uninf (autap); ***10**, 1 → 0, 33, 0; ***11**, 1 → 0, 66, 0; ***35**, 2 → 3, 100, 100;

Spalax: ***5**, 1 → 2, uninf (autap); ***9**, 0 → 1, uninf (autap); ***14**, 0 → 1, uninf (autap); ***15**, 2 → 3, 75, 50 (autap); ***21**, 1 → 2, 66, 50 (autap); ***25**, 0 → 1, uninf (autap); **26**, 0 → 1, 50, 50; **27**, 1 → 3, 75, 75 (autap); ***29**, 1 → 2, 66, 0; **30**, 0 → 1, 50, 0; ***39**, 0 → 1, 50, 0; ***45**, 1 → 2, 66, 0; **48**, 1 → 2, 66, 66; **52**, 1 → 2, 66, 80;

Tachyoryctoides: ***3**, 1 → 2, 66, 0; **16**, 1 → 2, 66, 50; ***18**, 2 → 3, 60, 33; ***40**, 2 → 1, 66, 66; ***41**, 2 → 1, 66, 50; ***47**, 2 → 3, 75, 66; ***50**, 2 → 1, 66, 66;

Pseudorhizomys: ***10**, 1 → 0, 33, 0; **16**, 1 → 2, 66, 50;

Pararhizomys: ***4**, 1 → 2, 100, 100 (autap); ***11**, 1 → 2, 66, 0 (autap); ***32**, 0 → 1, uninf (autap); ***38**, 0 → 1, uninf (autap); ***39**, 0 → 1, 50, 0.

根据上述 WinClada 所提供的特征状态的统计数值，我们可以得到若干对于分类较重要而有用的信息。这些信息可以简述如下。

在节点上所获得的信息：

1）节点 0 → 1，亦即除 *Paramys* 外，将所有其他 9 个分类单元聚合在一起的分支点上，共有特征为 13 个，其中 10 个是共近裔特征，3 个是同塑特征。特征中只有 2 个是头骨上的，即具宽大圆形的眶下孔（6-1）和小的颧弓板（7-1），其余的 11 个全为牙齿上的。3 个同塑特征（40-1，41-1，50-1）都与 *Tachyoryctoides* 相同，后者是反向演化的结果。节点 0 → 1 的 13 个共有特征表明，*Pappocricetodon* 和竹鼠及盲鼹鼠类的系统发育关系是很接近的。

2）节点 1 → 2，亦即除 *Pappocricetodon* 外，将所有其他 8 个内类群分类单元聚合在一起的分支点，共有 10 个特征，全为近裔特征。除特征 29-1（颞肌在下颌上的附着面增大）外，全为牙齿上的特征。这些近裔特征中有 5 个特征的 CI 和 RI 都是 100。这表明将 *Heramys* 和所有其他 7 个单元聚合在一起的支持度相当高。需要指出的是：*Heramys* 通常被认为是 *Spalax* 的祖先类型。*Heramys* 和 *Spalax* 具有 2 个同塑特征（29-2 和 45-2）也说明它们有较多的共同特征。*Heramys* 在本分支图中的位置显然与材料缺少有关。在材料较好的情况下，特别是有头骨标本时，不排除 *Heramys* 是盲鼹鼠祖先类型的可能。

3）节点 2 → 3 是将所有其余的 7 个分类单元聚合在一起的节点，共有特征 20 个，全为近裔特征。其中有 15 个是头骨和下颌上的，只有 5 个是牙齿上的。头骨上比较重要的特征是：头骨中 - 后部变高（1-1），鼻骨后端变窄（2-1），后伸到与前颌骨后端同一横线上（3-1），门齿孔短，其长约为齿隙的

1/3 左右（5-1），眶下孔具下裂隙（6-2），颧弓板较宽大（7-2），翼窝大而深（15-2），卵圆孔与中破裂孔愈合（17-1），外耳道短管状（18-2），位于关节窝后端下方（19-1），关节窝长大于宽（20-1）和下门齿在下颌后端处明显隆起（27-1）。总之，节点 2 → 3 是所有节点中得到特征支持最多的。这表明将 *Miorhizomys*, *Tachyoryctoides* 和副竹鼠类的两个属与现生竹鼠及盲鼹鼠类归为同一支系是比较合理的。

4）节点 3 → 4 是将 *Miorhizomys* 和现生竹鼠类（*Rhizomys* 和 *Tachyoryctes*）组合在一起的特征，只有 5 个。头骨和下颌上有 3 个特征，即头骨颧弓板很宽大（7-3），下颌髁突向上，而不是向后上方伸展（24-1），以及门齿后端在下颌骨上升支内侧的隆突很大（27-2）。此外，颊齿上的特征为：颊齿中等冠高，具单面高冠现象（35-2）和 m2 的下原凹退化（48-2）。在这些特征中，有 3 个是共近裔特征（24-1, 27-2 和 35-2），2 个是同塑特征：7-3 者与节点 6 → 7 者平行，48-2 者与 *Spalax* 者平行。这表明，这一节点的支持度相对较弱。有趣的是，在搜索中我们获得的次最简约树总是把 *Miorhizomys* 和 (*Spalax* (*Tachyoryctoides* (*Pseudorhizomys* + *Pararhizomys*))) 这一分支聚合在一起（见上）。这也从另一方面表明，*Miorhizomys* 的分类地位还需要更多的证据予以确定。

5）节点 3 → 6 是把 *Spalax* 和 (*Tachyoryctoides* (*Pseudorhizomys* + *Pararhizomys*)) 组合在一起的节点，共有 12 个特征，头骨 + 下颌和牙齿上的特征各为 6 个。头骨 + 下颌上较重要的特征是：咬肌 + 颞肌神经孔与颊肌神经孔分离，但它们的神经管短（16-1）、关节窝长远大于宽，接近或伸达项嵴（20-2）、两枕髁在腹面靠得很近（22-1，但 *Tachyoryctoides* 的这一特征不知）和下颌齿隙长于齿列长（31-1）。其中三个特征的 CI 和 RI 都为 100。头骨上有 2 个同塑特征，即吻部长远大于宽（0-1），与 *Tachyoryctes* 为平行演化；还有项面平，和项嵴同样斜向前上方（21-1），与 *Rhizomys* 为平行演化。在牙齿上，有 4 个特征的 CI 和 RI 都为 100，即颊齿全为脊形齿，中脊退化（36-3），上颊齿中凹呈褶沟状（42-2 和 43-2）和 m1-2 下中脊退化（53-2）。牙齿上只有一个同塑特征，即 i2 釉质层在两侧都很窄（33-1），与 *Rhizomys*+*Tachyoryctes* 为平行演化。总体上看，把 (*Spalax* (*Tachyoryctoides* (*Pseudorhizomys* + *Pararhizomys*))) 聚合在一起的共近裔特征共有 9 个，显然要比把 (*Miorhizomys* (*Rhizomys* + *Tachyoryctes*)) 聚合在一起的证据更强一些。

6）节点 6 → 7 是支持把 *Tachyoryctoides* 和副竹鼠类组合在一起的节点，只有 3 个共同特征，2 个为共近裔，1 个为同塑特征。2 个共近裔特征为：有副翼窝（13-1）和下颌咬肌嵴向外伸张，呈棚架状（28-1）。它们的 CI 和 RI 都为 100。这显示对组成 (*Tachyoryctoides* (*Pseudorhizomys* + *Pararhizomys*)) 分支的支持度也较低。

7）节点 7 → 8 是支持把副竹鼠类两个属作为一个分支的特征，共有 6 个特征，4 个共近裔，2 个平行演化的同塑特征。共近裔特征是：鼻骨后端变尖（2-2），前颌骨有侧背嵴（4-1），翼窝与中翼窝大小接近（15-1，亦为反向演化）和外耳道为雏形管状（18-1，为自节点 2 → 3 的 18-2 反向演化所得）。同塑特征为头骨中部高度低（1-0），为反向演化；m1-2 下后边凹退化，在 m3 中无（52-2）与 *Spalax* 为平行演化。总之，这一分支的支持强度为中等。

在末端分类单元上所获得的某些信息：

支持把 3 个化石属（*Heramys*, *Miorhizomys* 和 *Tachyoryctoides*）分别作为竹鼠类和盲鼹鼠类（包括副竹鼠类）的姐妹群的支持度都很弱。

Heramys 只有 2 个同塑特征：一个是下颌的颞肌附着面很大（29-2），另一个是 M3 内凹退化至无（45-2），都与 *Spalax* 为平行演化。这也许表明，将来如果有更多更好的材料发现，*Heramys* 会与 *Spalax* 有更近的系统关系。

Miorhizomys 没有区别于其姐妹群的自有特征。

Tachyoryctoides 的情况很特殊。其同塑特征较多，为 7 个。头骨上的特征是：鼻骨后端位置明显前于前颌骨后端（3-2），与 *Tachyoryctes* 为平行演化；颊肌、咬肌和颞深神经孔在同一深槽内（16-2），与 *Pseudorhizomys* 为平行演化，外耳道长管状（18-3），与 *Rhizomys* 为平行演化。牙齿上的特征是：M1-2 后脊与次尖前、后臂相连（40-1）与节点 0 → 1 相同；M3 后脊与次尖前臂相连（41-1）与

节点 0 → 1 相同；第一个下颊齿下后尖与下原尖为双连接（47-3）与节点 4 → 5 为平行演化；下臼齿下前边凹明显发育（50-1）与节点 0 → 1 相同。上述牙齿的特征中，除了 47-3 为平行演化外，其他的 3 个特征皆为反向演化所得，亦即它们都是近祖特征状态，与系统树中较早的节点 0 → 1 一致。这似乎表明，如果有更多更好的材料发现，不排除 *Tachyoryctoides* 在系统树中处于更基部位置的可能。目前已知，*Tachyoryctoides* 类化石的时代分布为晚渐新世—早中新世，而 *Pararhizomys* 类的时代分布为晚中新世早期—上新世早期。这两类时代分布期之间有一段时间（中中新世）的间隔。这似乎也支持了 *Tachyoryctoides* 处于较基部地位的可能性。

3. 副竹鼠类及鼠超科内相近科属级以上阶元的分类

当前分支系统学家依据所得最简约分支图建立分类的方法各异，但大体可以分为两类。一类是严格按照分支系统学的原理建立分类，拒绝采用林奈分类体系，例如"系统发育法规"（PhyloCode）学派。另一类则是遵循分支系统学的核心法则，采取若干约定折中措施以保留林奈分类的基本框架。本书作者认为后者更为合理和现实可行。Wiley 等（1991: 102–108）曾对后一派的观点进行了综述。其中比较符合我们的观点的有以下几点：

规则 1：只有单系类群可正式分类。

规则 2：分类必须能够反映分类单元之间的姐妹群关系。

约定 1：应用林奈等级体系。

约定 2：建立分类时尽量少改变现有分类。

约定 3：在分支树中构成不对称部分的单元可放于同一级别并依其分支顺序排列；列举时名单顺序代表分支顺序。

此外，似祖先单元（apparent ancestor，不必局限于种），作为约定 3 的个例，McKenna 和 Bell（1997: 28–30）建议，可以列在其高阶元姐妹群之前，不建立或归属于高阶单元。

依据我们所获得的在特征累加和迟转换优化条件下所产生的最简约分支图（图 77 B）和上一节的特征分析，我们至少可以得出以下几点分类信息：

1）节点 2 → 3，亦即除 *Pappocricetodon* 和 *Heramys* 外，将所有现生竹鼠类和盲鼹鼠类及其有关的化石属合并在一起的共近裔特征共 20 个，是这一分支图中含有共近裔特征最多的一个节点。这表明竹鼠类和盲鼹鼠类具有大量的共近裔特征，是 Muroidea 超科中系统关系非常接近的两个分支。当前大部分啮齿动物分类专家把它们当作两个独立的科。如将它们合并为一个更高一级的单元，则应该是一个超科。目前对 Muroidea 其他科的关系的研究还不很成熟，很难把 Muroidea 再提升一个级别（例如亚目等）而对其所有现在的科合并为几个大的超科。当前也许只能暂时保留 Muroidea 的超科级地位。

2）支持把 (*Spalax* (*Tachyoryctoides* (*Pseudorhizomys* + *Pararhizomys*))) 合并为一支的节点 3 → 6，共有 12 个特征，在数量上仅次于 2 → 3 和 0 → 1 两个节点。因此，把这一分支和竹鼠类分为两个分支是比较合理的选择。

3）*Spalax* 具有 8 个自近裔特征和 6 个同塑特征（2 个头骨上和 4 个牙齿上的）。这表明 *Spalax* 是节点 3 → 6 中形态分异程度最高的一个属，足以和 (*Tachyoryctoides* (*Pseudorhizomys* + *Pararhizomys*)) 分开，作为其姐妹群。

4）*Heramys* 由于目前所知的特征太少，其可编码特征状态仅占 43%（见表 20），这可能是造成其分类地位不可靠的主要原因。事实上，到目前为止该属一直被看作是盲鼹鼠类（不包括竹鼠类）的祖先类型。这都表明，*Heramys* 可能真和 *Spalax* 具有更近的系统关系。

5）把 *Miorhizomys* 当作 *Rhizomys*+*Tachyoryctes* 的姐妹群的证据也不够充分（只有 3 个共近裔和 2 个同塑特征），尚需对 *Miorhizomys* 有更多和更深入的研究才能予以证实。

6）如前所述，*Tachyoryctoides* 作为副竹鼠类的姐妹群的地位也不很可靠。它也可能应该处在更基部的位置。考虑到 *Tachyoryctoides* 主要出现在渐新世和早中新世，目前在分类上似乎把它和副竹鼠类分开也许更为恰当。

依据上述原则和约定以及对上述特征的分析和认识，我们可以提出如下分类方案（不包括 *Pappocricetodon*）：

Superfamily Muroidea
  Stem genus *Heramys*
  Family Rhizomyidae Winge, 1887
    Miorhizomys
    Rhizomys
    Tachyoryctes
  Family Spalacidae Gray, 1821
    Subfamily Spalacinae Gray, 1821
      Spalax
    Subfamily Tachyoryctoidinae Schaub, 1958
      Tribe Tachyoryctoidini Schaub, 1958
        Tachyoryctoides
      Tribe Pararhizomyini tribe nov.
        Pararhizomys
        Pseudorhizomys

上述方案是在遵循分支系统学的基本原则的前提下，较多地考虑到各个分类单元在形态特征上的差异程度，并采取了以顺序代表分支顺序的折中办法形成的。在这一方案中，我们仍然把 Rhizomyidae 和 Spalacidae 分为两个科，将 Spalacidae 科分为两个亚科，将副竹鼠类作为 Tachyoryctoidinae 亚科中的一个新建族——Pararhizomyini。这种方案既不同于 Bugge（1985）把所有的鼠形类整体上提高一个级别（亚目），把盲鼹鼠和竹鼠都提高到超科一级（见下），也不同于 McKenna 和 Bell（1997）把盲鼹鼠和竹鼠作为两个独立的亚科，和其他二十几个亚科一起归入鼠科内。总体上看，这种方案也许更适合目前化石材料不足的现状，而且对未来可能更好的分类方案留下了更多的改进余地。

上述的分析并没有包括 Bugge 根据头部动脉及供血演化模式所提出的关于鼠超科内科一级分类单元之间的系统发育的观点。这主要是由于软体解剖特征在化石中很难被保存和辨识，目前尚不具备作为可靠特征被纳入分支分析中。虽然如此，Bugge 的观点对我们了解副竹鼠类和盲鼹鼠类的分类地位仍然有很大的帮助。

Bugge 的一个重要的发现是，现生盲鼹鼠和竹鼠在头骨动脉分布及供血模式上比较接近，而和其他鼠形超科内其他各科有较明显的区别。在盲鼹鼠和竹鼠中蹬骨动脉近端部分（主要是在听泡内的部分）缺失，头部供血（主要是眶部和上颌部）的主要源动脉由颈内动脉远端部分（在脑颅内的部分）分出侧支与蹬骨动脉的远端部分衔接而供血（见图 3）。这和绝大部分传统意义上的鼠超科的鼠类（仓鼠类、小鼠类、沙鼠类及狭义的鼠类）都不同。在后者中蹬骨动脉是供给头部血液的主要源动脉。所以 Bugge 把盲鼹鼠和竹鼠作为两个科（Spalacidae 和 Rhizomyidae），并将这两个科合并提升为一个竹鼠超科（Rhizomyoidea），与鼠超科（Muroidea）、跳鼠超科（Dipoidoidea）及囊鼠超科（Geomyoidea）并列。

有意义的是，根据可见颈动脉通过的孔（切迹或沟）的位置、大小和形态，可以看出副竹鼠类化石似乎确实和现生盲鼹鼠及竹鼠更为接近。

1）副竹鼠类的蹬骨动脉可能在听泡内部已经退失。在所有副竹鼠和假竹鼠的标本中，听泡的内后角都有细长裂隙状的颈静脉孔（jugular for.），在其前部有时会看到向头骨内部伸展的细孔，可能是 IX–XI 神经的出口。在其外后角没有孔形明显的蹬骨动脉孔。在听泡内侧壁保存较好的标本（如

V 16287, V 16293–V 16299 和 V 16302 等）中也没有观察到有通向听泡内部的孔。因此，在副竹鼠类中，最大的可能是在听泡内没有蹬骨动脉。在这一点上，副竹鼠类只与鼠超科中的盲鼹鼠和竹鼠相同。

2）副竹鼠类内颈动脉如何与蹬骨动脉之远端（即眶上和眶下动脉支）相连接，目前尚不能确定。根据 Bugge 的观察，在盲鼹鼠中是 a5' 吻合支（anastomotic branch）自内颈动脉颅内部分与眶下动脉支（*ramus infraorbitalis*）在中破裂孔附近相连并供血（Bugge, 1985: 362–363, Fig. 4A；见本书图 3 C），这就使翼蝶管的后开孔很大并与颅腔相连通。在竹鼠中 则是 a4 吻合支在颅内与眶上动脉支在翼蝶管前开孔附近（Bugge, 1985: 362–363, Fig. 4B；见本书图 3 D）相连并供血，并不通过翼蝶管，所以翼蝶管也不发育。在副竹鼠类标本中，在外翼突后端之内有一个圆孔，向前开孔。这可能是翼蝶管的后开孔。由于该孔很小，即使 a5' 在此处与眶下动脉支相连，其供血量也一定很少，应该还有其他的供血渠道。至于是否有 a4、a1 或 a2 吻合支供血，由于是在前破裂孔内壁附近的结构，我们在化石上观察不到，因此不得而知。不过，这很可能反映了副竹鼠类乃是蹬骨动脉近端缺失类型（盲鼹鼠和竹鼠）的近祖类型的模式。

总之，对现有的副竹鼠类化石可能的头部动脉分布模式的推测与上述的分类方案，特别是将盲鼹鼠、竹鼠和副竹鼠类合并在一起是吻合的。

四、副竹鼠族若干生物学问题

（一）副竹鼠族部分肌肉复原

1. 头骨肌肉

啮齿类的头骨的肌肉很复杂，在此不能一一涉及。因副竹鼠族头骨的肌肉与鼠超科内其他科者相似，笔者在此参照 Greene（1935）、杨安峰和王平等（1985）对褐家鼠（*Rattus norvegicus domestrica*，又称大鼠）的解剖，以及 Howell（1926）对森林鼠（wood rat）的解剖，并对鼠超科的有关种类的头骨形态做了直接对比观察。本书只对副竹鼠族头骨上与咀嚼、咬啮和挖掘功能直接有关的一些肌肉作简要分析。

咬肌（*m. masseter*）　咬肌主要分为内、外、表三层。内层咬肌分前、后两部。前部主要起自上颌骨后缘及眶下窝，止于下颌骨咬肌嵴的前端。后部起自颧弓的腹缘和内面，止于下颌骨咬肌嵴。外层咬肌起自颧弓的腹 - 外面（包括颧弓板），止于下颌骨外侧下部和咬肌嵴。表层咬肌起自上颌骨颧弓板前内缘的咬肌结节，止于下颌角突的腹缘。副竹鼠类头骨的颧 - 咬肌结构属鼠型。它的眶下孔大，颧弓板面大而倾斜，上颌骨上的咬肌结节也较明显，它的下颌骨的咬肌窝大，咬肌嵴很发达。这些都表明副竹鼠类的三层咬肌都很发达。特别是它下颌上的咬肌嵴明显向外伸张，显然比竹鼠、鼢鼠和盲鼹鼠的都更发达，表明其内层咬肌可能也比后三者更为发达。副竹鼠类下颌骨与褐家鼠（*Rattus*）的相似，有明显向后延伸的角突，与竹鼠、鼢鼠和盲鼹鼠的明显不同。这表明副竹鼠类的表层咬肌和外层咬肌相对较长，其延伸的方向可能与后三类稍有不同。

颞肌（*m. temporalis*）　颞肌起自颞窝和颧弓内侧面，止于下颌骨冠状突前缘、内面及其与臼齿之间的沟中。副竹鼠类头骨的颞窝大，两侧的颞窝上缘相遇形成明显的矢状嵴。副竹鼠类的下颌骨的冠状突较发达，并稍高于髁突，其内侧下方供颞肌附着的区域（颞肌窝）为大而显著的凹面。这表明副竹鼠类的颞肌也较发达，能起到较快地关闭下颌的作用。

翼肌（*m. pterygoideus*）　翼肌分为翼外肌和翼内肌。翼外肌起自外翼突，止于下颌骨髁突内侧的翼外肌窝。翼内肌起自内翼突，止于下颌骨角突内面的翼内肌窝。副竹鼠类的内翼突较外翼突高大，下颌骨上的翼内肌窝也大而深，而髁突内侧供翼外肌附着的翼外肌窝的凹面较小而浅或不明显。可能副竹鼠类的翼内肌要比翼外肌发达。

副竹鼠类下颌的翼内肌窝与竹鼠和盲鼹鼠的不同：较大而深，并沿向后伸突的下颌角向后延伸较长。这表明副竹鼠类的翼内肌可能比竹鼠和盲鼹鼠等类的更发达，力臂也要长些。

二腹肌（*m. digastricus*）　褐家鼠的二腹肌起于副乳突，止于下颌联合的后面。副竹鼠类的副乳突很小，其大小与盲鼹鼠的相近，明显较竹鼠类的小。二腹肌的止点（下颌联合之后的二腹肌窝）也很小，这表明副竹鼠类的二腹肌可能比竹鼠类的细弱。二腹肌主要功能是张开口和拉下颌骨往后，表明副竹鼠类张开口和拉下颌骨往后的力量要比竹鼠类的小。

颊肌（*m. buccinator*）　副竹鼠类的头骨很特别的一点是，前颌骨的背侧有一明显的侧背嵴。该棱嵴的作用一直让笔者感到费解。根据该棱嵴的位置，笔者推测，该棱嵴及其下方的凹面可能是颊肌的附着处。依 Greene（1935）和杨安峰和王平等（1985）的记述，颊肌起自上颌骨和前颌骨，止于口轮匝肌，起提上唇的作用。而 Howell（1926）也记述了森林鼠的颊肌是起于前颌骨和上颌骨外面沿门齿齿根（从门齿齿槽端到眶下孔）形成的曲线。如果该侧背嵴的确是颊肌的起点的背缘的话，副竹鼠类的颊肌可能比较发达，提上唇的作用较大。

2. 前肢肌肉

有关副竹鼠类前肢肌肉的起、止点，笔者主要参考Greene（1935）、杨安峰和王平等（1985）对褐家鼠、Howell（1926）对森林鼠的描述，以及Vinogradov和Gambarian（1952）对查干鼠类（tsaganomyids）的肌肉复原，并参照了竹鼠、鼢鼠、山河狸等现生啮齿类的骨骼形态特征。

冈上肌（*m. supraspinatus*）　褐家鼠的冈上肌起自肩胛骨前缘、椎缘和冈上窝骨面，止于肱骨大结节，用以伸展肩关节。假竹鼠的冈上窝的形状与竹鼠和褐家鼠的相似，而且肱骨大结节的形态也与它们相似，其顶端具有明显的供冈上肌附着的凹面和粗糙面，表明副竹鼠类的冈上肌与它们的相似，也较发达。

冈下肌（*m. infraspinatus*）　褐家鼠的冈下肌起自肩胛冈、肩胛骨椎缘和冈下窝骨面，止于肱骨大结节，用以伸展肩关节。假竹鼠的冈下窝与竹鼠和山河狸的很相似，其后部有很发达的肌线，肩胛颈后面有粗隆，而假竹鼠肱骨的大结节也很发达，其外侧后面有明显的供冈下肌附着的凹面，很可能副竹鼠类的冈下肌与竹鼠和山河狸的相似，也很发达。

肩胛下肌（*m. subscapularis*）　褐家鼠的肩胛下肌起自肩胛下窝，止于肱骨小结节背缘，用以伸展肩关节。副竹鼠的肩胛下窝和肱骨小结节的形状均与竹鼠和山河狸的相似，可能它们的肩胛下肌也相似。

三角肌（*m. deltoideus*）　褐家鼠的三角肌有三个头，分别起自锁骨、肩峰、肩胛冈和肩胛骨的后缘等处，止于肱骨的三角肌粗隆。该肌三个头的作用分别是向内旋转前肢、外展和屈曲肩关节。尽管*Pseudorhizomys*的肩胛骨的肩峰和肩胛冈保存不好，但它的锁骨保存较好，特别是它的肱骨的三角肌粗隆很发达，明显向外伸，而且位置较低，约位于肱骨中部，这表明*Pseudorhizomys*的三角肌可能很发达。从三角肌粗隆的位置与山河狸的相当来判断，它的三角肌发达的程度可能与山河狸的相近。

大圆肌（*m. teres major*）　褐家鼠等的大圆肌起自肩胛骨后缘近端2/3处，止于肱骨体内侧缘的上部、小结节嵴的后面，用来屈曲肩关节。*Pseudorhizomys*的肩胛骨后缘近端已破损，但肱骨体内侧缘的上部，在小结节嵴的后面有明显的供大圆肌附着的、粗糙的凹面（＝圆肌隆起），可能*Pseudorhizomys*的大圆肌比较发达。

小圆肌（*m. teres minor*）　褐家鼠的小圆肌起自肩胛骨后缘远端1/3处，止于肱骨大结节，用来屈肩关节。*Pseudorhizomys*的肩胛骨后缘下部有发达的肌线，在肩胛颈的后面也有供肌肉附着的粗糙隆起，加上在肱骨较发达的大结节后侧外面有明显的供小圆肌附着的隆突，表明*Pseudorhizomys*的小圆肌也较发达。

胸浅肌（*m. pectoralis superficialis*）　褐家鼠的胸浅肌起于锁骨、胸骨和前部的6枚软肋，止于肱骨大结节嵴（或三角肌嵴）。其作用是：向内、向后拉动前肢。*Pseudorhizomys*的锁骨和肱骨的大结节嵴都较发达，较长，表明它的胸浅肌可能较发达。

胸深肌（*m. pectoralis profundus*）　褐家鼠和森林鼠的胸深肌均分为肩胛前部、肱骨部（或后部）和腹部三部分。分别起于第二至第五胸骨、剑胸骨和剑状软骨，而分别止于肩胛结节、肱骨大结节及肱骨嵴和肩胛骨喙突。用来向后拉动前肢。*Pseudorhizomys*虽未保存胸骨，但它的肩胛结节、喙突、肱骨大结节和肱骨嵴均较发达，表明胸深肌也较发达。

喙臂肌（*m. coracobrachialis*）　褐家鼠的喙臂肌起自肩胛骨的喙突，止于肱骨骨干远端1/2部分的内侧。其作用是伸展肩关节。*Pseudorhizomys*的肩胛骨的喙突较发达，肱骨骨干内缘中部、圆肌隆起的前下方有供其附着的粗糙面。这表明*Pseudorhizomys*的喙臂肌也较发达。

臂三头肌（*m. triceps brachii*）　褐家鼠的臂三头肌有长头、外头和内头三个头，它们分别起自肩胛骨腋缘远端1/3处，肱骨大结节和肱骨骨干近端2/3处，止于尺骨的肘突。其作用是伸展肘关节和屈曲肩关节。*Pseudorhizomys*的肩胛颈的后面有可供臂三头肌长头起点附着的较粗糙隆起，肱骨的大结节的外下方也有供其外头附着的隆突的粗糙面。虽然它的左、右尺骨的肘突均破损，但从保存的部分看，其肘突明显高出钩突，而且比较宽大，这表明肘突可能比较发达。这些都表明 *Pseudorhizomys* 的臂三头肌是很发达的。

臂二头肌（*m. biceps brachii*）　褐家鼠的臂二头肌有长、短两头。长头起自肩胛骨肩臼前缘，短头起自肩胛骨喙突。该肌腱经由结节间沟（臂二头肌沟）止于桡骨粗隆。其作用是屈肘伸肩。*Pseudorhizomys* 的肩胛骨的肩臼前缘有明显的肩胛结节和较发达的喙突，桡骨粗隆也很发达，而且供二头肌腱通过的肱骨的结节间沟也较宽大。这些都表明 *Pseudorhizomys* 的臂二头肌较发达。

臂肌（*m. brachialis*）　褐家鼠的臂肌起自肱骨大结节和肱骨颈，止于尺骨内侧面近冠状突下方，用以屈曲肘关节。*Pseudorhizomys* 的臂肌在肱骨颈和大结节附近的起点不清楚，但肱骨体前外侧的臂肌沟很宽阔，在桡骨骨体内缘近端有明显供臂肌远端肌腱通过的光滑面（桡骨臂肌沟），而且在尺骨冠状突下方有该肌止点形成的明显的粗糙凹面 [= 尺骨粗隆（tuul），Howell（1926）称其为臂肌嵴（brachial ridge）]。这表明 *Pseudorhizomys* 的臂肌较发达。

旋前圆肌（*m. pronator teres*）　褐家鼠的旋前圆肌起自肱骨内上髁，止于桡骨体内侧面中部。*Pseudorhizomys* 的肱骨内上髁有可能供旋前圆肌附着的凹面，而桡骨骨体前面中部近内缘处有供旋前圆肌附着的粗糙凹面。这表明 *Pseudorhizomys* 虽有旋前圆肌存在，但不很发达。

旋后肌（*m. supinator*）　褐家鼠的旋后肌起自肱骨外上髁，止于桡骨骨体内侧面旋前圆肌止点的下方；而 Howell（1926: 68）记述的森林鼠的旋后肌则起自桡骨环韧带，止于桡骨骨体外侧面上部。*Pseudorhizomys* 的肱骨外上髁上有供旋后肌附着的较小的凹面，与 *Rhizomys* 的相似，它的桡骨骨干前面中部近内缘处供旋后肌附着的粗糙面也较明显。这表明 *Pseudorhizomys* 的旋后肌也明显存在，而且其起、止点与褐家鼠的相似。

腕桡侧伸肌（*m. extensor carpi radialis*）　依 Greene（1935）、杨安峰和王平等（1985）的记述，褐家鼠的腕桡侧伸肌包括一长头和一短头。两者均起自肱骨外上髁。长头止于 McII 远端桡侧，而短头止于 McIII 远端桡侧。依 Howell（1926: 67–68, Fig. 28）的记述，森林鼠的腕桡侧伸肌的长头和短头均起于肱骨的外上髁嵴，分别止于 McII 和 McIII 的内背侧。依 Vinogradov 和 Gambarian［1952: Fig. 16（11, 12）］对鼢鼠、盲鼹鼠和查干鼠（原为 *Pseudotsaganomys*，现为 *Tsaganomys*）的复原，腕桡侧伸肌实际上是起于肱骨的外上髁嵴的上部。*Pseudorhizomys* 的肱骨外上髁嵴背侧有两个明显的凹面，上部者很可能供腕桡侧伸肌起点附着。这样，*Pseudorhizomys* 的腕桡侧伸肌的起点的位置很可能与鼢鼠、盲鼹鼠等的相似。而 *Pseudorhizomys* 的 McII 和 McIII 都有较发达的掌骨粗隆。其中，McIII 的掌骨粗隆为较发达的凹坑，位置与竹鼠、鼢鼠和山河狸的相似，位于骨体背侧近端近桡侧。而 McII 的掌骨粗隆在 *Pseudorhizomys* 的两个种（*Ps. indigenus* 和 *Ps. gansuensis*）中的形状和位置却有所不同：前者与竹鼠、鼢鼠、麝鼠（*Ondatra*）和山河狸的相近，为结节状的隆起，位于骨体背面中部近桡侧（见图 45 tmc），而后者则与 McIII 的掌骨粗隆相似，为较发达的凹坑，位于骨体背面上部的桡侧（见图 67 tmcII）。如果 *Pseudorhizomys* 的腕桡侧伸肌的长头止于 McII 的掌骨粗隆，而短头止于 McIII 的掌骨粗隆的话，这就表明 *Pseudorhizomys* 的腕桡侧伸肌可能很发达，而且其止点的位置有可能都要高于褐家鼠者。其中，腕桡侧伸肌短头在 McIII 的止点与森林鼠、竹鼠、鼢鼠和山河狸者相近。而腕桡侧伸肌长头在 McII 的止点的位置也与森林鼠、山河狸、鼢鼠、麝鼠（*Ondatra*）和竹鼠等相近。另外，因 *Ps. indigenus* 和 *Ps. gansuensis* 的 McII 掌骨粗隆的形状和位置彼此有所不同，这表明腕桡侧伸肌长头的形态结构和止点的位置在两个种中可能也有差别。

腕尺侧伸肌（*m. extensor carpi ulnaris*）　褐家鼠等的腕尺侧伸肌起自肱骨外上髁和尺骨肘突，止于 McV 近端。尽管 *Pseudorhizomys* 的尺骨肘突破损，但它的肱骨外上髁上有供腕尺侧伸肌附着的关节面；而且在它的 McV 近端外侧有很发达的结节状隆起和大的凹坑供腕尺侧伸肌附着。*Pseudorhizomys* 的腕尺侧伸肌很发达。

腕桡侧屈肌（*m. flexor carpi radialis*）　前人对褐家鼠等的腕桡侧屈肌的起点均记述为起自肱骨内上髁，但对其止点的记述并不相同：Greene（1935: 48, Fig. 88）、杨安峰和王平等（1985：图 59）表明该屈肌止于 McIII 近端；而杨安峰和王平等（1985: 56）则记述为止于 McII 近端；另外，Howell（1926:

65，Fig. 29）记述的森林鼠、王增涛等（2009：189）记述的 Wistar 鼠和 Hebel 和 Stromberg（1976：31）记述的实验鼠的腕桡侧屈肌也都止于 McII 近端。*Pseudorhizomys* 的肱骨内上髁很发达，其上可能供腕桡侧屈肌附着的面也较明显；不但其 McII 的掌面近端内侧缘有结节状隆起可能供腕桡侧屈肌附着，而且其 McIII 掌面近端内侧也有结节状隆起，也可能供腕桡侧屈肌附着。*Pseudorhizomys* 的腕桡侧屈肌可能比较发达。

腕尺侧屈肌（*m. flexor carpi ulnaris*） 褐家鼠的腕尺侧屈肌起自肱骨内上髁和尺骨肘突内侧，止于副腕骨（豌豆骨）。*Pseudorhizomys* 的尺骨肘突已破损，但其较发达的肱骨内上髁上有明显供腕尺侧屈肌附着的面，豌豆骨也较发达，表明 *Pseudorhizomys* 也有较发达的腕尺侧屈肌存在。

指总伸肌（*m. extensor digitorum communis*） 依 Greene（1935）、杨安峰和王平等（1985）的记述，褐家鼠的指总伸肌起自肱骨外上髁，止于第二至第五指的第三指节骨的基部。而依 Vinogradov 和 Gambarian［1952：Fig. 16（13）］对鼢鼠、盲鼹鼠和查干鼠的复原，指总伸肌起于肱骨的外上髁嵴腕桡侧伸肌的下方。*Pseudorhizomys* 的肱骨外上髁嵴背侧的两个明显的凹面中，其下部者很可能供指总伸肌起点附着。*Pseudorhizomys* 不但第二至第五指的第三指节骨的伸腱突都很发达，而且第二至第五指的第一指和第二指的背面都有供指总伸肌肌腱附着或滑过的伸腱结节。我们在文献中发现，在有些哺乳动物（如人类和大熊猫等）中指总伸肌可以止于第二至第五指的第一至第三指节骨的背侧。很可能 *Pseudorhizomys* 的指总伸肌也可以止于第二至第五指的第一至第三指节骨的背侧。这表明，*Pseudorhizomys* 的指总伸肌很可能非常发达。

指侧伸肌（*m. extensor digitorum lateralis*） 褐家鼠的指侧伸肌起自肱骨外上髁，指总伸肌的外侧处，在腕附近分成二腱，分别止于第四指和第五指的第三指节骨的基部。*Pseudorhizomys* 的肱骨外上髁上有三个凹面，其外上者可能是指侧伸肌的起点。*Pseudorhizomys* 的第四指和第五指的第三指节骨上的伸腱突都很发达，它们很可能供指侧伸肌的附着。*Pseudorhizomys* 的指侧伸肌也较发达。

拇长伸肌（*m. extensor pollicis longus*） 褐家鼠的拇长伸肌起自尺骨的伸肌面，止于拇指的末指节骨（＝爪指骨）的基部。*Pseudorhizomys* 的拇指的爪指骨的伸腱突虽比第二至第五指爪指骨者弱小，但仍明显存在。*Pseudorhizomys* 可能是具有拇长伸肌的。

拇短伸肌（*m. extensor pollicis brevis*） 褐家鼠的拇短伸肌起自桡骨骨体的背外侧，止于拇指的第一指节骨的近端。*Pseudorhizomys* 的桡骨骨体前面下 3/4 近外缘处有很明显的粗糙隆凸，而拇指的第一指节骨近端有明显的伸腱结节。这些表明 *Pseudorhizomys* 有拇短伸肌存在。

拇展肌（*m. abductor pollicis*） 依 Howell（1926：68）和 Greene（1935：51）的记述，森林鼠和褐家鼠的拇外展肌起自镰状骨，止于拇指的第一指节骨基部的桡侧，而杨安峰和王平等（1985）认为褐家鼠的拇外展肌起自桡骨和尺骨的骨体，止于第一掌骨远端内侧。*Pseudorhizomys* 有较发达的镰状骨，而且拇指的第一指节骨基部的桡侧有明显可能供拇展肌附着的结节状隆起，表明 *Pseudorhizomys* 有较发达的拇展肌存在。

指浅屈肌（*m. flexor digitorum superficialis*） 褐家鼠的指浅屈肌起自肱骨内上髁，止于第二至第五指的第二指节骨近端两侧。*Pseudorhizomys* 的肱骨内上髁很发达，而且它的第二至第五指的第二指节骨的近端掌面两侧都有供指浅屈肌附着的区域。这些都表明 *Pseudorhizomys* 的指浅屈肌也可能较发达。

指深屈肌（*m. flexor digitorum profundus*） 褐家鼠的指深屈肌有四个头，分别起自肱骨内上髁，尺骨肘突下方的尺骨体和桡骨骨体后面。止于第二至第五指的第三指节骨。依 Howell（1926：64–65）研究，森林鼠的指深屈肌远端有五条肌腱，分别止于第二至第五指的第三指节骨和拇指的爪指骨。*Pseudorhizomys* 的肱骨内上髁很发达，桡骨后面和尺骨后内面有较宽的供指深屈肌附着的面。*Pseudorhizomys* 与褐家鼠的不同，而与森林鼠、竹鼠和大熊猫等相似：不但第二至第五指的爪指骨掌侧有很发达的爪指骨粗隆（tuberosity of ungual phalanx），而且拇指的爪指骨也有同样发达的爪指骨粗隆，甚至比竹鼠的还要发达。可能在这一点上，*Pseudorhizomys* 与森林鼠、大熊猫（北京动物园等，1986：144）的相似：指深屈肌远端形成五个较大的肌腱，除了分别止于第二至第五指的爪指骨的四个肌腱外，而且还有很发达的第五条肌腱，止于拇指的爪指骨。

拇短屈肌（*m. flexor pollicis brevis*） 褐家鼠的拇短屈肌起自腕掌深韧带，止于拇指第一指节骨基部外侧。*Pseudorhizomys* 的拇指的第一指节骨掌面近端外侧有明显的肌肉附着的结节，表明 *Pseudorhizomys* 有拇短屈肌存在。

拇收肌（*m. adductor pollicis*） 依 Greene（1935: 520）研究，褐家鼠的拇收肌起自头状骨和 McII 和 McIII 近端，止于拇指第一指节骨近端外侧。*Pseudorhizomys* 的 McII 和 McIII 掌侧面近端内侧缘有脊状隆起供拇收肌附着。它的拇指的第一指节骨掌面近端的外侧也有明显的肌肉附着的结节。这表明 *Pseudorhizomys* 的拇收肌可能存在。

（二）副竹鼠族某些骨骼和肌肉机能－习性分析

啮齿类的适应性强，能生活在多种多样的生态环境中，其中大多数为穴居性的，在地下营掘土或半掘土生活；也有树栖或半树栖的；甚至有半水栖的。副竹鼠和假竹鼠的生活方式究竟属于哪一类，是个值得探讨的问题。因假竹鼠保存的材料较好、较多，下面首先对该属的特征进行分析。

1. 假竹鼠的机能－习性分析

假竹鼠头骨的背侧较平直，眼眶都很小（说明它们的眼很小），上、下齿隙都较长，门齿孔较小；下颌具发达的咬肌嵴；i2 后端伸达髁突下方，并在上升支颊侧形成明显的隆突；颈椎和四肢较粗短，具较发达的供肌肉附着的结节或突起等。这些特征都与在地下营掘土生活的啮齿类很相似（见 Stein，2000），表明假竹鼠可能是穴居、在地下营掘土生活的啮齿类。

Hildebrand（1982）、Stein（2000）、Hildebrand 和 Goslow（2001）在研究脊椎动物的掘土方式时认为啮齿类主要以三种方式掘土：①前肢交替屈伸以爪刨土的爪刨型（scratch digging），例如地松鼠；②以门齿凿土的齿凿型（chisel-tooth digging），例如囊鼠等；③少数啮齿类具有强壮的颈部肌肉和前肢伸肌以扬头的办法造穴，可称之为头拱型（head-lift digging），例如盲鼹鼠。此外，不少啮齿类也并非只采用一种方法掘土。Stein（2000）对近 20 种穴居啮齿类的掘土方式进行了深入研究和系统总结。Stein 采纳了 Hildebrand 所提出的三种主要掘土方式，并对其进行了重新解释和定义。**爪刨型**：前肢交替屈伸，以爪破土（break up soil）和松土（loosen soil），以前掌推土（move soil），代表性动物是囊鼠类（Geomyidae）、梳鼠（*Ctenomys*）、滨鼠（*Bathyergus*）和鼢鼠（*Myospalax*）等；**齿凿型**：以平伏门齿破土，在强壮头部和颌部肌肉的配合下以门齿松土，以头和脚推土，代表性动物是所有的竹鼠类（Rhizomyidae）、鼢足鼠（*Spalacopus*）、除滨鼠属外的所有滨鼠类，如隐鼠（*Cryptomys*）和裸鼢鼠（*Heterocephalus*）等；**头拱型**：在头的协助下以门齿形成钻和铲破土、松土和推土，代表性动物是鼹形田鼠（*Ellobius*）和小盲鼹鼠（*Nannospalax*）等。

如果假竹鼠是穴居的、在地下营掘土生活的啮齿类，它们用何种方法掘土呢？

（1）假竹鼠的头骨形态结构的机能－习性分析

假竹鼠的头骨约为三角形，颞窝较大，颧弓明显地往外张；矢状嵴和项嵴较发达；项面较宽，并稍向前倾（可提供强有力的颈肌附着），前颌骨前端向前伸，超过鼻骨的前端；I2 较少弯曲，属前伸型齿，其后端起于近 M1 处，并在 M1 之前形成明显的隆突；上、下齿隙较长；i2 较长等。这些都与 *Rhizomys* 相似。而且假竹鼠的 I2 和 i2 的弯曲度甚至比 *Rhizomys* 还小，I2 的前端远远地伸到吻部的前方，而 i2 明显伸长等。假竹鼠的头骨的上述特点与齿凿型的啮齿类 *Rhizomys*, *Cryptomys* 和 *Heterocephalus* 很相似，而与爪刨型的 *Myospalax* 和 *Geomys* 等不同（后者的头骨无明显的矢状嵴，I2 较明显弯曲，i2 较短等），也不同于头拱型的 *Spalax* 和 *Ellobius* 等（后者的头骨项面强烈向前倾斜，I2 较明显地弯曲；

此外，*Spalax* 的吻部很窄，使头骨形成三角形）。这表明，假竹鼠主要是用门齿和有力的下颌骨和颈部的肌肉来松土和挖掘土，属主要用齿凿型方法掘土的啮齿类。

Flynn 等（1987）认为门齿的釉质层的外层较厚和施氏明暗带的倾斜度低等与挖掘有关，也与所掘土的性质有关。尽管假竹鼠的门齿的釉质层的外层要比 *Rhizomys* 的薄很多，但内层的施氏明暗带的倾斜度是变化的，这表明假竹鼠在门齿挖土受力时可能更有利于防止破裂。另外，假竹鼠的门齿的釉质层的外层较薄也可能与它们所掘的土层较松软有关。Hua 等（2015）认为门齿的微细结构与动物所食的食物的坚硬程度有关。假竹鼠的门齿的微细结构与竹鼠的不同，可能与它们的食材不同有关。竹鼠主要以抗破裂较强的竹子为食。而假竹鼠的门齿结构表明，它可能不食竹子，而主要是以比竹子抗破裂较小的植物为食。从产假竹鼠的地层岩性和共生的植物孢粉组合和三趾马动物群来看，假竹鼠当时生活的环境是较干燥的开阔草原，并不适于竹子生长，不可能有竹子供假竹鼠食用。

假竹鼠的牙齿咀嚼方式和 *Rhizomys* 者明显不同。*Rhizomys* 的下颌骨较高，咬肌也很发达，表明它们的上、下咬合能力和掘土能力都很强。*Rhizomys* 的内、外翼突同等发达（内、外翼肌同等发达），表明其下颌骨左右移动和牙齿横向咀嚼的能力较强。竹鼠类的臼齿为颊 - 舌侧单面高冠，其冠面不水平，上臼齿冠面向外倾，下臼齿冠面向近中方向倾斜。这也反映了竹鼠类的咀嚼方法应以横向的咀嚼为主。但假竹鼠的臼齿为前 - 后侧高冠，其冠面近于水平，表明假竹鼠的咀嚼方法与 *Rhizomys* 的不同，是以前 - 后向的咀嚼为主。事实上，假竹鼠的内、外翼突明显地不对称，外翼突明显地小于内翼突（翼外肌明显小于翼内肌），这也证明了假竹鼠的下颌骨的横向移动的能力较弱，因此其横向咀嚼的能力要弱于 *Rhizomys* 者。另外，假竹鼠的下颌骨水平部的高度较 *Rhizomys* 的低，其下颌角较 *Rhizomys* 明显更向后伸，表明假竹鼠的外层咬肌和表层咬肌延伸方向与 *Rhizomys* 的不完全相同，它们除了有关闭下颌的作用外，还有较强的向前拉动下颌的作用。假竹鼠的关节窝伸长，向后延伸到项嵴，表明允许下颌骨前后移动的距离较 *Rhizomys* 的长。这说明，它们的掘土能力更强；另外也表明，相对于 *Rhizomys* 而言，假竹鼠的前后咀嚼的能力较强、咀嚼的距离也较长。

（2）假竹鼠前肢的机能 - 习性分析

假竹鼠前肢较短，其中，肱骨（32.6 mm）长于桡骨（28 mm），桡骨长于前脚（约 25 mm）（详见表 14–17）。假竹鼠的肩胛骨的肩臼为长卵圆形；喙突很发达，强烈突出（表明喙臂肌等肌肉的附着区较大，而且减少了肩关节的侧向运动）。肩胛骨后部有很发达的供冈下肌附着的肌线，在肩胛颈后面有粗隆，肱骨大结节后面供冈下肌附着处也很发育。这都说明假竹鼠冈下肌也很发达，伸展肩关节的能力较强。这些部分的形态结构与 *Rhizomys* 的较相似，而与鼢鼠和麝鼠的区别较大。假竹鼠的肱骨较粗壮，肱骨头为卵圆形。肱骨的三角肌粗隆很发达，并向远方移至骨体中部。其位置也与 *Rhizomys* 者相近（见表 21）。这表明假竹鼠的三角肌很发达，用以内旋前肢、外展和屈曲肩关节的能力较强。肱骨的内上髁和外上髁嵴都较发达。桡骨近端与肱骨的关节面为卵圆形。尺骨具深凹的半月切迹。McII 和 McIII 有很发达的掌骨粗隆、指节骨有很发达的伸腱突或结节，掌面有供屈肌通过的沟或附着的粗糙面或结节，爪指骨具很发达的爪指骨粗隆。特别是假竹鼠的爪指骨较粗壮而长。这些都表明假竹鼠有很发达的腕和指的伸肌和屈肌，可以用前肢挖和刨土或抛土。

Lehmann（1963: 64）指出用爪刨型方法掘土的啮齿类（如 *Geomys* 和 *Ctenomys*），在肢骨上的特点是：肱骨有较粗大的三角肌粗隆和较宽的上髁；尺骨有较长的肘突，较大的冠状突和较深的骨体侧窝，脚爪特别增大、伸长成"鹰爪"形等。与现生用爪刨型方法掘土的 *Myospalax* 和 *Geomys* 等比较，假竹鼠的肱骨的三角粗隆向下移的程度不如 *Myospalax* 和 *Geomys* 那么远，肱骨远端的宽度与长度之比显然小于后者，内上髁发达的程度也较后者的小些（详见表 21）；尺骨冠状突较小，骨体的侧窝的凹入明显较浅和侧棱较弱小。此外，假竹鼠前脚的五个爪虽有较明显的长爪，但仍属正常类型，不特别增大、也不弯曲成"鹰爪"型。显然，假竹鼠不可能像 *Myospalax* 和 *Geomys* 那样强有力地用爪来刨土，即不属于完全依靠脚爪刨型掘土的啮齿类。假竹鼠的形态特点都与 *Rhizomys* 的较相近，其主要的掘土的方

表 21　部分穴居啮齿类肱骨一些部位比值比较

种（Species）	肱骨上部长 *** / 肱骨全长 UL/L	内上髁长 / 肱骨全长 LME/L	内上髁宽 / 肱骨全长 WME/L
Pseudorhizomys indigenus	0.51	0.14	0.14
Rhizomys sp. (OV 1055)	0.52–0.55	0.11–0.12	0.11–0.12
*Myospalax myospalax**	0.63	0.17	0.17
*Spalax leucodon**	0.59	0.058	
*Geomys bursarius**	0.54	(0.16–0.17**)	(0.17–0.18**)
*Ellobius lutescens**	0.51	0.076	

* 依 Vinogradov 和 Gambarian (1952, Table 1)（after Vinogradov and Gambarian, 1952, Table 1）。

** 依 *Geomys bursarius*（OV 432）测量。

*** Vinogradov 和 Gambarian（1952）表 1 中称其为肱骨大结节长，实际可能是肱骨上部长（= UL：肱骨头顶至三角肌粗隆下端的距离）[The length of the greater tuberosity in Table 1 of Vinogradov et Gambarian (1952) may be, in fact, the length of the upper part of the humerus (= UL: distance from top of the head of humerus to the lower end of the deltoid tuberosity)；笔者依 *Myospalax myospalax* (OV 478) 测量的 UL 值计算的该比值为 0.65 (The ratio of UL of *Myospalax myospalax* is 0.65 based on measurement of OV 478)]。

法是属于齿凿型的。

Rhizomys 虽主要用齿凿型的方法掘土，但也用脚爪辅助掘土（见 Stein, 2000: Tab. 1.1）。假竹鼠与 *Rhizomys* 的肢骨比较，虽然肩胛骨、肱骨和尺骨等总的形态很相似，但也有一些区别：假竹鼠的肱骨的内上髁发育的程度比 *Rhizomys* 的大些，而比头拱型的 *Spalax* 和 *Ellobius* 的发达得多（表 21）；假竹鼠的第二至第五指的爪指骨的远端横向较狭窄（尽管狭窄的程度不如 *Myospalax* 的那样明显），呈尖锐的锥形，具较发达的纵向的沟和棱（而 *Rhizomys* 的相对较短而宽扁，约呈尖铲形，两侧呈薄片状向外延伸，具很少的纵向的沟和棱）等。假竹鼠的爪指骨的远端表面具有较发达的纵向的沟棱，表明假竹鼠的爪指骨可能具有较发达的角质爪，显然它的脚爪比 *Rhizomys* 的更有利于掘土。另外，假竹鼠的胸浅肌、胸深肌和前肢的伸肌和屈肌都比较发达。这表明，假竹鼠的前肢除了有运土、抛土的功能外，它的第二至第五指还有一定的掘土功能，以辅助门齿掘土。加之，假竹鼠可能具有拇短伸肌、拇展肌、拇长伸肌，这些肌肉的作用主要是用来伸和外展拇指的，这也进一步表明假竹鼠的拇指具有相当的伸展功能，其拇指可能也有抛土或刨土的功能。

总之，根据头骨和前肢的解剖特征，假竹鼠可能也是营地下生活的啮齿类，它们挖土的方法可能主要属齿凿型，即主要以有力的头骨、下颌骨和颈椎的肌肉协助门齿来挖土。它们的前肢除了具有运土的功能外，其前脚爪可能还有一定的辅助掘土的功能。综上所述，假竹鼠应主要是用齿凿型方法掘土，并辅以脚爪刨土的啮齿类。

2. 副竹鼠的机能 - 习性分析

副竹鼠和假竹鼠的头骨在基本形态特征上很接近：头骨总的形态近三角形，吻部较长，颧弓明显向两侧扩张；其背面较平；眼眶很小；关节窝伸长至项嵴；矢状脊和项嵴较发达；项面较低宽，并稍向前方倾斜；外翼突较内翼突短小；上、下颌齿隙较长；下颌骨咬肌嵴很发达，并向外扩展呈棚架状；下颌角突明显的向后伸突；i2 特别伸长，匍匐前伸，后端伸达髁突下方，并在上升支颊侧形成明显的隆突等。这些都表明副竹鼠应该和假竹鼠一样也是营地下生活，属于主要用齿凿型方法挖土的啮齿类。

但副竹鼠与假竹鼠在头骨上也有明显的区别（详见表 9）。最主要的区别是：副竹鼠的 I2 弯曲度较大，其后端仅达门齿孔附近，其前端弯向下后方，唇侧具明显的纵棱；前颌骨前端和鼻骨的前端约在同一垂线上；颧弓板的位置较靠后，前颌骨侧背嵴较长，颧骨较短，不伸达泪骨等。副竹鼠的这些特特征表明，它不大可能用 I2 挖土。

副竹鼠的颅后骨骼保存得很少，而且其中的绝大多数都保存不完全。从现有的材料来看。副竹鼠的颈椎、肩胛骨、肱骨和掌骨都与假竹鼠者相似。

　　虽然副竹鼠可能和假竹鼠一样，也属主要用齿凿型方法掘土的啮齿类，但它们的挖土方式可能并不完全相同：假竹鼠可能用上、下门齿挖土，而副竹鼠则可能只用下门齿挖土，而上门齿则主要用于切割植物的根、茎和叶等。不过到目前为止，我们对副竹鼠的颅后骨骼所知甚少，至今还未发现它的尺骨和指节骨，特别是爪指骨。对副竹鼠的生活方式的推论还需要更多更好的标本来证明。

五、副竹鼠族的地史和古地理分布

临夏盆地的副竹鼠和假竹鼠化石产自晚中新世灞河期柳树组的下 - 中部。副竹鼠属包括三个种：三趾马层副竹鼠（*Pa. hipparionum*）、华夏副竹鼠（*Pa. huaxiaensis*）和陇副竹鼠（*Pa. longensis*）。它们分别产自柳树组的三个不同的层位（见表 2）。其中，陇副竹鼠产于柳树组下部，属郭泥沟哺乳动物群，其时代被认为是晚中新世灞河早期，距今约 11.1 Ma（Deng, 2005；Qiu et al., 2013；Deng et al., 2013）。三趾马层副竹鼠产自柳树组中部下层，属大深沟哺乳动物群。其时代被认为是晚中新世灞河中期，距今约 9.5 Ma（Deng, 2005；Qiu et al., 2013；Deng et al., 2013）。华夏副竹鼠产自柳树组的中部上层，属杨家山哺乳动物群，较大深沟动物群的时代稍晚，其时代被认为是晚中新世，晚灞河期，距今约 8.3 Ma（Deng, 2005；Qiu et al., 2013；Deng et al., 2013）。

临夏盆地的假竹鼠已知包括四个不同的种：土著假竹鼠（*Ps. indigenus*）、甘肃假竹鼠（*Ps. gasuensis*）、平齿假竹鼠（*Ps. planus*）和原始假竹鼠（*Ps. pristinus*）。尽管原始假竹鼠具有许多较原始的特征，但是由于该标本是从柳树组地层表面捡拾到的，目前还很难确定其确切的时代。平齿假竹鼠的地点和地史分布也不清楚。只有两个种的产出层位清楚：土著假竹鼠产于柳树组的中部上层，属杨家山动物群，甘肃假竹鼠产于柳树组中部，属大深沟动物群和杨家山动物群的两个层位中，其时代为晚中新世灞河中 - 晚期，距今约 9.5-8.3 Ma。总之，临夏盆地从郭泥沟动物群到杨家山动物群都有副竹鼠类化石发现。其生存的时代为晚中新世灞河期，距今约 11-8 Ma。

临夏盆地以外地区产的副竹鼠类化石，目前已知有三趾马层副竹鼠、秦副竹鼠（*Pa. qinensis*）、华夏副竹鼠和黑河假竹鼠（属存疑）（*Ps.? hehoensis*）四个种，以及副竹鼠属的两个未定种（*Pararhizomys* spp.）。三趾马层副竹鼠发现于我国陕西省的府谷和秦安，内蒙古的宝格达乌拉和蒙古西部的 Altan Teli。其时代分布为晚中新世灞河晚期至保德期，距今约 8-6 Ma（Kowalski, 1968；Zhang et al., 2005；Li, 2010）。秦副竹鼠产于我国陕西省蓝田坝河组的下部。其时代原被认为是晚中新世灞河较早期，距今约 10 Ma（Zhang et al., 2005），后被改为晚中新世灞河中期，距今约 9.37 Ma（Zhang et al., 2013）。如果德日进（Teilhard de Chardin, 1942）记述的产自庆阳地区的晚中新世保德期的砖红色的泥岩中的"桑氏原鼢鼠（*Prosiphneus licenti*）"确是华夏副竹鼠的话（见本书第 43-44 页），这个种的时代也可能延伸到晚中新世保德中期（距今约 7-6 Ma）。*Ps.? hehoensis* 仅发现于西藏比如县的布隆盆地，其时代最初被认为是上新世早期（郑绍华，1980），后被改为晚中新世灞河中期（邱占祥、邱铸鼎，1990；Qiu et Qiu, 1995；Deng, 2006；Qiu et al., 2013；Deng et Ding, 2015）。副竹鼠的两个未定种中，一个产自青海柴达木盆地的深沟，其时代原被认为是晚中新世灞河早期（Qiu et Li, 2008；Li, 2010），现被改为晚中新世灞河中期（Qiu et al., 2013），笔者暂称产自青海深沟的未定种为 *Pararhizomys* sp. I；另一个产自内蒙古的高特格地区，属下高特格动物群，其时代为上新世的早期，距今约 4.18-4.63 Ma（Teilhard de Chardin et Young, 1931；Li, 2010；Qiu et al., 2013），笔者暂称其为 *Pararhizomys* sp. II。

总之，世界上目前已知的副竹鼠类化石包括 2 属 9 种（其中一个为属存疑种）和 2 个未定种。它们绝大多数仅发现于我国（图 78），其中又以甘肃临夏盆地发现的最多，化石最丰富。在已知的副竹鼠类中，以产于临夏盆地柳树组下部的陇副竹鼠为出现最早的代表，其产出时代为灞河早期。秦副竹鼠、三趾马层副竹鼠、甘肃假竹鼠、黑河假竹鼠（？）和青海柴达木盆地深沟的 *Pararhizomys* sp. I 大致都出现于晚中新世灞河中期（Qiu et Li, 2008；Li, 2010）。此外，原始假竹鼠也可能在晚中新世灞河中期或更早出现。三趾马层副竹鼠分布的范围最广，在我国陕西府谷和秦安、甘肃临夏盆地、内蒙古中部，以及蒙古国西部等地都发现了该种的化石。其时代分布也最长，从灞河中期一直延续到保德期（Teilhard de Chardin et Young, 1931；Kowalski, 1968；Zhang et al., 2005；Li, 2010）。甘肃假竹鼠在地理分布上目前仅限于临夏盆地，其时代则由灞河中期延续到灞河晚期。而产于陕西蓝田灞河组下部的秦副竹鼠和发现于西藏布隆盆地布隆组的黑河假竹鼠（？），其时代则均限于晚中新世灞河中期。华夏副竹鼠和土著假竹鼠出现得更晚，时代为灞河晚期。其中土著假竹鼠在地理上也仅限于临夏盆地，其时代则仅限于晚中新世灞河晚期。而华夏副竹鼠可能分布的范围稍广些，在甘肃东部的庆阳地区可能出现过，其延续的时间可能较长，一直到保德中期。内蒙古高特格的 *Pararhizomys* sp. II 出现得最晚，其时代为上新世的早期。

　　综上所述，副竹鼠类目前已知的地理和地史分布都很有限。地理上仅限于北纬 31° 至 48°，东经 92° 至 115° 的东亚中部地区（图 78），时代为晚中新世灞河早期至上新世早期。其中副竹鼠属（包括 2 未定种）目前已知的地理分布最广，东自我国内蒙古的高特格，西至蒙古国西部的 Altan Teli 和我国青海省的深沟。它们的时代分布也最长，从晚中新世早期到上新世早期的地层中都发现了它们的痕迹。假竹鼠属目前已知的地理和地史分布范围都较小，除了可能归入该属的黑河假竹鼠（？）发现于西藏外，其他种的化石均仅发现于临夏盆地的晚中新世，中 - 晚灞河期的柳树组中部的地层中。

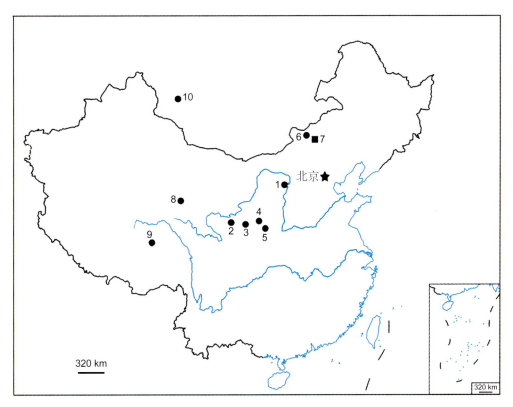

图 78　副竹鼠族的地理分布

Fig. 78　Geographic distribution of the pararhizomyines

1. 陕西府谷（Fugu, Shaanxi），2. 甘肃临夏（Linxia, Gansu），3. 甘肃秦安（Qin'an, Gansu），4. 甘肃庆阳（Qingyang, Gansu），5. 陕西蓝田（Lantian, Shaanxi），6. 内蒙古宝格达乌拉（Baogeda Ula, Nei Mongol），7. 内蒙古高特格（Gaotege, Nei Mongol），8. 青海深沟（Shengou, Qinghai），9. 西藏布隆（Bilung, Xizang），10. 蒙古 Altan Teli（Mongolia）

●中新世地点（Miocene locality），■上新世地点（Pliocene locality）

地图据国家测绘地理信息局网站，审图号：GS（2016）1569 号

六、副竹鼠族生存时期临夏盆地的古环境

目前已知，营地下生活的啮齿类绝大多数都居住在有足够可食用植物的较开阔地区的地下土层中，土壤可干可湿，只要透气性或排水性好即可（Lacey et al., 2000: 2；Bushet al., 2000: 185）。临夏盆地的副竹鼠类化石均产于晚中新世柳树组。柳树组产有丰富的三趾马动物群。该动物群中大量种类都是适于开阔草原的类型，表明临夏盆地在晚中新世时为温带较干旱的开阔草原环境（邓涛，2011）。侯素宽等对中国三趾马动物群中有蹄动物化石牙齿釉质中碳同位素组成进行了分析，指出中国西部（包括临夏盆地）"在晚中新世和早上新世时是以 C3 草本植物为特征的开阔温带草原"（侯素宽等，2006：85）。马玉贞等（1998: 303）认为临夏盆地"8.5–6.0 Ma"的阶段 C（带 VI）（= 柳树组所产孢粉）的孢粉组合中以草本，特别是旱生、半旱生草本为优势分子，含一定量的温带、暖温带阔叶树；其植被类型为干旱环境中生长的干草原。显然，临夏盆地在晚中新世时为开阔的干旱或半干旱的温带草原，不但能为动物（包括地下生活的啮齿类）提供广阔的生活场地，而且能提供较大量的可食用植物。副竹鼠类的臼齿为较高冠的脊形齿，也能适应于摄食草原地区的植物的茎、叶和根。这种生态环境较适合穴居的副竹鼠类生活。

在临夏盆地发现的副竹鼠类还有一个特点，即发现的个体很分散。除了陇副竹鼠曾有三个个体发现于同一地点的同一层位中以外，其他属、种的 14 件标本都是在一个地点只发现一个个体的标本。这与三趾马动物群中大部分大型哺乳动物在同一地点经常发现大量个体的情况完全相反。这种现象至少可有两种解释：或是采集的偏好，或是副竹鼠类的生活习性为独居，而非群居。事实上，在临夏盆地以外的 7 个发现副竹鼠类化石地点，也是一个地点只发现一个个体的标本（Teilhard de Chardin et Young, 1931；Kowalski, 1968；郑绍华，1980；Zhang et al., 2005；Qiu et Li, 2008；Li, 2010）。这表明，副竹鼠类可能像现生的竹鼠（*Rhizomys*）一样，是独居者，每个洞穴内仅居住一只成年的副竹鼠。对于上面提到的三件陇副竹鼠的标本产于同一地点的例外情况，可以动物死后再搬运来解释。因为其中有两件（V 16292.2 和 V 16292.3）均只保存有吻部，显然并非原地埋藏。

Flynn（1985, 1990: 174）曾指出，竹鼠类穴居的习性使得其成年个体活动性变低，导致居群隔离，从而有利于土著特性的快速分化，分异成不同穴居的生态灶。化石记录也表明穴居者的分异速率要比非穴居者高。副竹鼠类目前全部已知 2 属 9 种和 2 个未定种的时代分布也只有约 700 万年（从 11 Ma 至 4 Ma）。同样，副竹鼠类在临夏盆地也经历了在短时期内高速分化的历史。在晚中新世的约 300 万年期间（从 11 Ma 至 8 Ma）就出现了 2 属 7 种。反之，临夏盆地的副竹鼠类的快速分化也从另一个角度证实了副竹鼠类是地下穴居生活者。

七、副竹鼠族在亚洲产生和发展的原因

副竹鼠族（Pararhizomyines）是一类在晚中新世—早上新世时仅分布于东亚中部中纬度地区的土著啮齿动物，其分布区正好位于青藏高原北部和东北缘及其以北地区。新生代以来全球的气候和亚洲的古地理都发生了明显的变化，而这些变化都与青藏高原的隆升有密切关系。副竹鼠类的起源和发展显然也与此有关。

从早始新世晚期到早渐新世，全球的气候有一渐进的和阶段性的变冷、变干的过程（Wolfe, 1978; Berggren et Prothero, 1992; Zachos et al., 2001）。始新世大部分时间里北半球的气候在冬季仍然温暖潮湿，而在夏季则炎热多雨，适于各种常绿和阔叶落叶林生长。渐新世时北半球的气温变得比始新世时要干而凉。孢粉资料显示，此时喜暖植物减少，松柏类增多，草本被子植物得到发展（李浩敏、郑亚惠，1995: 365）。

始新世—渐新世期间的全球气候的明显转变是与亚洲大陆的气候变冷、变干相关联的（Dupont-Nivet et al., 2007）。始新世以前，喜马拉雅地区还处于古地中海地区。在始新世—渐新世时，印度板块与古亚洲板块进入初步拼合阶段，海水从亚洲大陆逐渐并最终完全退出（马宗晋等，2003: 342–344）。强小科等（2010）指出，晚渐新世时青藏高原南部的隆起阻挡了来自南方大洋的水气，导致了亚洲内陆干旱化的增强。在临夏盆地晚渐新世地层中发现的孢粉组合（阶段 A）代表干旱气候的疏（森）林 - 草原植被类型（马玉贞等，1998）。适于林地 - 草原生活的巨犀动物得到发展（邱占祥、王伴月，2007）。但青藏高原在渐新世时隆升得并不太高，其高度并未阻止青藏高原南、北两侧的哺乳动物的迁徙和交流。例如，在青藏高原以北的亚洲内陆和以南的印巴次大陆都有巨犀动物群存在（Qiu et al., 2001; 邱占祥、王伴月，2007; Deng et Ding, 2015）。在亚洲内陆与巨犀共生的大哺乳动物主要是适应较干旱气候的动物（邓涛，2004a, 2011）。适于较干旱气候的啮齿动物，如梳趾鼠类（Ctenodactylidae）、查干鼠类（Tsaganomyidae）和拟速掘鼠（Tachyoryctoides）等也陆续出现，并繁衍、分化和发展（Bohlin, 1937, 1946; Wang, 1997, 2001; Wang et Qiu, 2012）。

从晚渐新世晚期（27–26 Ma）到中中新世（~15 Ma）全球的气候又逐渐回暖（Zachos et al., 2001）。在早中新世晚期—中中新世早期（22–14 Ma）东亚的气候变得较温暖、潮湿（刘裕生、郑亚惠，1995: 392）。与此同时，由于青藏高原抬升到一定的高度（约 2000 m 或更高），使亚洲的气候由全球的带状模式转变为以季风为主的模式，东南和西南季风与从亚洲内陆来的冬季季风交汇，而副地中海退缩到土库曼斯坦西部克孜勒卡亚地区，进一步加强了东亚季风，分别增加了季风区和我国西北部的潮湿和干燥程度（Zhang et al., 2007）。较强的东南和西南季风从海洋带来的潮湿气候使临夏盆地等受季风影响的地区变得更温暖、潮湿（Shi et al., 1999; 李吉均等，2001; Guo et al., 2008）。临夏盆地这一阶段的孢粉组合（阶段 B）属于森林植被阶段。这一阶段乔木发育，森林茂盛（马玉贞等，1998）。在早中新世晚期的山旺期，亚洲内陆的哺乳动物群似乎仍处于过渡阶段，哺乳动物化石群中保留有适应干旱气候的种类（如梳趾鼠类和拟速掘鼠类），而适应潮湿的长鼻类和牛科在临夏盆地也开始出现并发展。亚洲内陆的中中新世通古尔期（15–11 Ma），正好与全球的中中新世气候最佳期（Zachos et al., 2001）大致相当。这时亚洲内陆相当潮湿（Liu et al., 2009），出现了以森林型为主的哺乳动物群。该动物群种类繁多。其中，生活在水边、以水生植物为食的铲齿象和半水栖的啮齿动物河狸类等都很繁盛（邱占祥、关键，1986; 邱铸鼎，1996; 叶捷等，2001; 邓涛，2004a, b, 2011）。很可能该地区在当时是森林环境，并有较丰富的水体。而适应于较干燥环境的啮齿类，有的（如查干鼠类和拟速掘鼠）相继绝灭；有的（如梳趾鼠类）被迫向南迁徙至南亚，并由南亚进一步向西迁徙，最后到达非洲，而在亚洲内陆从此完全绝灭（Wang, 1997）。

中中新世以后，全球的气候又逐渐变冷、变干旱，南极冰盖再次形成（Zachos et al., 2001）。青藏高原进一步隆升（邱占祥等，1997；安芷生等，2006），不但使亚洲内陆进一步干旱化，而且加强了亚洲季风系统（Kutzbach et al., 1993）。与此同时，副特提斯海的逐渐退缩，更使亚洲内陆大陆化，也加强了季风环流（汪品先，2005）。东亚季风将水分从海洋带到中国东部，使东部变得更加湿热，并使亚洲内陆的干旱带由古近纪的纬向分布到中新世时退缩到我国西北和蒙古等周边地区（Wang, 1990；Shi et al., 1999；汪品先，2005；Sun et Wang, 2005）。

在晚中新世时（11–6 Ma）青藏高原可能又发生了一次重要的隆升（安芷生等，2006）。这促使了季风进一步加强（Prell et Kutzbach, 1992；汪品先，2005；Sun et Wang, 2005）。在此期间，在临夏盆地沉积的柳树组的岩性主要为褐黄色粉砂岩、粉砂质泥岩和泥岩，并含有大量的钙质结核，表明当时的气候较干燥（方小敏等，1997）。亚洲内陆的生态环境变为以 C3 植物为特征的温带草原（侯素宽等，2006）。临夏盆地在此期间的孢粉组合（阶段 C）被称为草原植被阶段。孢粉组合中草本，特别是旱生、半旱生草本量大增，含一定量的温带、暖温带阔叶树。植被类型为干旱环境中生长的干草原（马玉贞等，1998）。

在这种环境下，亚洲内陆的哺乳动物也发生了很大的变化。那些喜水的动物有的完全绝灭（如笨河狸，*Anchitheriomys*），有的虽残存到晚中新世早期，但之后也很快绝灭（如铲齿象，*Platybelodon*，豪狸，*Hystricops* 和单沟河狸，*Monosaulax* 等）。仅在部分地区有林栖动物（如爪兽）生存。代之以适应较干燥环境的开阔草原型动物为主的三趾马动物群的出现（邓涛，2004a，2011）。在这种环境下，副竹鼠类和鼢鼠类得以在亚洲内陆起源，并迅速演化和分异发展，占据了拟速掘鼠等啮齿动物原来生活的生态灶。

在青藏高原北部柴达木盆地发现的 *Pararhizomys* sp. I 化石产于深沟地点的上油沙山组。在同一地点和层位中还发现了原始等级裂腹鱼类化石（Chang et al., 2010）。原始等级裂腹鱼类的现生种类现聚居于青藏高原周围，海拔 1250–2500 m 一带（曹文宣等，1981；Chang et al., 2010；张弥曼、Miao，2016）。西藏布隆盆地产三趾马动物群化石地层的孢粉组合中的植物现在分布区的海拔一般不超过2000 m（吴玉书、于浅黎，1980）。这样，根据与副竹鼠类同一地点和同一层位中发现的植物化石和鱼化石推测，晚中新世时青藏高原北部（包括布隆盆地和柴达木盆地）产化石的地区的海拔通常可能仍在 2000 m 左右（黄万波等，1980；吴玉书、于浅黎，1980；Chang et al., 2010）。这一方面表明，青藏高原北部在晚中新世时海拔可能与临夏盆地现在的海拔大致相当；另一方面也表明，副竹鼠类在海拔 1250–2500 m 的环境下也可以生存。换句话说，副竹鼠类当时产生和生存的环境有可能与现生的原始等级裂腹鱼类生活的青藏高原周边地区环境类似。

就地理位置而言，产副竹鼠类化石的地区正好位于现在我国西北干旱区（或蒙新高原区、或东干草原）的自然区内（徐惟诚等，1999: 207；张荣祖，1999: 8），也位于亚洲内陆中新世的干旱带内（Sun et Wang, 2005；汪品先，2005）。在该地区内干草原植被得到发展，温度和湿度变低。这种较干冷的环境似乎正好适于副竹鼠类和鼢鼠类生活。因此，副竹鼠类和鼢鼠类在这一时期在该地区得以起源，并迅速发展。然而，在北纬 48° 以北的地区，可能气候太寒冷，不适于副竹鼠类生活。大兴安岭以东及其以南的地区属东部季风区（或潮湿区），潮湿程度都很高；而阿尔泰山脉以西的中亚属西干草原，其气候也要比东干草原温暖而潮湿，显然都不适于副竹鼠类生活。因此，副竹鼠类只能分布在这较窄小的区域内。相比之下，鼢鼠类分布的范围要广些，往北可到西伯利亚，往东可到兴安岭以东地区（Zheng, 1994）。这似乎表明鼢鼠类的适应能力可能要比副竹鼠类的强些。

上新世时青藏高原又发生了较强烈的隆升，被称为青藏运动（李吉均等，1996）。不仅青藏高原北部在此期间强烈隆升，西藏以北或东北的亚洲中部的山脉也经历了大幅度的隆升（Zheng et al., 2000；汪品先，2005）。这表现在青藏高原北缘党河流域的上新统顶部巨厚砾岩层（王晓鸣等，2008）和酒西盆地巨厚的玉门砾岩，以及临夏盆地的巨厚的积石砾岩（李吉均等，1995）的出现。青藏高原和蒙古高原的隆升，直接导致亚洲北缘的西伯利亚河系改向北流。通过注入北冰洋的淡水的增加，促使海水结冰，导致北冰洋的冰盖形成（Zachos et al., 2001；汪品先，2005），使亚洲内陆的气候变得更寒冷和干燥，雨量普遍减少，日温差变大。亚洲内陆的酒西盆地的孢粉系列显示，在上新世时，从5.67 Ma开始，该地区旱化逐渐加强，植被变为荒漠草原型（马玉贞等，2004）。由于持续的极度干旱化，在柴达木盆地上新世狮子沟组地层中还出现了骨骼非常粗的一种鱼——献文鱼（*Hsianwenia*，见 Chang et al., 2008；张弥曼、Miao，2016）。塔克拉玛干沙漠也在5.3 Ma左右形成（Sun et Liu, 2006）。安芷生等（2006: 680–682）指出，3.6–2.6 Ma期间东亚冬、夏季风同时增强，内陆干旱化急剧扩展；2.6 Ma以后，北半球进入大冰期时代，全球冰量增加的趋势和冰期旋回有助于亚洲内陆的干旱趋势的加大。显然，上新世及其以后，亚洲内陆变得既干旱又严寒，环境变得相当恶劣。

当环境条件发生急剧变化时，生活在该环境中的动物可能至少会出现三种不同的应对变化：①全部或部分成员通过产生变异，经自然选择而适应新的环境，仍在原地区生存；②全部或部分成员迁移到与它们原生活环境接近或类似的地区，逐渐适应并得以生存和繁衍；③无法适应变化了的新环境，又无迁徙可能，最终走向绝灭。鼢鼠类可能属于第①种情况。较原始的属 *Prosiphneus* 虽已绝灭，但有的种类演变为新的属种（如 *Chardina* 和 *Mesosiphneus* 等），适应了新环境。后来尽管经历了第四纪冰期这样恶劣的气候环境，鼢鼠类仍能在原起源地，甚至在更广大的地区生存了下来（Zheng, 1994；张荣祖等，1997: 222–225；张荣祖，1999: 208–210）。梳趾鼠类可能属于第②种情况。尽管在中中新世以后已在亚洲内陆绝灭，但在早中新世时，青藏高原南、北两侧的大陆的气候环境相近，而且青藏高原抬升的高度也还不能阻止哺乳动物的迁徙，因此，梳趾鼠类的部分成员得以往南迁到南亚，后经南亚、西亚，逐渐迁徙到北非，并在北非一直生存到现在（见 Wang, 1997）。副竹鼠类则属于第③种情况。其本身的适应能力可能就较弱，在环境变恶劣的情况下，不能随环境的变化产生适应新环境的变异。青藏高原的持续隆升，使副竹鼠类分布区以南的地区，海拔更高，变为高寒地区，副竹鼠类无法生存，也无法向南逾越。副竹鼠类分布区以北的地区，气候更寒冷，也无法向北迁移。其以东和东南地区当时仍属湿热的潮湿区，而以西的中亚地区，属西干草原，其气候有明显的季节性变化，也较湿热，显然也都不适于副竹鼠类生存。这表明当时周边地区没有适合副竹鼠类生活的环境供其迁徙并继续生存，最终导致副竹鼠类在上新世晚期在亚洲完全绝灭。

八、结　论

1）研究表明，临夏盆地在晚中新世时已知共有 2 属 7 种副竹鼠类（包括 1 新属和 6 新种）生存过。它们产自三个不同的层位和时代。其中陇副竹鼠（*Pa. longensis*）出现得最早，产于晚中新世灞河早期的柳树组下部。晚中新世灞河中期的柳树组上部下层产有三趾马层副竹鼠（*Pa. hipparionum*）和甘肃假竹鼠（*Ps. gansuensis*），可能还有原始假竹鼠（*Ps. pristinus*）。晚中新世灞河晚期的柳树组上部上层产有华夏副竹鼠（*Pa. huaxiaensis*）、甘肃假竹鼠和土著假竹鼠（*Ps. indigenus*）。还有一种平齿假竹鼠（*Ps. planus*），其产出地点和层位尚不清楚。在临夏盆地以外地区还发现有秦副竹鼠（*Pa. qinensis*）和黑河假竹鼠（？）（*Ps.? hehoensis*），以及副竹鼠的 2 个未定种，这样副竹鼠类目前已知共有 2 属 9 种和 2 个未定种。

2）基于分支系统学的原则和约定，本书中拟采用的分类方案是：竹鼠科（Rhizomyidae）和盲鼹鼠科（Spalacidae）为鼠超科（Muroidea）中的两个科。盲鼹鼠科包含两个亚科：盲鼹鼠亚科（Spalacinae）和拟速掘鼠亚科（Tachyoryctoidinae）。后一亚科包含两个族，即拟速掘鼠族（Tachyoryctoidini）和副竹鼠族（Pararhizomyini）。本书所研究的全部材料均属于副竹鼠族。

3）根据头骨和头后骨骼的形态和肌能分析，本书作者认为，副竹鼠族是一类穴居的啮齿类。它们主要用门齿，即齿凿型的方法掘土，辅以用爪刨型的方法协助掘土和运土。但副竹鼠和假竹鼠两属在掘土的方法上并不完全相同。前者可能仅主要用下门齿掘土，而后者可能用上、下门齿一起掘土。

4）副竹鼠族生活的地区正好位于现在的东干草原的自然区内。对其生存时期气候及与其共生的动物和植物群的分析表明，副竹鼠族是一类适应于较干燥环境下的开阔草原的啮齿动物。副竹鼠族在东亚中部中纬度地区的产生、发展和绝灭的历史，与晚中新世—早上新世时东亚中部的地理、气候环境的变化有关，特别与青藏高原隆升有密切关系。

参 考 文 献

An Z S, Zhang P Z, Wang E Q, Wang S M, Qiang X K, Li L, Song Y G, Chang H, Liu X D, Zhou W J, Liu W G, Cao J J, Li X Q, Shen J, Liu Y, Ai L. 2006. Changes of the monsoon-arid environment in China and growth of the Tibetan Plateau since the Miocene. Quat Sci, 26 (5): 678–693 (in Chinese with English abstract) [安芷生 , 张培震 , 王二七 , 王苏民 , 强小科 , 李力 , 宋友桂 , 常宏 , 刘晓东 , 周卫健 , 刘卫国 , 曹军骥 , 李小强 , 沈吉 , 刘禹 , 艾莉 . 2006. 中新世以来我国季风 - 干旱环境演化与青藏高原的生长 . 第四纪研究 , 26 (5): 678–693]

Beijing Zoo, Peking University, Beijing Agricultural University et al. 1986. Morphology of the Giant Panda—Systematic Anatomy and Organ-history. Beijing: Science Press. 1–641 (in Chinese) [北京动物园 , 北京大学 , 北京农业大学等 . 1986. 大熊猫解剖——系统解剖和器官组织学 . 北京 : 科学出版社 . 1–641]

Bendukidze O G, de Bruijn H, van den Hoek Ostende L W. 2009. A revision of late Oligocene association of small mammals from the Aral Formation (Kazakhstan) in the National Museum of Georgia, Tbilissi. Palaeodiversity, 2: 343–377

Berggren W A, Prothero D R. 1992. Eocene–Oligocene climatic and biotic evolution: An overview. In: Berggren W A, Prothero D R (eds). Eocene–Oligocene Climatic and Biotic Evolution. Princeton: Princeton Univ Press. 1–28

Blanga-Kanfi S, Miranda H, Penn O, Pupko T, DeBry R W, Huchon D. 2009. Rodent phylogeny revised: analysis of six nuclear genes from all major rodent clades. BMC Evolutionary Biology, 9: 71

Bleefeld A R, McKenna M C. 1985. Skeletal integrity of *Mimolagus rodens* (Lagomorpha, Mammalia). Am Mus Novit, 2806: 1–5

Bohlin B. 1937. Oberoligozäne Säugetiere aus dem Shargaltein-Tal (Western Kansu). Palaeont Sin, New Ser C, (3): 1–66

Bohlin B. 1946. The fossil mammals from the Tertiary deposit of Taben-buluk, western Kansu. Palaeont Sin, New Ser C, (8b): 1–259

Bugge J. 1970. The contribution of the stapedial artery to the cephalic arterial supply in muroid rodents. Acta Anat, 76: 313–336

Bugge J. 1971. The cephalic arterial system in mole-rats (Spalacidae), bamboo rats (Rhizomyidae), jumping mice and jerboas (Dipodoidea) and dormice (Gliroidea) with special reference to the systematic classification of rodents. Acta Anat, 79: 165–180

Bugge J. 1974. The cephalic arterial system in insectivores, primates, rodents and lagomorphs, with special reference to the systematic classification. Acta Anat, 87 (Sup. 62): 1–160

Bugge J. 1985. Systematic value of the carotid arterial pattern in rodents. In: Luckett W P, Hartenberger J-L (eds). Evolutionary Relationships among Rodents—A Multidisciplinary Analysis. New York: Plenum Press. 355–379

Busch C, Antinuchi C D, del Valle J C, Kittlein M J, Malizia A I, Vassallo A I, Zenuto R R. 2000. Population ecology of subterranean rodents. In: Lacey E A, Patton J L, Cameron G N (eds). Life Underground—The Biology of Subterranean Rodents. Chicago: Univ Chicago Press. 183–226

Cao W X, Chen Y Y, Wu Y F, Zhu S Q. 1981. Origin and evolution of schizothoracine fishes in relation to the upheaval of the Qinghai-Xizang Plateau. In: The Comprehensive Scientific Expedition to the Qinghai-Xizang Plateau, Chinese Academy of Sciences (ed). Studies on the Period, Amplitude and the Type of the Uplift of the Qinghai-Xizang Plateau. Beijing: Science Press. 118–131 (in Chinese with English abstract) [曹文宣 , 陈宜瑜 , 武云飞 , 朱松泉 . 1981. 裂腹鱼类的起源和演化及其与青藏高原隆起的关系 . 见 : 中国科学院青藏高原综合科学考察队 (编). 青藏高原隆起的时代、幅度和形式问题 . 北京 : 科学出版社 . 118–131]

Carleton M D, Musser G G. 1985. 11. Muroid rodents. In: Anderson S, Jones J K (eds). Orders and Families of Recent Mammals of the World. New York: John Wiley et Sons. 289–379

Chang M M, Miao D S. 2016. Review of the Cenozoic fossil fishes from the Tibetan Plateau and their bearing on paleoenvironment. Chinese Sci Bull, 61(1): 1–15 (in Chinese with English abstract) [张弥曼 , Miao D S. 2016. 青藏高原的新生代鱼化石及其古环境意义 . 科学通报 , 61 (1): 1–15]

Chang M M, Wang X M, Liu H Z, Miao D S, Zhao Q H, Wu G X, Liu J, Li Q, Sun Z C, Wang N. 2008. Extraordinarity thick-bones fish linked to the aridification of the Qaidam Basin (northern Tibetan Plateau). PNAS, 105 (36): 13246–13251

Chang M M, Miao D S, Wang N. 2010. Ascent with modification: Fossil fishes witnessed their own group's adaptation to the uplift of the Tibetan Plateau during the Late Cenozoic. In: Long M Y, Gu H Y, Zhou Z H (eds). Darwin's Heritage Today: Proceeding of the Darwin 200 Beijing International Conference. Beijing: Higher Education Press. 60–75

Constantinescu G M, Schaller O (eds). 2012. Illustrated Veterinary Anatomical Nomenclature (3rd edition). Stuttgart: Enke Verlag. 1–620

Cooper G, Schiller A L. 1975. Anatomy of the Guinea Pig. Cambridge: Hrvard Univ Press. 1–417

Dawson M R, Tong Y S. 1998. New material of *Pappocricetodon schaubi,* an Eocene rodent (Mammlia: Cricetidae) from the Yuanqu Basin, Shanxi Province, China. Bull Carnegie Mus Nat Hist, 34: 278–285

Daxner-Höck G, Badamgarav D, Maridet O. 2015. Evolution of Tachyoryctoidinae (Rodentia, Mammalia): evidences of the Oligocene and Early Miocene of Mongolia. Ann Naturhist Mus Wien, Serie A, 117: 161–195

de Bruijn H, Marković Z, Wessels W. 2013. Late Oligocene rodents from Banović (Bosnia and Herzegovina). Palaeodiversity, 6: 63–105

de Bruijn H, Bosma A A, Wessels W. 2015. Are the Rhizomyinae and the Spalacinae closely related? Contradistinctive conclusions between genetics and palaeontology. Palaeobio Palaeoenv, 95: 257–269

Deng T. 2004a. Evolution of the late Cenozoic mammalian faunas in the Linxia Basin and its background relevant to the uplift of the Qinghai-Xizang Plateau. Quat Sci, 24 (4): 413–420 (in Chinese with English abstract) [邓涛 . 2004a. 临夏盆地晚新生代哺乳动物群演替与青藏高原隆升背景 . 第四纪研究 , 24 (4): 413–420]

Deng T. 2004b. Establishment of the middle Miocene Hujialiang Formation in the Linxia Basin of Gansu and its features. Jour Stratigr, 28 (4): 307–312 (in Chinese with English abstract) [邓涛 . 2004b. 临夏盆地中中新统虎家梁组的建立及其特征 . 地层学杂志 , 28 (4): 307–312]

Deng T. 2005. Character, age and ecology of the Hezheng biota from northwestern China. Acta Geologica Sinica, 79 (6): 739–750

Deng T. 2006. Chinese Neogene mammal biochronology. Vert PalAs, 44 (2): 143–163

Deng T. 2009. Late Cenozoic environmental changes in the Linxia Basin (Gansu, China) as indicated by Cenograms of fossil mammals. Vert PalAs, 47 (4): 282–298

Deng T. 2011. Diversity variations of the Late Cenozoic mammals in the Linxia Basin and their response to the climatic and environmental backgrounds. Quat Sci, 31 (4): 577–588 (in Chinese with English abstract) [邓涛 . 2011. 临夏盆地晚新生代哺乳动物的多样性变化及其对气候环境背景的响应 . 第四纪研究 , 31 (4): 577–588]

Deng T, Ding L. 2015. Paleoaltimetry reconstructions of the Tibetan Plateau: progress and contradiction. Nat Science Rev, 2 (4): 417–437

Deng T, Wang X M, Ni X J, Liu L P, Liang Z. 2004. Cenozoic stratigraphic sequence of the Linxia Basin in Gansu, China and its evidence from mammal fossils. Vert PalAs, 42 (1): 45–66 (in Chinese with English summary) [邓涛 , 王晓鸣 , 倪喜军 , 刘丽萍 , 梁忠 . 2004. 临夏盆地的新生代地层及其哺乳动物化石证据 . 古脊椎动物学报 , 42(1): 45–66]

Deng T, Qiu Z X, Wang B Y, Wang X M, Hou S K. 2013. Late Cenozoic biostratigraphy of the Linxia Basin, northwestern China. In: Wang X M, Flynn L, Fortelius M (eds). Fossil Mammals of Asia—Neogen Biostratigraphy and Chronology. New York: Columbia Univ Press. 243–273

Dupont-Nivet G, Krijgsman W, Langereis C G, Abels H A, Dai S, Fang X M. 2007. Tibetan Plateau aridification linked to global cooling at the Eocene-Oligocene transition. Nature, 445: 635–638

Ellerman J R. 1940. The Families and Genera of Living Rodents. Volume I. Rodents other than Muridae. London: British Museum (Nature Hist). 1–689

Ellerman J R. 1941. The Families and Genera of Living Rodents. Volume II. Family Muridae. London: British Museum (Nature Hist). 1–690

Evans H E, Christensen G C. 1979. Miller's Anatomy of the Dog. Philadelphia: W B Saunders Company. 1–1181

Fang X M, Li J J, Zhu J J, Chen H L, Cao J X. 1997. Division and age dating of the Cenozoic strata of the Linxia Basin in Gansu. Chinese Sci Bull, 42 (14): 1457–1471 (in Chinese) [方小敏 , 李吉均 , 朱俊杰 , 陈怀录 , 曹继秀 . 1997. 甘肃临夏盆地新生代地层绝对年代测定与划分 . 科学通报 , 42 (14): 1457–1471]

Flynn L J. 1982. Systematic revision of Siwalik Rhizomyinae (Rodentia). Geobios, 15 (3): 327–389

Flynn L J. 1985. Evolutionary patterns and rates in Siwalik Rhizomyidae (Rodentia). Acta Zool Fennica, 170: 141–144

Flynn L J. 1990. The natural history of rhizomyid rodents. In: Nevo E, Reig O (eds). Evolution of Subterranean Mammals at the Organismal and Molecular Levels. New York: Wiley-Liss. 155–183

Flynn L J. 2009. The antiquity of *Rhizomys* and independent acquisition of fossorial traits in subterranean Muroid. Bull Am Mus Nat Hist, (331): 128–156

Flynn L J, Nevo E, Heth G. 1987. Incisor enamel microstructure in blind mole rats: addative and phylogenetic significance. Jour Mammal, 68 (3): 500–507

Gervais P. 1846. Observations sur diverse espèces de mammifères fossils du Midi de la France. Ann des sc nat, Zool, ser 3e, 5: 248

Greene E C. 1935. Anatomy of the rat. Trans Amer Philos Soc, New Ser, 27: 1–370

Gromova V. 1959. Giant rhinoceroses. Trav Paleont Inst AS USSR, 71: 1–164 (in Russian)

Guo Z T, Sun B, Zhang Z S, Peng S Z, Xiao G Q, Ge J Y, Hao Q Z, Qiao Y S, Liang M Y, Liu J F, Yin Q Z, Wei J J. 2008. A major reorganization of Asian climate by the early Miocene. Clim Past 4: 153–174

Guthrie D A. 1963. The carotid circulation in the Rodentia. Bull Mus Comp Zool, 128 (10): 455–481

Guthrie D A. 1969. The carotid circulation in *Aplodontia*. Jour Mammal, 50 (1): 1–7

Hebel R, Stromberg M W. 1976. Anatomy of the Laboratory Rat. Baltimore: The Williams et Wilkins Company. 1–173

Hildebrand M. 1982. Analysis of Vertebrate Structure (2nd ed.). New York: John Wiley et Sons Inc. 1–654

Hildebrand M, Goslow G E. 2001. Analysis of Vertebrate Structure (5th ed.). New York: John Wiley et Sons Inc. 1–635

Hill J E. 1935. The cranial foramina in rodents. Jour Mammal, 16 (2): 121–129

Hofmeijer G K, de Bruijn H. 1985. The mammals from the Lower Miocene of Aliveri (Island of Evia, Greece). Part 4: The Spalacidae and Anomalomyidae. Proc Kon Ned Akad Wetensch, Ser. B, 88 (2): 185–198

Hou S K, Deng T, Wang Y. 2006. Stable carbon isotopic evidence of tooth enamel for the late Neogene habitats of the *Hipparion* fauna in China. In: Dong W (ed). Proc of Tenth Annual Meeting of the Chinese Soc. Vert Paleontol. Beijing: China Ocean Press. 85–94 (in Chinese with English abstract) [侯素宽 , 邓涛 , 王杨 . 2006. 中国新近纪晚期三趾马动物群生活环境的化石稳定碳同位素证据 . 见 : 董为主编 . 第十届中国古脊椎动物学术年会论文集 . 北京 : 海洋出版社 . 85–94]

Howell A B. 1926. Anatomy of the Wood Rat. Comparative Anatomy of the Subgenera of the American Wood Rat (Genus *Neotoma*). Monograph Amer Soc Mammalogy, no.1. Baltimore: Williams et Wilkins Company. 1–225

Hua L C, Ungar P S, Zhou Z R, Ning Z W, Zheng J, Qian L M, Rose J C, Yang D. 2015. Dental development and microstructure of bamboo rat incisors. Biosurface and Biotribology, 1 (2015): 263–269

Huang W P, Ji H X, Chen W Y, Hsu C Q, Zheng S H. 1980. Pliocene stratum of Guizhong and Bulong basin, Xizang. Palaeontology of Xizang, 1. Beijing: Science Press. 4–17 (in Chinese with English abstract) [黄万波 , 计宏祥 , 陈万勇 , 徐钦琦 , 郑绍华 . 1980. 西藏吉隆、布隆盆地的上新世地层 . 西藏古生物 , 第一分册 . 北京 : 科学出版社 . 4–17]

International Commission on Zoological Nomenclature. 2000. International Code of Zoological Nomenclature (Fourth Edition). Padova: Tipografia la Garangola. 1–126

Jacobs L L, Flynn L J, Li C K. 1985. Comments on rodents from the Chinese Neogene. Bull Geol Inst University of Uppsala, N S, 11: 59–78

Jansa S A, Weksler M. 2004. Phylogeny of muroid rodents: relationship within and among major lineages as determined by IRBP gene sequences. Molecula Phylogenetics and Evolution, 31: 256–276

Jansa S A, Giarla T C, Lim B K. 2009. The phylogenetic position of the rodent genus *Typhlomys* and the geographic origin of Muroidea. Jour Mammal, 90 (5): 1083–1094

Ji H X, Hsu C Q, Huang W P. 1980. The *Hipparion* Fauna from Guizhong Basin, Xizang. Palaeontology of Xizang, 1. Beijing:

Science Press. 18–32 (in Chinese with English abstract) [计宏祥 , 徐钦琦 , 黄万波 . 1980. 西藏吉隆沃马公社三趾马动物群 . 西藏古生物 , 第一分册 . 北京 : 科学出版社 . 18–32]

Koken E. 1885. Ueber fossile Säugethiere aus China. Palaeontologische Abhandlungen, Dritter B. Heft 2: 1–85

Kowalski K. 1968. *Pararhizomys hipparionum* Teilhard & Young, 1931 (Rodentia) from Pliocene of Altan Teli, Western Mongolia. Palaeont Polonica, (19): 163–168

Kowalski K. 1974. Middle Oligocene rodents from Mongolia. Palaeont Polonica, (30): 147–178

Kutzbach J E, Prell W L, Ruddiman W F. 1993. Sensitivity of Eurasian climate to surface uplift of the Tibetan Plateau. Jour Geol, 101: 177–190

Lacey E A, Patton J L, Cameron G N (eds). 2000. Life Underground—the Biology of Subterranean Rodents. Chicago: Univ Chicago Press. 1–449

Lehmann W H. 1963. The forelimb architecture of some fossorial rodents. Jour Morphol, 113 (1): 59–76

Li C K, Qiu Z D. 1980. Early Miocene Mammalian fossils of Xining Basin, Qinghai. Vert PalAs, 18 (3): 198–214 (in Chinese with English summary) [李传夔 , 邱铸鼎 . 1980. 青海西宁盆地早中新世哺乳动物化石 . 古脊椎动物学报 , 18 (3): 198–214]

Li C K, Zheng J J, Ting S Y. 1989. An Early Eocene ctenodactyloid rodent of Asia. In: Black C C, Dawson M R (eds). Papers on Fossil Rodents in Honor of Albert Elmer Wood. Nat Hist Mus Los Angeles County. Sci Ser. 33: 179–192

Li H M, Zheng Y H. 1995. Palaeogene flora. In: Li X X (ed). Fossil Floras of China through the Geological Ages. Guangzhou: Guangdong Science and Technology Press. 455–505 [李浩敏 , 郑亚惠 . 1995. 早第三纪植物群 . 见 : 李星学 (主编). 中国地质时期植物群 . 广州 : 广东科技出版社 . 345–382]

Li J J, Fang X M, Zhu J J, Zhong W, Cao J X, Wang J L, Zhang Y C, Wang J M, Kang S C. 1995. Paleomagnetic chronology and type sequence of the Cenozoic stratigraphy of the Linxia Basin in Gansu Province of China. Annual Study on Formation, Evolution, Environmental Change and Ecology of Tibetan Plateau (1994). Beijing: Science Press. 41–54 (in Chinese with English abstract) [李吉均 , 方小敏 , 朱俊杰 , 钟魏 , 曹继秀 , 王建力 , 张叶春 , 王建民 , 康世昌 . 1995. 临夏盆地新生代地层古地磁年代与模式序列 . 青藏高原形成演化、环境变迁与生态系统研究学术论文年刊 (1994). 北京 : 科学出版社 . 41–54]

Li J J, Fang X M, Ma H Z, Zhu J J, Pan B T, Chen H L. 1996. Geomorphological and environmental evolution in the upper reaches of the Yellow River during the late Cenozoic. Science in China, Ser D, 39 (4): 380–390 [李吉均 , 方小敏 , 马海洲 , 朱俊杰 , 潘保田 , 陈怀录 . 1996. 晚新生代黄河上游地貌演化与青藏高原隆起 . 中国科学 , D 辑 , 26 (4): 316–322]

Li J J, Fang X M, Pan B T, Zhao Z J, Song Y G. 2001. Late Cenozoic intensive uplift of Qinghai-Xizang Plateau and its impacts on environments in surrounding area. Quart Sci, 21 (5): 381–391 (in Chinese with English abstract) [李吉均 , 方小敏 , 潘保田 , 赵志军 , 宋友桂 . 2001. 新生代晚期青藏高原强烈隆起及其对周边环境的影响 . 第四纪研究 , 21 (5): 381–391]

Li Q. 2010. *Pararhizomys* (Rodentia, Mammalia) from the Late Miocene of Baogeda Ula, Central Nei Mongol. Vert PalAs, 48 (1): 48–62

Lin G H, Wang K, Deng X G, Nevo E, Zhao F, Su J P, Guo S C, Zhang T Z, Zhao H B. 2014. Transcriptome sequencing and phylogenomic resolution within Spalacidae (Rodentia). BMC Genomics, 15: 32

Liu L P, Eronen J T, Fortelius M. 2009. Significant mid-latitude aridity in the middle Miocene of East Asia. Palaeogeogr Palaeoclimatol Palaeoecol, 279: 201–206

Liu Y S, Zheng Y H. 1995. Neogene floras. In: Li X X (ed). Fossil Floras of China through the Geological Ages. Guangzhou: Guangdong Science and Technology Press. 506–551 [刘裕生 , 郑亚惠 . 1995. 晚第三纪植物群 . 见 : 李星学 (主编). 中国地质时期植物群 . 广州 : 广东科技出版社 . 383–416]

Ma Y Z, Li J J, Fang X M. 1998. Records of the climatic variation and pollen flora from the red beds at 30.6–5.0 Ma in Linxia district. Chinese Sci Bull, 43 (3): 301–304 (in Chinese) [马玉贞 , 李吉均 , 方小敏 . 1998. 临夏地区 30.6~5.0 Ma 红层孢粉植物群与气候演化记录 . 科学通报 , 43 (3): 301–304]

Ma Y Z, Fang X M, Li J J, Wu F L, Zhang J. 2004. Vegetational and environmental changes during late Tertiary–early Quaternary in Jiuxi Basin. Sci China, Ser. D, Earth Sci, 34 (2): 107–116 (in Chinese) [马玉贞 , 方小敏 , 李吉均 , 吴福莉 , 张军 . 2004.

酒西盆地晚第三纪—第四纪早期植被与气候变化. 中国科学, D 辑, 地球科学, 34 (2): 107–116]

Ma Z J, Du P R, Hong H J (eds). 2003. Structure and dynamics of the earth. Guangzhou: Guangdong Science and Technology Press. 1–564 (in Chinese with English abstract) [马宗晋, 杜品仁, 洪汉净 (编著). 2003. 地球构造与动力学. 广州 : 广东科技出版社. 1–564]

McKenna M C, Bell K S. 1997. Classification of Mammals above the Species Level. New York: Columbia University Press. 1–631

Meng J, Hu Y M, Li C K. 2003. The osteology of *Rhombomylus* (Mammalia, Glires) implications for phylogeny and evolution of Glires. Bull Am Mus Nat Hist, (275): 1–247

Michaux J, Reyes A, Catzeflis F. 2001. Evolutionary history of the most speciose mammals: molecular phylogeny of muroid rodent. Molecular Biology and Evolution, M18 (11): 2017–2031

Musser G G, Heaney L R. 1992. Philipine rodents: definitions of *Tarsomys* and *Limnomys* plus a preliminary assessment of phylogenetic patterns among native Philippine Murines (Murinae, Muridae). Bull Am Mus Nat Hist, (211): 1–138

Nixon K C. 1999–2002. WinClada ver. 1.0000 Published by the author, Ithaha, NY, USA

Norris R W, Zhou K, Zhou C, Yan G, Kilpatrick C W, Honeycutt R L. 2004. The phylogenetic position of zakors (Myospalacinae) and comments on the families of muroids (Rodentia). Molecular Phylogenetics and Evolution 31: 972–978

Pan Q H, Wang Y X, Yan K (eds). 2007. A Field Guide to the Mammals of China. Beijing: China Forestry Publishing House. 1–420 (in Chinese) [潘清华, 王应祥, 岩昆 (主编). 2007. 中国哺乳动物彩色图鉴. 北京 : 中国林业出版社. 1–420]

Prell W L, Kutzbach J E. 1992. Sensitivity of the Indian monsoon to forcing parameters and implications for its evolution. Nature, 360: 647–652

Qiang X K, An Z S, Song Y G, Chang H, Sun Y B, Liu W G, Ao H, Dong J B, Fu C F, Wu F, Lu F Y, Cai Y J, Zhou W J, Cao J J, Xu X W, Ai L. 2011. New eolian red clay sequence on the western Chinese Loess Plateau linked to onset of Asian desertification about 25 Ma ago. Science China, Earth Sci, 54 (1): 136–144 [强小科, 安芷生, 宋友桂, 常宏, 孙有斌, 刘卫国, 敖红, 董吉宝, 符超峰, 吴枫, 卢凤艳, 蔡演军, 周卫健, 曹军骥, 徐新文, 艾莉. 2010. 晚渐新世以来中国黄土高原风成红粘土序列的发现：亚洲内陆干旱化起源的新记录. 中国科学, 地球科学, 40 (11): 1479–1488]

Qiu Z D. 1996. Middle Miocene Micrommalian Fauna from Tunggur, Nei Mongol. Beijing: Science Press. 1–216 (in Chinese with English summary) [邱铸鼎. 1996. 内蒙古通古尔中新世小哺乳动物. 北京 : 科学出版社. 1–216]

Qiu Z D, Li Q. 2008. Late Miocene micromammals from the Qaidam Basin in the Qinghai-Xizang Plateau. Vert PalAs, 46 (4): 284–306

Qiu Z D, Wang X M, Li Q. 2013. Neogene faunal succession and biochronology of central Nei Mongol (Inner Mongolia). In: Wang X M, Flynn L, Fortelius M (eds). Fossil Mammals of Asia—Neogene Biostratigraphy and Chronology. New York: Columbia Univ Press. 155–186

Qiu Z X, Guan J. 1986. A lower molar of *Pliopithecus* from Tongxin, Ningxia Hui Autonomous Region. Acta Anthropol Sinica, 5 (3): 201–207 (in Chinese with English summary) [邱占祥, 关键. 1986. 宁夏同心发现的一颗上猿牙齿. 人类学学报, 5 (3): 201–207]

Qiu Z X, Qiu Z D. 1990. Neogene local mammalian fauna: Succession and ages. J Stratigr, 14 (4): 241–260 (in Chinese) [邱占祥, 邱铸鼎. 1990. 中国晚第三纪地方哺乳动物群的排序及其分期. 地层学杂志, 14 (4): 241–260]

Qiu Z X, Qiu Z D. 1995. Chronological sequence and subdivision of Chinese Neogene mammalian faunas. Palaeogeogr Palaeoclimatol Palaeoecol, 116: 41–70

Qiu Z X, Wang B Y. 2007. Paracerathere Fossils of China. Palaeont Sin, New Ser C, (29): 1–396 (in Chinese with English summary) [邱占祥, 王伴月. 2007. 中国的巨犀化石. 中国古生物志, 新丙种第 29 号 : 1–369]

Qiu Z X, Wang B Y, Qiu Z D, Xie G P, Xie J Y, Wang X M. 1997. Recent advances in study of the Xianshuihe Formation in Lanzhou Basin. In: Tong Y S, Zhang Y Y, Wu W Y, Li J L, Shi L Q (eds). Evidence for Evolution—Essays in Honor of Prof. Chungchien Young on the Hundredth Anniversary of His Birth. Beijing: Ocean Press. 177–192 (in Chinese with English abstract) [邱占祥, 王伴月, 邱铸鼎, 颉光普, 谢骏义, 王晓鸣. 1997. 甘肃兰州盆地咸水河组研究的新进展. 见 : 童永生, 张银运, 吴文裕, 李锦玲, 史立群 (编). 演化的实证——纪念杨钟健教授百年诞辰论文集. 北京 : 海洋出版社.

177–192]

Qiu Z X, Wang B Y, Qiu Z D, Heller F, Yue L P, Xie G P, Wang X M, Engesser B. 2001. Land mammal geochronology and magnetostratigraphy of mid-Tertiary deposits in the Lanzhou Basin, Gansu Province, China. Eclogae Geol Helv, 94: 373–385

Qiu Z X, Qiu Z D, Deng T, Li C K, Zhang Z Q, Wang B Y, Wang X M. 2013. Neogene Land Mammal Stages/Ages of China—Toward the goal to establish an Asian land mammal stage/age scheme. In: Wang X M, Flynn L, Fortelius M (eds). Fossil Mammals of Asia—Neogene Biostratigraphy and Chronology. New York: Columbia Univ Press. 29–90

Repenning C A. 2003. *Mimomys* in North America. In: Flynn L J (ed). Vertebrate Fossils and Their Context - Contributions in Honor of Richard H. Tedford. Bull Am Mus Nat Hist, (279): 469–512

Rose K D, DeLeon V B, Missiaen P, Rana R S, Sahni A, Singh L, Smith T. 2008. Early Eocene lagomorph (Mammalia) from Western India and the early diversification of Lagomorpha. Proc Royal Soc B, 275: 1203–1208

Sarica N, Sen S. 2003. Spalacidae (Rodentia). In: Fortelius M, Kappelman J, Sen S et al. (eds). 2003. Geology and Paleontology of the Miocene Sinap Formation, Turkey. New York: Columbia Univ Press. 141–162

Schaub S. 1958. Simplicidenta (=Rodentia). In: Piveteau (ed). Traité Paléontologie. Tom 6. L'Origine des Mammifères et les aspects fondamentaux de leur Évolution, Vol. 2. Paris: Masson et Ci, 793–815

Shi Y F (施雅风), Tang M C (汤懋苍), Ma Y Z (马玉贞). 1999. Linkage between the second uplifting of the Qinghai-Xizang (Tibetan) Plateau and the initiation of the Asian monsoon system. Sci China, Ser D, 42 (3): 303–312

Simpson G G. 1945. The principles of classifications and a classification of mammals. Bull Am Mus Nat Hist, 85: 1–350

Stein B R. 2000. Morphology of Subterranean Rodent. In: Lacey E A, Patton J L, Cameron G N (eds). Life Underground—the Biology of Subterranean Rodents. Chicago: Univ Chicago Press. 19–61

Steppan S J, Adkins R M, Anderson J. 2004. Phylogeny and divergence—Date estimates of rapid radiations in muroid rodents based on multiple nuclear genes. Systematic Biology, 52 (4): 533–553

Sun J M, Liu T S. 2006. The age of the Taklimakan Desert. Sicence, 312: 1621

Sun X J, Wang P X. 2005. How old in the Asian monson system?—Palaeobotanical records from China. Palaeogeogr Palaeoclimatol Palaeoecol, 222: 181–222

Szalay F S. 1985. Rodent and lagomorph morphotype adaptations, origins, and relationships: some postcranial attributes and analyzed. In: Luckett W P, Hartenberger J-L (eds). Evolutionary Relationships among Rodents—A Multidisciplinary Analysis. New York: Plenum Press. 83–132

Teilhard de Chardin P. 1942. New rodents of the Pliocene and lower Pleistocene of North China. Publication of Institute de Géo-Biologie Paper, (9): 1–101

Teilhard de Chardin P, Young C C. 1931. Fossils mammals from the Late Cenozoic of northern China. Palaeont Sin, Ser C, 9 (1): 1–89

Tong Y S. 1992. *Pappocricetodon*, a pre-Oligocene cricetid genus (Rodentia) from Central China. Vert PalAs, 30 (1): 1–16 (in Chinese with English summary) [童永生 . 1992. 中国中部中晚始新世仓鼠类一新属——祖仓鼠 (*Pappocricetodon*). 古脊椎动物学报 , 30 (1): 1–16]

Tong Y S. 1997. Middle Eocene small mammals from Liguanqiao Basin of Henan Province and Yuanqu Basin of Shanxi Province, Central China. Palaeont Sin, New Ser C, (26): 1–256 (in Chinese with English summary) [童永生 . 1997. 河南李官桥和山西垣曲盆地始新世中期小哺乳动物 . 中国古生物志 , 新丙种第 26 号 : 1–256]

Ünay E. 1999. Family Spalacidae. In: Rössner G E, Heissig K (eds). The Miocene Land Mammals of Europe. München: Verlag Dr. Friedrich Pfeil. 421–425

Vinogradov B C, Gambarian P P. 1952. The cylindrodonts of the Oligocene of Mongolia and Kasakhstan (Cylindrodontidae, Glires, Mammalia). Trudy Paleont Inst, 41: 13–42 (in Russian)

Voss R S. 1988. Systematics and ecology of ichthyomyine rodents (Muroidea): patterns of morphological evolutionh in a small adaptive radiation. Bull Am Mus Nat Hist, 188 (2): 259–493

Wahlert J H. 1974. The cranial foramina of protrogomorphous rodents; an anatomical and phylogenetic study. Bull Mus Compara

Zool, 146 (8): 363–410

Wahlert J H. 1983. Relationships of the Florentiamyidae (Rodentia, Geomyoidea) based on cranial and dental morphology. Am Mus Novitates, 2769: 1–23

Wahlert J H. 1985a. Skull morphology and relationships of geomyoid rodents. Am Mus Novitates, 2812: 1–2

Wahlert J H. 1985b. Cranial foramina of rodents. In: Luckett W P, Hartenberger J-L (eds). Evolutionary Relationships among Rodents. A Multidisciplinary Analysis. NATO SCI Ser, Ser A: Life Sciences 92: 311–332

Wang B Y. 1997. The mid-Tertiary Ctenodactylidae (Rodentia, Mammalia) of Eastern and Central Asia. Bull Am Mus Nat Hist, (234): 1–88

Wang B Y. 2001. On Tsaganomyidae (Rodentia, Mammalia) of Asia. Am Mus Novitates, (3317): 1–50

Wang B Y, Dawson M R. 1994. A primitive cricetid (Mammalia: Rodentia) from the Middle Eocene of Jiangsu Province, China. Ann Carnegie Mus, 63 (3): 239–256

Wang B Y, Qiu Z X. 2012. *Tachyoryctoides* (Muroidea, Rodentia) fossils from Early Miocene of Lanzhou Basin, Gansu Province, China. Swiss J Palaeontol, (2012) 131: 107–126

Wang P X. 1990. Neogene stratigraphy and paleoenvironments of China. Palaeogeogr Palaeoclimatol Palaeoecol, 77: 315–334

Wang P X. 2005. Cenozoic deformation and history of sea-land interaction in Asia. Earth Science, 30 (1): 1–18 (in Chinese with English abstract) [汪品先 . 2005. 新生代亚洲形变与海陆相互作用 . 地球科学 , 30 (1): 1–18]

Wang X M, Wang B Y, Qiu Z X. 2008. Early explorations of Tabenbuluk region (western Gansu Province) by Birger Bohlin— Reconciling classic vertebrate fossil localities with modern stratigraphy. Vert PalAs, 46 (1): 1–19

Wang Z T, Hao L W, Li G S, Li C H (eds). 2009. Atlas of Anatomy of Wistar Rat. Jinan: Shandong Science and Technology Press. 1–255 (in Chinese) [王增涛 , 郝丽文 , 李桂石 , 李常辉 (编). 2009. Wistar 大鼠解剖图谱 . 济南 : 山东科学技术出版社 . 1–255]

Wesselman H B, Black M T, Asnake M. 2009. Small mammals. In: Haile-Selassie Y, Woldegabriel G (eds). *Ardipithecus kadabba* Late Miocene Evidence from the Middle Awash, Ethiopia. Berkeley: Univ California Press. 105–133

Wible J R, Wang Y Q, Li C K, Dawson M R. 2005. Cranial anatomy and relationships of a new ctenodactyloid (Mammalia, Rodentia) from the Early Eocene of Hubei Province, China. Ann Carnegie Mus, 74 (2): 91–150

Wiley E O, Siegel-Causey D, Brooks D R, Funk V A. 1991. The Compleat Cladist. A Primer of Phylogenetic Procedures. Univ Kansas Mus Nat Hist Special Publ, No. 19: 1–158

Wolfe J A. 1978. A paleobotanical interpretation of Tertiary climates in Northern Hemisphere. Am Sci, 66: 694–703

Wood A E. 1955. A revised classification of the rodents. Jour Mammal, 36 (2): 165–187

Wood A E. 1962. The early Tertiary rodents of the Family Paramyidae. Trans Am Philos Soc, New Ser, 52 (1): 1–261

Wu Y S, Yu Q L. 1980. Pollen-spores assemblages from localities of *Hipparion* Fauna in Xizang and its significance. Palaeontology of Xizang, 1. Beijing: Science Press. 76–82 (in Chinese with English abstract) [吴玉书 , 于浅黎 . 1980. 西藏高原含三趾马动物群化石地点孢粉组合及其意义 . 西藏古生物 , 第一分册 . 北京 : 科学出版社 . 76–82]

Xu W C (ed). 1999. Encyclopaedia Britannica (International Chinese Edition), 16. Beijing: China Encyclopaedia Press. 1–548 [徐惟诚 (总编辑). 1999. 不列颠百科全书 (国际中文版) 16. 北京 : 中国大百科全书出版社 . 1–548]

Yang A F, Wang P et al. (eds). 1985. Anatomys and Histology of Rat. Beijing: Science Press. 1–241 (in Chinese) [杨安峰 , 王平等 (编). 1985. 大鼠的解剖和组织 . 北京 : 科学出版社 . 1–241]

Ye J, Wu W Y, Meng J. 2001. The age of Tertiary strata and mammal faunas in Ulungur River area of Xinjiang. Jour Strati, 25 (4): 283–287 (in Chinese with English abstract) [叶捷 , 吴文裕 , 孟津 . 2001. 新疆乌伦古河地区第三纪哺乳动物群初析及地层年代确定 . 地层学杂志 , 25 (4): 283–287]

Young C C, Liu P T. 1950. On the mammalian fauna at Koloshan near Chungking, Szechuan. Bull Geol Soc China, 30: 43–90

Zachos J, Pagani M, Sloan L, Thomas E, Billups K. 2001. Trends, rhythms, and aberrations in global climate 65 Ma to present. Science, 292: 686–693

Zhang R Z. 1999. Zoogeography of China. Beijing: Science Press. 1–502 (in Chinese) [张荣祖 . 1999. 中国动物地理 . 北京 : 科

学出版社 . 1–502]

Zhang R Z et al. 1997. Distribution of mammalian species in China. Beijing: China Forestry Publishing House. 1–280 (in Chinese and English) [张荣祖等 . 1997. 中国哺乳动物分布 . 北京 : 中国林业出版社 . 1–280]

Zhang Y L. 1983. Palaeontological Latin in Nomenclature. Beijing: Science Press. 1–429 (in Chinese) [张永辂 . 1983. 古生物命名拉丁语 . 北京 : 科学出版社 . 1–429]

Zhang Z Q, Flynn L J, Qiu Z D. 2005. New materials of *Pararhizomys* from northern China. Palaeont Electronica, 8 (1), 5A: 1–9

Zhang Z Q, Kaakinen A, Liu L P, Lunkka J P, Sen S, Gose W A, Qiu Z D, Zheng S H, Fortelius M. 2013. Mammalian biochronology of Late Miocene Bahe Formation. In: Wang X M, Flynn L, Fortelius M (eds). Fossil Mammals of Asia—Neogene Biostratigraphy and Chronology. New York: Columbia Univ Press. 187–202

Zhang Z S, Wang H J, Guo Z T, Jiang D B. 2007. What triggers the transition of palaeoenvironmental pattern in China, the Tibetan Plateau uplift or the Paratethys Sea retreat? Palaeogeogr Palaeoclimatol Palaeoecol, 245: 317–331

Zheng H B, Powell C M, An Z S, Zhou J, Dong G R. 2000. Pliocene uplift of the northern Tibetan Plateau. Geology, 28 (8): 715–718

Zheng S H. 1980. The *Hipparion* fauna of Bulong Basin, Biru, Xizang. Palaeontology of Xizang, 1. Beijing: Science Press. 33–47 (in Chinese with English abstract) [郑绍华 . 1980. 西藏比如布隆盆地三趾马动物群 . 西藏古生物 , 第一分册 . 北京 : 科学出版社 . 33–47]

Zheng S H. 1993. Quaternary Rodents of Sichuan-Guizhou Area, China. Beijing: Science Press. 1–270 (in Chinese with English summary) [郑绍华 . 1993. 川黔地区第四纪啮齿类 . 北京 : 科学出版社 . 1–270]

Zheng S H. 1994. Classification and evolution of the Siphneidae. In: Tomida Y, Li C K, Setoguchi T (eds). Rodent and Lagomorph Families of Asian Origins and Diversification. National Sci Mus Monographs, (8): 57–76

英/拉（缩写）- 汉颅后骨骼解剖名词索引

fmcV　facet for McV　与第五掌骨关节面　80, 85, 217, 219

fmg　facet for magnum　与头状骨关节面　29, 80, 84

fna　facet for navicular　与跗舟骨关节面　31, 32

fodf　facet for odontoid fossa of the atlas　枢椎齿突与寰椎
　的齿窝关节面　40

fphI　facet for Ph1　与第一指节骨关节面　81

fprzy　facet on prezygapophysis　前关节突上关节面　65,
　67, 107, 109

fpzy　facet on postzygapophysis　后关节突上关节面　40,
　63–65, 106, 107, 109, 231

fsc-lu　facet for scapho-lunar　与腕舟 - 月骨关节面　75,
　76, 79

ftr　*foramen transversarium*（= transverse foramen, trf）
　横突孔　40, 198, 199

ftR1　facet for tubercle of 1st rib　与第一肋骨肋骨结节关
　节面　65, 68

ftR4　facet for tubercle of 4th rib　与第四肋骨肋骨结节关
　节面　109

ftrd　facet for trapezoid　与小多角骨关节面　79, 82

ftrm　facet for trapezium　与大多角骨关节面　81, 82

ftro　facet for trochlea of humerus　与肱骨滑车关节面　74, 75

fts　facet for tibial sesamoid　与胫侧籽骨关节面　31, 32

ful　facet for ulna　与尺骨关节面　74, 75

fun　facet for unciform　与钩骨关节面　29, 84–86, 218

G

gbr　groove for *m. brachialis*　臂肌沟　75

glc　glenoid cavity　肩臼　41, 42, 71, 110, 199, 213, 233

gras　groove of astragalus　距骨沟　31, 32, 193

gtu　greater tuberosity　大结节　42, 43, 72, 110, 199, 213, 234

gva　groove for vertebral artery　椎动脉沟　61, 62, 105,
　209, 230

H

hR1　partial head of R1　第一肋骨的部分肋骨头　65, 66,
　69, 211

Hu　humerus　肱骨　35, 43, 60, 72, 111, 199, 213, 233

I

ilvf　internal lateral vertebral foramen　内侧椎孔　38, 61,
　198, 209, 230

invf　intervertebral foramen　椎间孔　64

ispf　infraspinous fossa　冈下窝　71, 110, 213, 233

L

ldvc　*lamina dorsalis vertebrae cervicalis* VI　第六颈椎横
　突背支　64, 65, 107

lep　lateral epicondyle　外上髁　72, 74, 214

lepcr　lateral epicondylar crest　外上髁嵴　72, 74, 214

lf　lateral foramen　侧壁孔　88, 89

lm　*linea muscularis*　肌线　71, 110

lr　lateral ridge　滑车外嵴　31, 103, 193, 230

ltu　lesser tuberosity　小结节　42, 43, 72, 73, 110, 199,
　213, 234

lvaf　lateral vertebro-alar foramen　侧椎 - 翼孔　61, 198,
　209, 230

lvf　lateral vertebral foramen　侧椎孔　38, 61, 198, 209, 230

lvvc　*lamina ventralis vertebrae cervicalis* VI　第六颈椎横
　突腹板　64, 65, 107

M

Mc　metacarpus 或 metacarpal　掌骨　80, 83, 114, 217, 235

mep　medial epicondyle　内上髁　72, 74, 214

mepcr　medial epicondylar crest　内上髁嵴　72, 74, 214

Mg　magnum　头状骨　112, 113, 235

mr　medial ridge　滑车内嵴　30, 31

Ms　manus　前脚　77, 111, 216, 234

msg　musculo-spiral groove　臂肌沟　72, 74, 214

N

nas　neck of astragalus　距骨颈　31

nhu　neck of humerus　肱骨颈　72

nra　neck of radius　桡骨颈　74, 75, 215

nsc　neck of scapula　肩胛颈　42, 71, 110, 199, 213, 233

O

odf　odontoid fossa　齿窝　38, 39, 61, 62, 105, 209, 230

odpr　odontoid process　齿突　40, 63, 106, 198, 210, 231

ol　olecranon　肘突　76, 77, 216

olf　olecranon fossa　肘窝　72, 74, 214

P

pcf　proximal calcaneal facet　近端跟骨关节面　31, 193

Ph　phalange　指节骨，趾节骨　86, 115, 220, 237

pran　*processus anconaenus*　钩突　76, 77, 216

przy　prezygapophysis　前关节突　40, 65, 67, 107, 109, 199

Ps　pisiform　豌豆骨　111, 112, 234

pva　pedicle of vertebral arch　椎弓根　40, 41, 65

pzy   postzygapophysis   后关节突   40, 41, 63, 65, 67, 106, 107, 109, 199, 210, 231

R

R   rib   肋骨   69, 108, 211, 233

Ra   radius   桡骨   27, 60, 74, 111, 112, 214, 234

rano   radial notch   桡切迹   76, 77, 216

S

sacr   sagittal crest   矢状崎   74, 75, 215

Sc   scapula   肩胛骨   35, 41, 60, 71, 104, 110, 199, 213, 233

Sc-lu   scapho-lunar   腕舟 - 月骨   27, 111, 112, 234

sctu   scapular tuber   肩胛结节   42, 43, 71, 199, 213

semno   semilunar notch   半月切迹   76, 77, 216

sp   spinous process   棘突   40, 41, 65, 67, 106, 109, 198, 199, 210, 231–233

spsc   spine of scapula   肩胛冈   42, 43, 71, 110, 213, 233

sscf   subscapular fossa   肩胛下窝   71, 110, 213, 233

sspf   supraspinous fossa   冈上窝   71, 110, 213, 233

stypr   styloid process of radius   桡骨茎突   27, 28, 75, 192

stypu   styloid process of ulna   尺骨茎突   77

suf   sustentacular facet   载距关节面   32, 193

sup   sustentacular process   载距突   31

T

T   thoracic vertebra   胸椎   66, 108, 211, 232

tmc   tuberosity of metacarpal   掌骨粗隆   29, 30, 82, 84, 204, 218, 219

To   toe   脚趾［如 ToIII（third toe, 第三趾）等］   93, 94, 223

tosf   tongue-shaped facet   舌状面   31, 193

Trd   trapezoid   小多角骨   111, 112

trf   transverse foramen（= *foramen transversarium*, ftr）横突孔   41, 210, 230, 231

Trm   trapezium   大多角骨   61, 77, 78, 111, 112

tro   trochlea   滑车   31, 72, 193, 214

trpr   transverse process   横突   40, 41, 63, 65, 67, 106, 107, 109, 199, 210, 231, 232

tt   teres tuberosity   圆肌隆起   72, 74

tunph   tuberosity of ungual phalanx   爪指骨粗隆   87, 89–91, 118, 221, 222, 238

tur   tubercle of rib   肋骨结节   69, 108, 109, 212, 233

tura   tuberosity of radius   桡骨粗隆   75, 76

tuul   tuberosity of ulna   尺骨粗隆   76, 77, 216

U

Ul   ulna   尺骨   60, 76, 111, 112, 215, 234

Un   unciform   钩骨   61, 78, 79, 112, 113, 217, 235

V

va   ventral arch   腹弓   38, 39, 61, 62, 105, 198

vb   vertebral body   椎体   40, 198, 199

vc   vertebral canal   椎管   38

vf   vertebral foramen   椎孔   39, 61, 198

vfs   vertebral fossa   椎窝   40, 41, 63, 65, 106, 107, 109, 199, 210, 231, 232

vn   vertebral notch   椎切迹   64, 65, 210

vof   volar foramen   掌孔   87, 89, 91, 118

vos   volar sulcus   掌沟   87, 89–91, 118

vt   ventral tubercle   腹结节   38, 39, 61, 62, 105, 198, 209, 230

W

w   wing   寰椎翼   39, 40, 62, 105, 198

PALAEONTOLOGIA SINICA

Whole Number 200, *New Series C, Number* 31

Edited by

Nanjing Institute of Geology and Palaeontology

Institute of Vertebrate Paleontology and Paleoanthropology

Chinese Academy of Sciences

Late Miocene Pararhizomyines from Linxia Basin of Gansu, China

by

Wang Banyue   Qiu Zhanxiang

(*Institute of Vertebrate Paleontology and Paleoanthropology, Chinese Academy of Sciences*)

With 78 Figures and 21 Tables

SCIENCE PRESS

Beijing, 2018

LIST OF PUBLICATIONS "PALAEONTOLOGIA SINICA"
NEW SERIES C

Late Miocene Pararhizomyines from Linxia Basin of Gansu, China

Wang Banyue   Qiu Zhanxiang

(Institute of Vertebrate Paleontology and Paleoanthropology, Chinese Academy of Sciences)

Summary

Contents

PREFACE

In May, 2000, the Institute of Vertebrate Paleontology and Paleoanthropology, the Chinese Academy of Sciences (hereafter IVPP), initiated a cooperative program with the Hezheng Paleozoological Museum (hereafter HPM) to search and extensively collect fossils from the Cenozoic deposits of the Hezheng area. Hezheng is a county of the Linxia Hui Autonomous Prefecture of Gansu Province, well known for its richness of "dragon bones" (mainly late Cenozoic mammalian fossils) since the late 20th century. The majority of the collected materials are skulls and jaws of large Miocene mammals (various proboscideans, carnivorans, chilotheres, hipparionine horses, giraffes, pigs, bovids, etc). Small mammal fossils, though highly desired by paleontologists for their important role in geologic time calibration, were largely neglected by local "dragon bone" collectors. In May, 2008, while prospecting the collected materials in HPM, we found considerable number of well-preserved skulls, jaws, and even postcranial bones belonging to *Pararhizomys* and some closely related new forms. From 2009 on, the senior author (Wang) of the present monograph started to concentrate on study of this group of animals. Two papers describing the referred specimens to the type species (*Pararhizomys hipparionum*) and two new species of this genus of the Hezheng material had been submitted to the *Vertebrata PalAsiatica* in 2011. The editor of the journal, Dr. Shi Liqun, while reviewing the manuscripts, realizing that a large amount of pararhizomyine specimens were being studied and waiting for publication, kindly advised the senior author to prepare a monograph to encompass all the pararhizomyine materials so far found from the Hezheng area rather than publishing them separately. This proposal soon won hearty support from Prof. Li Chuankui, the leading rodent authority of IVPP. This encouraged the senior author to finally make up her mind to prepare such a monograph.

With the increase of the studied materials and consequently the widening of the scope of problems to be treated, new challenges emerged. Firstly, few anatomical studies on skull had so far been done, not only for fossils, but also for extant forms of closely related rodent groups. Sometimes, in order to solve some anatomical uncertainties, in addition to the literatures on anatomy of *Rattus* and guinea pig, one has to resort to anatomy of humans, domestic animals, and some carnivores, which are anatomically better studied than rodents in general. Secondly, the better preserved skulls, jaws and skeleton parts, often of the same individuals, provide rare opportunities to apply cladistic analysis to test phylogenetic position of *Pararhizomys* and the like in the superfamily Muroidea. Under such circumstances, the junior author (Qiu) has been invited to join the work since August, 2014.

During the course of writing, the authors have received great and multifarious help from colleagues in and out of IVPP and abroad. Among them, notably, are: Profs. Li Chuankui, Qiu Zhuding, Zheng Shaohua, Zhang Zhaoqun, Wu Wenyu, Deng Tao, Ni Xijun, Li Qiang, and Bi Shundong of IVPP, and Dr. L. Flynn of Peabody Museum, Harvard University. Mr. He Wen, Director, and Mr. Chen Shanqin, Vice-director of the HPM, kindly placed all the HPM specimens at the authors' disposal whenever necessary for the completion of the present monograph. Profs. Qiu Zhuding, Zhang Zhaoqun, and Ni Xijun helped to review the Chinese manuscript and made valuable suggestions. Dr. Wang Xiaoming of Natural History Museum of Los Angeles County helped to review and improve the English summary of the monograph. To all of them the authors would like to express their sincerest thanks.

I. INTRODUCTION

1. Research history of *Pararhizomys* and significance of the new findings from Linxia Basin

Pararhizomys is one of the extinct endemic rodents lived in Late Miocene-Early Pliocene in central area of East Asia. The genus *Pararhizomys* (type species *Pa. hipparionum*) was established by Teilhard de Chardin and Young in 1931, based on a partial left hemimandible collected from Fugu, Shaanxi Province. They supposed that *Pararhizomys* might have closer relationship with *Rhizomys*, but only referred it to the superfamily Muroidea without familiar affiliation. Since then, *Pararhizomys* has not been studied again until 1968, about 37 years later, when Kowalski described an incomplete skull with mandible from Altan Teli of Mongolia as *Pa. hipparionum* and referred it to the family Rhizomyidae. Since then until 2005, for another 37 years, news about findings of *Pararhizomys* only sporadically appeared in literatures. Since 2005, Zhang et al. (2005) described two *Pararhizomys* species from Shaanxi and Gansu provinces: *Pa. hipparionum*, based on partial maxilla and left mandibular ramus, and *Pa. qinensis* sp. nov., based on an anterior part of skull and two isolated molars. Then, Qiu and Li (2008) reported *Pararhizomys* sp., based on a right m2 from Shengou, Qaidam Basin of Qinghai Province. Recently, Li (2010) reported *Pa. hipparionum*, based on a partial left hemimandible, some isolated teeth and several postcranial bones from Baogeda Ula of Nei Mongol Autonomous Region. Thus, up to now only two species of *Pararhizomys* have been reported based on comparatively poor material.

Systematic position of *Pararhizomys* has long been problematic. It was referred either uncritically to Muroidea (Teilhard de Chardin and Young, 1031), or to Rhizomyidae (Simpson, 1945; Young et Liu, 1950; Kowalski, 1968), Rhizomyinae of the Muridae (McKenna et Bell, 1977), or to Spalacidae (Flynn, 1982, 1990). Although trying hard to untangle this complex problem, Zhang et al. (2005) and Li (2010) failed to solve the problem, mainly because of paucity of the available specimens, and they were only able to refer *Pararhizomys* to family incertae sedis in Muroidea.

Recently, a fairly good number of specimens belonging to the *Pararhizomys*-group, including skulls (some with mandibles) and partial postcranial skeletons, were collected from Hezheng County of the Linxia Basin, Gansu Province. Most of them are provided with provenance information. As described in the systematic description chapter, this material contains seven species of two genera. This is the best material of the *Pararhizomys*-group so far known. The Hezheng material does not only show much more anatomical characters of skulls, allowing us to apply cladistic analysis, but also show more complete skeletal morphological features, enabling us to carry out some functiono-behavioral analysis of the *Pararhizomys*-group animals.

2. Materials

1) Fossil specimens studied

The specimens of the pararhizomyine fossils from the Linxia Basin including skulls, mandibles and postcranial bones belong to 22 individuals. Their contents are systematically enumerated in the systematic description chapter (Tabs. 2 and 10). They are collected from 15 localities of Hezheng, Guanghe and Dongxiang counties (Fig. 1 and Tab. 1).

2) Specimens used for comparison purpose

A. Specimens (mainly skulls with mandibles and postcranial bones) of extant rodents kept in IVPP: *Rhizomys sinensis*, *R. pruinosus*, *Rattus norvegicus*, *Apodemus agrarius*, *Myospalax myospalax*, *Ondatra zibethicus*, *Microtus fortis*, *M. clackei*, *M. mongolicus*, *Tscheskia triton*, *Cynomys ludovicianus*, *Marmota monax*, *M. himalayana*, *Tamias sibricus*, *Erethizon dorsatum*, *Thomomys bottae*, *Geomys bursarius*, *Myocastor coypus*, *Meriones meridianus*, and *Allactaga sibirica*.

B. Specimens on loan from personal collections:

Undescribed fossil skulls of *Tachyoryctoides* from Prof. Qiu Zhuding, IVPP;

Skeletons of extant *Aplodontia rufa* and *Eospalax fontanierii* from Prof. Ni Xijun, IVPP;

Skull of extant *Castor* from Prof. Li Chuankui, IVPP;

Skull of extant *Spalax* from Prof. Zhang Zhaoqun, IVPP;

Skeleton of extant *Homo sapiens* from Prof. Wu Xiujie, IVPP;

Casts of skull with mandible of *Pa. hipparionum* from Mongolia, donated by Dr. L. Lindsay, University of Arizona, USA.

3. Methodology

1) Subdivision of skeletal elements and anatomical terms

Hildebrand and Goslow's system of subdivision of skeletal elements (2001) is adopted here. There are two major divisions: head skeleton (= cranial part) and postcranial skeleton (= postcranial part). From the point of view of evolution and ontogeny, the head skeleton of mammals, or the skull, consists of cranium and splanchnocranium. The latter is composed of the face bones including maxillae and mandible, and the hyoid bones. Since in mammals the maxillae are always welded together with the other facial bones, but the mandible and the hyoid bones invariably remain separate, in practical use, only the maxillae are included in the skull (*sensu stricto*). As noted by Hildebrand and Goslow (2001: 115), the skull is used "for the single unit that forms the braincase and upper jaw and houses the nose and ear." Hildebrand and Goslow admitted that "The word is useful, but inexact." The mandible is composed of two halves, often called hemimandibles by some paleontologists. In turn, each hemimandible is composed of horizontal ramus and ascending ramus (note the meaning of the ramus here is different from that used in human anatomy). The postcranial skeleton consists of vertebral column, ribs, sternum and limb bones.

The anatomical terms of the skull and postcranial bones used in this monograph is mainly taken from Greene (1935), Yang et al. (1985), with references to Cooper and Schiller (1975), Hebel and Stromberg (1976), Wang et al. (2009), Bugge (1970, 1971, 1974, 1985), and Wahlert (1974), in cross-reference with the "Illustrated Veterinary Anatomical Nomenclature" (Constantinescu et Schaller, 2012).

2) Positions and functional implications of foramina in skull

(1) Cephalic arterial systems and blood-supply patterns in rodents

The terms used by Greene (1935) for the cephalic arterial system and its related foramina were primarily taken from Emmel's Basle Anatomical Nomenclature (B. A. N., 1927). Some of them are different from those commonly used later. Fig. 2 shows the terms used by Greene (1935), while Fig. 3 shows those of Bugge commonly used since the 1970s–1980s.

Towards the end of the 1970s knowledge of the cephalic arterial systems in rodents had been greatly advanced. The advancement had been almost uniquely made by Danish anatomist J. Bugge in his six original papers published in 1970 to1974. Bugge dissected 478 skulls of 78 extant rodent species. He established the primitive basic pattern of the cephalic arterial system in rodents, as "comprising the internal-external carotid artery system and the stapedial artery with its supraorbital, infraorbital and mandibular branches." The modifications then occur in different rodent groups as a result of the obliteration of certain parts of the arteries and emerging of new anastomoses at various positions and in various combinations.

(2) Terms and positions of the cranial foramina

Based on previous works (Hill, 1935, Guthrie, 1963, 1969 and Bugge, 1970–1974), Wahlert (1974: 363) made an explicit statement that "Presence or absence, relative position, number, and relative size of foramina are useful characters in determining relationships." Using protrogomorphous rodents as representative, Wahlert listed 44 foramina, each with an explanation of its position, size and form, and function. Later, Voss (1988) and Musser and Heaney (1992) described these foramina in more detail based on Ichthyomyini materials. Of Wahlert's 44 foramina 26 are well applicable to our fossils, and are chosen in the present monograph, with some amendments as to their explanations (in detail in Chinese text). We added 4 new terms of foramina, which are shown in our fossil specimens, but cannot be clearly recognized in protrogomorphous rodents.

The abbreviations of the terms of the cranial foramina used in this monograph are listed as follows (* new terms of foramina proposed here; numbers in () are those of Wahlert):

1 (2). inf — incisive foramen;

2 (5). iof — infraorbital foramen;

3 (8). nlf — nasolacrimal foramen;

*4. psaf — posterior superior alveolar foramen, situated on the posterior half of the floor of the infraorbital canal;

5 (11). spf — sphenopalatine foramen;

6 (16). dpf — dorsal palatine foramen;

7 (12). etf — ethmoidal foramen;

8 (14). opf — optic foramen;

9. aafi — anterior alar fissure, created by Wahlert (1983: 2), a substitute for Hill's sphenoidal fissure (Hill, 1935: 124);

9a (19). sf — sphenoidal fissure (*sensu stricto*);

9b (20). fr — *foramen rotundum*;

10. fap— *foramen alare parvum*, a small foramen situated at the lower end of the external wall of the anterior alar fissure, nearby the *foramen rotundum* and the anterior opening of the alisphenoid canal;

11 (3). ppf — posterior palatine foramen;

12 (4). pmf — posterior maxillary foramen or notch;

13 (23). mf — masticatory foramen;

14 (23p). mfp — posterior foramen of masseteric nerve canal;

15 (24). bf — buccinator foramen;

16 (24p). bfp — posterior foramen of buccinator nerve canal;

*17. dtf — deep temporal foramen, situated above and posterior to the masticatory foramen, serving as the exit of the deep temporal nerve;

*18. dtfp — posterior foramen of deep temporal nerve canal;

19 (26). fo — *foramen ovale*;

20 (28). mlf — middle lacerate foramen;

21. petf — petrotympanic fissure, located between the squamosal and the auditory bulla;

*22. pcpf — posterior foramen of pterygoid canal;

23 (21). asc — alisphenoid canal, and ascp – posterior foramen of alisphenoid canal;

24 (29). eucf — foramen of Eustachian canal;

25 (30). icf — internal carotid foramen;

26 (32). juf — jugular foramen;

27 (40). chuf — foramen of canal of Huguier;

28 (33). hyf — hypoglossal foramen;

29 (34). pgf — postglenoid foramen;

30 (37). styf — stylomastoid foramen;

31 (38). msf — mastoid foramen.

(3) Terms used for dentition

The hypsodonty of the molars in *Pararhizomys* is of a particular type: the mesial side of the crown is higher than its distal side in the upper molars, and *vice versa* in the lower ones. This kind of hypsodonty is here called mesial and distal hypsodonty for upper and lower molars respectively, or mesiodistal hypsodonty in abbreviated form. In higher crowned *Pararhizomys* species the crowns on lingual sides of both upper and lower molars are higher than those on the buccal sides. This kind of hypsodonty, named here as lingual hypsodonty, is essentially different from "unilateral hypsodonty," where the lingual crown is higher than the buccal crown in the upper molars but *vice versa* in the lower molars (= linguobuccal hypsodonty).

The terms of the crown elements are shown in Fig. 4.

In order to know the enamel microstructure of incisors in the pararhizomyines, the lower incisors of two species of two genera (*Pa. hipparionum* and *Ps. gansuensis* gen. et sp. nov.) are chosen for microstructure investigation. The incisors were cut from the mandibles. They are imbedded in artificial resin (Tech 7200) for about 12 hours, then sectioned transversely and longitudinally with EXAKT 300CP. After abrading and polishing, they were etched for about 60 seconds with 0.1 mol/L phosphoric acid. Having been rinsed and dried, the specimens were coated with gold and examined with a scanning electronic microscope (S-3700N).

3) Methods of measurements

(1) Cranial measurements

The system of the cranial measurements is mainly taken from Pan et al. (2007: 4–8, fig. 3), with minor amendments by the present authors. The measurements used in this monograph are as follows:

1. Condylonasal length (CNL)：Anterior end of nasal–posterior border of occipital condyle.

2. Condylobasal length (CBL)：Anterior end of premaxilla–posterior border of occipital condyle.

3. Length of anterior part of skull (LAS)：Anterior end of premaxilla–anterior border of M1 alveolus.

4. Length of posterior part of skull (LPS)：Anterior border of M1 alveolus–posterior border of occipital condyle.

5. Palatal length (PL)：Anterior end of premaxilla–posterior border of hard palate.

6. Length of maxillary diastema (LMXD): Posterior border of I2 alveolus–anterior border of M1 alveolus.

7. Length of palatal bridge (LPB): Posterior border of incisive foramen–posterior border of hard palate.

8. Anterior width of palatal bridge (AWPB): Minimal distance between left and right M1 alveoli.

9. Posterior width of palatal bridge (PWPB): Minimal distance between left and right M3 alveoli.

10. Length of incisive foramen (LIF): Maximal length of incisive foramen.

11. Rostrum length (RL): Anterior end of premaxilla–anterior border of zygomatic arch at premaxillo-maxillary suture.

12. Anterior width of rostrum (AWR): Maximal width of anterior part of rostrum.

13. Posterior width of rostrum (PWR): Distance between anterior borders of left and right zygomatic arches measured at premaxillo-maxillary sutures.

14. Rostrum height (RH): Minimal distance from top of nasal to lower border of premaxilla measured immediately behind I2.

15. Height of middle part of skull (HMS): Height of skull measured immediately before M1.

16. Nasal length (NL): Maximal length of nasal.

17. Anterior width of nasals (AWN): Maximal width of anterior parts of left and right nasals.

18. Posterior width of nasals (PWN): Width of posterior parts of left and right nasals measured at lateral ends of naso-frontal sutures.

19. Interorbital width (IOW): Distance between lateral sides of left and right lacrimal tubercles.

20. Width of postorbital constriction (WPOC): Minimal width of postorbital constriction.

21. Zygomatic width (ZW): Maximal distance of lateral borders of left and right zygomatic arches.

22. Width of cranial part at mastoid processes (WCM): Maximal width between lateral borders of left and right mastoid processes.

23. Length of auditory bulla (LAB): Maximal length of auditory bulla excluding anterior spine.

24. Width of auditory bulla (WAB): Maximal transverse width of auditory bulla.

25. Distance between two auditory bullae (DAB): Minimal distance between left and right auditory bullae.

26. Nuchal height (NH): Vertical distance from inion to lower border of occipital condyle.

27. Height of *foramen magnum* (HFM): Maximal vertical distance of *foramen magnum*.

28. Width of *foramen magnum* (WFM): Maximal transverse width of *foramen magnum*.

29. Mandibular length (MDL): Anterior end of mandible excluding incisor–posterior end of angular process.

30. Mandibular height (MDH): Vertical distance from top of coronoid process to lower border of ascending ramus.

31. Height of condyloid process of mandible (HCPM): Vertical distance from top surface of condyle to lower border of ascending ramus.

32. Depth of mandibular notch (DMN): Vertical distance from top of coronoid process to bottom of mandibular notch.

33. Length of mandibular diastema (LMD): Posterior border of i2 alveolus–anterior border of m1 alveolus.

34. Height of mandiblar diastema (HMD): Minimal height of mandible at diastema.

35. Height of horizontal ramus of mandible (HHR): Height of horizontal ramus measured between m1 and m2 on lateral side.

(2) Measurements of teeth

In his paper on *Pararhizomys*, Li Qiang set the "index H" to represent the relative height of the molar crown (see Li, 2010: 50–51, Fig. 3). In the Chinese abstract he stated that "the larger is the index H, the lower is the crown" (Li, 2010: 48). According to Li (2010, 55–57, Fig. 8), the deposits yielding *Pararhizomys hipparionum* in Baogeda Ula is younger than those yielding V 14178 in Lamagou and the holotype in age. Accordingly, the "index H" of the former should be smaller than those of the latter two. Unfortunately, this cannot be borne out by his actual measurements. For example, in Li's Table 4 the LNH (representing Li's "index H" on lingual side of the molar) of M2 and m2 of V 16306 are 1.7 mm and 0. 90–0.95 mm respectively, while the LNH of M2 of V 14178 is 1.15 mm and that of the holotype m2 is 0.8 mm. The formers (1.7 mm and 0.90–0.95 mm) are larger than the latters (1.15 mm and 0.8 mm). Apparently, Li's "index H" represents only the minimal distance between the bottom of the reentrants and the dental tract (= boundary between crown and root of molar), but cannot adequately express the variation of the crown height in this particular case of mesiodistal hypsodonty in *Pararhizomys*. Besides, Li used the lateral and medial border lines of the occlusal surface as his datum lines. Since the occlusal surface becomes more and more oblique relative to the crown axis in mesiodistally hypsodont teeth in pararhizomyines with wearing, Li's measurements based on the datum lines of the occlusal surface would become more and more inaccurate in reflecting the crown height.

The method of measurements applied by Zheng to *Rhizomys* (Zheng, 1993: 102, Fig. 49) seems to be more appropriate in expressing hypsodonty. In the present monograph the method of measuring the molars are mainly taken from Zheng (1993), partly amended by the present authors (see Fig. 5). The main measurements are explained in the following lines:

1. Molar length (= mesiodistal length, L)：Maximal antero–posterior distance measured perpendicular to the posterior wall in upper molar/anterior wall in lower molar.
2. Molar width (W)：Maximal width of molar crown (vertical to the mesiodistal axis of molar).
3. Length of reentrant of molar on occlusal surface (LRM)：Transverse extension of reentrant on occlusal surface.
4. Width of reentrant of molar on occlusal surface (WRM)：longitudinal extension of reentrant on occlusal surface.
5. Depth of reentrant of molar on lateral sides (DRM)：Extension toward roots of reentrant on lateral sides.
6. Hypsodonty proxy (Hp)：Longest distance from bottom of reentrant to the lowest point of the dentine tract (= DT, Repenning, 2003) on sides, parallel to the posterior side on upper molar/anterior side on lower molar (see Fig. 5).
6a. Hypsodonty proxy on buccal side (Hpb), including Hypsodonty proxy on anterior buccal reentrant (Hpb1) and Hypsodonty proxy on posterior buccal reentrant (Hpb2).
6b. Hypsodonty proxy on lingual reentrant (Hpl).

(3) General terms used in postcranial bones

The general anatomical and topographic terms of the postcranial bones and their measurements used in this monograph are mainly taken from Qiu and Wang (2007). The only difference is that the axes of the manus and pes are defined in a slightly different way. In the present monograph the median plane of the vertebral column is considered as the directional reference for all the limb bones, the manus and pes included. Thus, for all the manus and pes bones the side closer to the median plane of the body, rather than the axes of the manus and pes themselves, is defined as the medial side and the opposite side the lateral.

Some commonly used abbreviations in measurements are: L=length, W=width, H=height, D=distance, APD=anteroposterior distance, PE=proximal end, DE=distal end, Max=maximum, Min=minimum; Ⓛ, left; Ⓡ, right; "+" indicates slightly larger and "~" indicates *circa*.

All the measurements of skull and postcranial bones are taken with a vernier caliper to the nearest 0.05 mm. The teeth are measured with a Wild Heerbrugg microscope to the nearest 0.01 mm.

Abbreviations   HMV: prefix of catalogue numbers of vertebrate fossils of the Hezheng Paleozoological Museum, Gansu Province; IVPP Loc. LX (abbreviated as LX in text): prefix of field localities of the Linxia Basin of IVPP; IVPP V (abbreviated as V in text): prefix of catalogue numbers of vertebrate fossils of IVPP; OV: prefix of catalogue numbers of the Osteological Division of Vertebrates in IVPP.

II. SYSTEMATIC DESCRIPTION OF PARARHIZOMYINES OF LINXIA BASIN

Superfamily Muroidea Illiger, 1811
Family Spalacidae Gray, 1821
Subfamily Tachyoryctoidinae Schaub, 1958
Tribe Pararhizomyini tribe nov.

Type genus   *Pararhizomys* Teilhard de Chardin et Young, 1931.

Other included genus   *Pseudorhizomys* gen. nov.

Geographic distribution and geologic age   Middle north latitudinal region of East Asia, including Nei Mongol, Shaanxi, Gansu, Qinghai and Xizang of China, and Mongolia; Late Miocene–Early Pliocene.

Diagnosis   Mid-large sized muroids. Skull is low, with myomorphous zygomasseteric structure. Facial and cranial parts are subequal in length, with a narrow and long rostrum, but a wider cranial part. Zygomatic plate extends anterosuperiorly and widens laterally, with its lower surface facing anteroinferiorly and slightly externally. The attachment area of lateral masseter muscle is confined to maxilla and bordered anteriorly by a distinctly curved ridge. Posterior end of zygomatic arch is separated from nuchal crest by a notch. Maxillary diastema is long owing to the backward shifting of the upper molars. Narrow and pointed posterior ends of nasals are located roughly at the same transverse line as those of premaxillae. Premaxilla possesses laterodorsal crest. Premaxillo-maxillary suture intersects incisive foramen, which is small, about 1/4–1/3 the length of the diastema. Infraorbital foramen is large in form, with a distinct ventral slit. Naso-lacrimal fossa is situated within the infraorbital canal. Orbit is small. Parapterygoid fossa is present on posterior part of hard palate. Pterygoid fossa is narrow and shallow. Both masticatory and buccinator nerve canals are short. Glenoid fossa is very long, extending to the nuchal crest, and adjoined by the auditory bulla at its posterointernal corner. Embryonic external auditory meatus is pushed downwards. *Foramen ovale* is confluent with middle lacerate foramen. No stapedial foramen is seen. Sagittal and nuchal crests are well developed. No interparietal bone is observed. Occipital condyles are short and wide, situated close to each other on ventral side. Nuchal surface slightly slants forward toward its top. Petromastoid portion is relatively large, but the paramastoid process is very small.

Mandible is sciurognathous, with short and thick horizontal ramus. Diastema is longer than or subequal to the lower cheek tooth row in length. Masseteric fossa extends anteriorly below m1/m2, with well-developed and flaring masseteric ridge. Mental foramen is located below anterior margin of m1. Distinct temporal fovea extends to the base of the medial side of the coronoid process. Condyloid process extends posterosuperiorly. Posterior end of lower incisive alveolus forms a prominent bulge on buccal surface of ascending ramus. Angular process is well-developed and pointing backwards, and posterior border of ascending ramus is concave. Internal pterygoid fovea is large and deep.

Dental formula is 1·0·0·3/1·0·0·3. Molars are mesiodistally hypsodont and/or subhypsodont with roots, lophodont, without mesoloph(id)s: upper molars without posterosinus; M1 and m1–2 with two buccal reentrants and three transverse loph(id)s, M2–3 and m3 with 1 or 2 buccal reentrant(s), upper molars with one lingual reentrant and lower molars with 1 or 2 lingual reentrant(s), each reentrant extending to the middle on occlusal surface; M3/m3 tend to be reduced. Incisors are strong and enamels covering on mesial and lateral sides are very narrow; i2 has one or two longitudinal ridges on labial surface.

Pararhizomys Teilhard de Chardin et Young, 1931

Type species *Pararhizomys hipparionum* Teilhard de Chardin et Young, 1931.

Other included species *Pararhizomys qinensis* Zhang, Flynn et Qiu, 2005 (Lantian, Shaanxi), *Pa. huaxiaensis* sp. nov. and *Pa. longensis* sp. nov. In addition, there are two indeterminate species: *Pararhizomys* sp. I (Qaidam, Qinghai) and *Pararhizomys* sp. II (Gaotege, Nei Mongol).

Geographic distribution and geologic age Shaanxi, Gansu, Qinghai and Nei Mongol of China, and Western Mongolia; Late Miocene–Early Pliocene.

Emended diagnosis Anterior ends of premaxillae and nasals align in about the same vertical line. Dorsal surface of nasal is convex longitudinally and transversely. Ventral part of premaxillo-maxillary suture extends anterolaterally from the posterior part of the incisive foramen. Premaxilla bears a remarkable and long laterodorsal crest, which is subparallel with the anterior part of naso-premaxillary suture, rendering the dorsal surface of the premaxilla band-formed. Dorsal part of premaxillo-maxillary suture is straight, extending posteriorly to meet the anteriormost point of the maxillo-frontal suture, forming an obtuse angle. Infraorbital foramen is wide in shape. Anterior root of zygomatic arch and the zygomatic plate are located more posteriorly: the anterior border of orbit roughly aligns with the posterior ends of nasals and premaxillae transversely, and the anteriormost point of the emarginated posterior border of zygomatic plate is located at the level posterior to the incisive foramen. Anterior border of attachment area of the lateral masseter muscle forms a curved ridge. Maximal width of skull is across the middle parts of the strongly curved zygomatic arches. The jugal bone is very short, with its anterior end lying far away from the anterior border of orbit. Parietal bones form an irregular rectangle, longer than wide, with an anterior central spine inserted between the two frontals. Two pairs of posterior palatine foramina are located in deep palatine sulci. Parapterygoid fossa is large and deep. Posterior border of the hard palate is located posterior to M3. Mesopterygoid fossa is much wider than pterygoid fossa. The paired internal pterygoid processes converge posteriorly and slant laterally. Deep temporal foramen is confluent with the masticatory foramen, but separated from the buccinator one. The posterior apertures of the three above mentioned foramina are also confluent. Petromastoid portion appears relatively large on nuchal surface. Posterior part of the mandibular diastema is deeply concave. Lower border of the horizontal ramus is rather straight. Mental foramen is located lower than the anterior end of masseteric fossa. Mandible notch shallow, less than 1/3 of mandibular height. I2 is orthodont, with its anterior end turning slightly backwards in direction, and has one prominent longitudinal labial ridge. The posterior end of the I2 originates at the level of the posterior end of the incisive foramen. The i2 has one or two longitudinal labial ridges. Molars are mesiodistally hypsodont, each with one lingual reentrant. Each of M2–3 and m3 is with only one buccal reentrant, without anterosinus(id). The m3 is subquadrate, with transverse and directly opposite mesosinusid and sinusid.

Pararhizomys hipparionum Teilhard de Chardin et Young, 1931
(Figs. 6–15, 75; Tabs.1–5, 15, 16)

Holotype V 412 [former No. c/90 of C. L. G. S. C. (Cenozoic Laboratory, Geological Survey of China)], partial horizontal ramus of left hemimandible with m1–3, from Loc. 10 of Xinminzhen (formerly Chen-chiang-pao), Fugu County, Shaanxi Province; "Pontian Red Clay," now considered Baodean ALMA.

Described specimens 1) V 16286.1, skull with mandible, V 16286.2, partial left radius and V 16286.3, right femur, all from LX 200047 (Qiaojiacun); 2) HMV 1923, incomplete skull, from LX 200207 (Lima); 3) V 16287, incomplete skull, from Linxia Basin (locality and horizon unknown), Gansu Province (for details see Tabs. 1, 2).

Locality and horizon (China) Shaanxi: Xinminzhen and Lamagou of Fugu, Baode Fm.; Gansu: Qin'an, "Red Clay"; Qiaojiacun (Guanghe) and Lima (Hezheng), lower part of middle Liushu Fm.; Nei Mongol: Baogeda Ula (Abag Banner), Baogeda Ula Fm.; Upper Miocene, medial Bahean–Baodean ALMA/S.

Diagnosis Premaxilla has a shallow groove on upper part of its lateral surface. Anteriormost point of the emarginated posterior border of zygomatic plate is situated anterior to M1. Mandibular diastema is subequal to the lower cheek tooth row in length. Molars are mesiodistally hypsodont. Buccal and lingual reentrants are subequally deep on M1–2 and m1–3 and mesosinus deeper than sinus on M3. On occlusal surface, sinus of M3 usually extends anterobuccally, slightly overlapping mesosinus. Mesosinusid of m1 bends strongly anteriorly, with its buccal part extending anteriorly beyond the protosinusid.

Differential diagnosis *Pa. hipparionum* differs from *Pa. qinensis* in having longer maxillary diastema and longer M2.

Comments The correctness of the species name *hipparionum* is to be further verified. Whether it is the genitive case of plural form, or an adjective derived from *Hipparion* is difficult to decide. In the first case, *hipparionum* should be retained, while in the latter case, it should be changed into *hipparionus* in accordance with the Code's rule of agreement in gender.

1) Description

(1) **Skull** (Figs. 6–9; Tab. 3)

Large-sized muroid rodent, with myomorphous zygomasseteric structure. Like *Rhizomys* and *Spalax* the skull has a wide cranial part and anteriorly slanting nuchal surface. Different from that of *Rhizomys*, the skull is lower and narrower; with relatively narrower and longer rostrum, longer maxillary diastema, shorter cranial part, and having a distinct sagittal crest. Unlike *Spalax*, the degree of slanting of the nuchal surface is lesser, and the parietal bone is longer than wide in the described skulls.

Dorsal view (Figs. 6 A, 8 C) The rostrum is narrow and long, with nearly parallel lateral sides. The nasal (N) is wedge-shaped, forming a pointed posterior end. The naso-premaxillary sutures are mostly straight and convergent posteriorly. The surface of the nasal is convex longitudinally and transversely. The transverse convexity is more prominent on the anterior part, becoming less so or flat on the posterior part. The posterior end of the nasal extends close to the posteriormost part of the strongly zigzag premaxillo-frontal suture, which extends anterolaterally. The dorsal part of the premaxilla (Pm) forms a long band, bordered by a distinct laterodorsal crest (pmdlc, see Figs. 6 A, 7 A, and 8 A), which extends from the anterosuperior corner of the infraorbital foramen (iof) to the anterolateral border of the upper incisor alveolus, almost parallel to the naso-premaxillary suture on its anterior part. The dorsal part of the premaxillo-maxillary suture extends roughly longitudinally. It links with the maxillo-frontal suture, which goes posterolaterally to the lacrimal (L). The lacrimal is probably mainly located in the orbit and a hardly noticeable lacrimal tubercle appears on the dorsal surface of the skull. In V 16287 the left infraorbital foramen (iof) is better preserved and clearly shown in Fig. 6 A and 6 B. It is within the maxillary bone, partly visible from dorsal side, but more clearly seen from the anterior view (Fig. 6 B). Its upper part is a large oval opening, with its long axis stretching in inferolateral direction. At its inferointernal corner there is a distinct nasolacrimal fossa (nlf), separated from the infraorbital canal by a bony septum. Below the nasolacrimal fossa is the distinct ventral slit of the infraorbital foramen. The zygomatic arches are strongly expanded, making the middle part of the skull round in shape. The posterior border of the upper branch of the anterior root of the zygomatic arch is slightly concave, with its anteriormost point aligned at the same transverse line with the naso-frontal suture (Fig. 8 C). The middle part of the zygomatic arch is triangular in cross-section, with sharply crested upper border; flat and wide lateral surface, facing superolaterally; wide medial surface and narrow bottom. The posterior end of the zygomatic arch does not reach to the nuchal crest (nc), and separated from the latter by a notch. The jugal bone (Fig. 8 A, J) is short, and its anterior end does not extend to the anterior border of the orbit. The postorbital constriction is prominent. The weak frontal crest (fc) originates from the lacrimal tubercle. The left and right frontal crests converge posteriorly and unite at the level anterior to the postorbital constriction to form a distinct sagittal crest (sc). The parietal bone (P) is roughly rectangular in outline, longer than wide, with an anterior middle spine inserting in the frontals. The parietal has a smooth surface, slightly concave transversely near the sagittal crest. The temporal fossa is large. The temporal crest (tc) is short and forms a semicircular concavity. Its posterior part becomes gentle. No interparietal bone is visible.

Lateral view (Figs. 7 A, 8 A)　The dorsal profile is slightly wavy, with the middle part above the postorbital contribution slightly concave, and the parts anterior and posterior to that constriction weakly convex. The anterior ends of the nasal and premaxilla are almost at the same vertical line (Fig. 8 A). The maxillary diastema is much longer than the upper cheek tooth row. The upper cheek tooth row shifts posteriorly so that the posterior part of the skull is subequal to the anterior part of the skull in length (see Tab. 3). The lateral surface of the premaxilla is broad. On its upper part there is a gentle longitudinal bulge, showing the relief of the I2, which passes through the medial wall of the infraorbital foramen. Between the dorsolateral crest and the bulge is a very shallow longitudinal groove. The premaxillo-maxillary suture runs roughly vertically in front of the infraorbital foramen. The insertion of lateral masseter muscle is confined within maxilla. The anterior rim of this muscle insertion forms a distinct and anteriorly convex ridge (araml). The orbit shifts more anteriorly relative to the upper cheek tooth row. The sphenopalatine foramen (spf) is situated above the M1. The dorsal palatine foramen (dpf) is situated posteroinferior to spf and above the anterior part of M2. The ethmoidal foramen (etf) is better preserved on the left side in V 16287. It is small, situated anterosuperior to the optic foramen. The optic foramen (opf) is small and located above the M2, posterosuperior to the sphenopalatine foramen. The anterior alar fissure (aafi) is large and located posteroinferior to the optic foramen and above the M2 and M3. On the right side of V 16287 it is clearly seen that the anterior alar fissure is subdivided in two parts: the upper part should be the sphenoidal fissure (*sensu stricto*, sf) and lower one the *foramen rotundum* (fr). The masticatory foramen (mf) is separated from the buccinator foramen (bf) and is located dorsal to the latter. From the masticatory foramen a distinct furrow extends to the temporal area passing by the anterior border of the posterior root of the zygomatic arch, which may serve as the passage of the deep temporal nerve. It seems that the deep temporal foramen (dtf) is confluent with the masticatory foramen in *Pa. hipparionum*. The posterior foramen of the pterygoid canal (pcpf) is reduced and located in the posteromedial side of the pterygoid fossa, near the base of the internal pterygoid process. The glenoid fossa (glf) is very long, extending posteriorly to the nuchal crest, pushing the external auditory meatus (eam) downwards. The anterior part of the glenoid fossa is flat or slightly convex and formed by the squamosal bone only. Its posterior part is large and transversely deeply concave and composed of squamosal bone and lateral side of the auditory bulla. The postglenoid foramen (pgf) is fissure-like and located in the posterior part of the glenoid fossa (glf), extending along the squamoso-tympanic suture. The petrotympanic fissure (petf) is located anterolateral to the auditory bulla and its posterolateral end meets the postglenoid foramen (pgf). The mastoid process (msp) is very small. The stylomastoid foramen (styf) is squeezed into the lower part of the area between the external auditory meatus (eam) and the mastoid process. The nuchal surface slightly slants forwards. The occipital condyle (occ) is located behind the nuchal crest and bulla, and its lower border is higher than the lower border of the bulla.

Ventral view (Figs. 7 B, 8 B, 9 A) No interpremaxillary foramen is seen. The incisive foramen (inf) is located against the middle of the maxillary diastema and is about 1/4–1/3 length of the diastema (see Tab. 3). The weakly zigzag premaxillo-maxillary suture intersects the incisive foramen at around the middle of the posterior half. The zygomatic plate is broad and concave, extending anterodorsally and facing lateroventrally. Its anterior border is bounded by a well-developed curved rim. The anteriormost point of the insertion of the lateral masseteric muscle is nearly opposite the anterior end of the incisive foramen in position. The anteriormost point of the emarginated posterior border of the zygomatic plate is located at the level between the incisive foramen and the M1. The masseteric tubercle (mt) is located on the anterior part of the zygomatic plate, which may provide insertion for the superficial masseter muscle. The latter is located in about the same transverse line as the posterior end of incisive foramen, closer to incisive foramen than to infraorbital foramen. The left and right cheek tooth rows are parallel to each other or slightly converge anteriorly. Along the intermaxillary and interpalatine sutures a sagittal ridge extends from the incisive foramen to the posterior border of the palate, forming a posterior spine, making the posterior border of the hard palate roughly omega-shaped. On both sides of the sagittal ridge the two palatine sulci (ps) deepen near the M1–M2, and widen posteriorly forming wider and shallower concavities. Two pairs of posterior palatal foramina (ppf) are present in the palatine sulci. The anterior pair lies medial to the posterior part of M1 within the maxilla, while the posterior pair lies medial to the anterior part of M2 within the palatine (Pl). The parapterygoid fossa (pptf) forms a triangular depression on the posterior part of the hard palate. There are also some minor nutrient foramina in the parapterygoid fossa. The posterior maxillary foramen (pmf) is located at the posterolateral corner of the parapterygoid fossa, and almost in the same transverse line as the posterior border of the hard palate, behind the M3. The mesopterygoid fossa (mptf) is relatively wide, much wider than the pterygoid fossa (ptf). The large and high internal pterygoid process (ipp) slants laterally, and extends posteromedially. The external pterygoid process (epp) is short and low and extends slightly posterolaterally. The posterior aperture of the masseteric nerve canal (mfp) is confluent with those of the deep temporal nerve canal and the buccinator nerve canal (dtfp and bfp). The posterior foramen of the alisphenoid canal (ascp) is located anteroventral to the *foramen ovale* (fo) and medial to the confluent posterior apertures of the deep temporal nerve, masseteric nerve and buccinator nerve canals. The *foramen ovale* is confluent with the middle lacerate foramen (mlf).

Only V 16287 preserves the basisphenoid and basioccipital bones, which are slightly skewed. Based on the preserved part, the basisphenoid (Bs) and basioccipital (Bo) form a trapezoid, narrowing anteriorly. The basilar tubercle (bt) is faint. A sagittal ridge and a pair of foveae can be seen on the basioccipital. The hypoglossal foramen (hyf) is located anterolateral to the occipital condyle. The auditory bulla (ab) is moderately inflated, oval in outline, with its major axis extending anteromedially and a spine in the same direction. The foramen of canal of Huguier (chuf) is situated anteromedial to the external auditory meatus. The foramen of Eustachian canal (eucf) is located anterior to the spine of the bulla. No bony septum is seen in the broken bulla. The internal carotid foramen (icf) is located at the anterior 1/3 of medial border of the bulla. The jugular foramen (juf) is narrow. The two foramina mentioned above are connected by a groove on the medial margin of the bulla. On the medial side of the bulla near the jugular foramen no distinct foramen for stapedial branch is visible. The occipital condyle is wide and short in ventral side, and the two condyles are close to each other.

Posterior view (Fig. 6 C) The nuchal surface is about semicircular in outline. The arched nuchal crest (nc) slightly protrudes at the inion, where the sagittal crest meets the nuchal crest. The nuchal surface is roughly flat, without distinct external nuchal crest. The petromastoid portion (Pms) exposed on the nuchal surface forms an irregular lozenge, occupying the lateroventral part of the latter. It is relatively large, with its upper part being slightly wider than the lateral part of the occipital bone lateral to the *foramen magnum*. The large mastoid foramen (msf) is situated slightly higher than the upper border of the *foramen magnum* (fm). The *foramen magnum* is large, about 1/3 the nuchal height and roughly circular in shape, higher than wide. The posterior surface of the occipital condyle (occ) is relatively narrow.

(2) Mandible (Fig. 9 B, C; Tab. 3)

Only V 16286.1 skull is preserved with its mandible. The left and right hemimandibles are united at the symphysis. The mandible is sciurognathous. The horizontal ramus is short and thick, with a rather straight lower border below the molars and a vertically concave lingual surface. The symphysis extends forwards and upwards, forming an angle of about 40° with the lower border of the horizontal ramus. The chin process (cp) forms a distinct crest along the lower border. On the medial side of the crest there is a long ovoid concavity, which may provide insertions for *m. digastricus*, *m. transverse mandibularis* and *m. geniohyoideus*, etc. Here we call it digastric fovea (df). The mandibular diastema is subequal to the lower cheek tooth row in length. Viewed laterally, the diastema is roughly S-shaped, forming a deep concavity in front of m1. The masseteric fossa (msf) extends anteriorly below m1/m2, with its masseteric ridge (mr) prominently flaring. The mental foramen (mtf) is located below the anterior margin of m1, at the level near the lower 1/3 of the horizontal ramus, hence lower than the anterior end of the masseteric fossa.

The ascending ramus of the mandible is rather long. The coronoid process (crp) curves slightly posteriorly at its upper part. Its anterior border rises from near the anterior part of m2, forming an angle of about 105°– 110° with the alveolar border of the horizontal ramus. On the lower part of the medial surface of the coronoid process there is a distinct concavity, which may be the attachment area of the temporal muscle, and is here called temporal fovea (tf). The mandibular foramen (mdf) is small and located below the posterior margin of the coronoid process and posterolateral to m3, at about the same level as the alveolar border of molars. The depth of mandibular notch is less than 1/3 height of mandible. The condyloid process (cdp) is situated slightly lower than the coronoid process (crp). It stretches dorsoposteriorly and slightly bends lingually. On the medial side of the condyloid process the attachment area of the external pterygoid muscle is not distinct. The articular facet of the condyle of mandible is ovoid-hemispherical in shape, longer than wide. The internal pterygoid fovea (iptf) is large and deep, with its anterior end extending below the coronoid process and its lower margin rolling up lingually. The bulge (i2b) formed by the posterior end of the lower incisor alveolus is prominent, situated below the mandibular notch on the buccal side. The angular process is broken.

(3) Teeth (Fig. 10; Tab. 4)

The dental formula is 1·0·0·3/1·0·0·3. The molars are mesiodistally hypsodont. On the lateral sides the lower limits of the enamel [= dentine tract (DT); Repenning, 2003] are oblique relative to the occlusal surface. The occlusal surfaces of the molars have 2 or 3 reentrants and 2 or 3 loph(id)s. As a rule the reentrants become closed basins or disappear when moderately or heavily worn. The three specimens described here are of middle- and old-aged individuals, therefore only a few of them remain open.

The M1 is the largest of the upper molars. The occlusal surface is trapezoid in outline, longer than wide, with slightly shorter lingual side. There are two buccal reentrants (anterosinus and mesosinus) and one lingual one (sinus) and 2 or 3 lophs. The mesosinus is the longest and its lingual end turns slightly posteriorly. The anterosinus is nearly transverse. The sinus extends towards the paracone and inserts between the anterosinus and mesosinus. In M2 the crown is subquadrate in outline. It has one buccal (mesosinus) and one lingual reentrant (sinus). The mesosinus is longer than the sinus, and bends slightly posteriorly at its lingual part. The sinus extends slightly anterobuccally in front of the mesosinus, and partially overlaps the latter. The M3 is trapezoid in outline, with narrower and slightly convex posterior side. Like M2, it also has one buccal (mesosinus) and one lingual reentrant (sinus). The sinus is long, extending anterobuccally. The mesosinus is shorter, but deeper than the latter, extending transversely or anterolingually to meet or overlap the latter partially.

The I2 is orthodont, bending strongly so that its anterior end even turns slightly backwards (see Fig. 8 A). Its posterior end originates from the anterior margin of the maxillary, externoposterior to the incisive foramen, far away from the M1. The cross section of I2 is triangular in form, with a rounded lingual angle and an almost flat labial side. The enamel layer covers the whole labial side, less so on the medial and lateral sides. On the labial side there is a prominent longitudinal ridge, located nearer to the medial side than to the lateral one (see Fig. 8 B).

The lower molars of V 16286.1 are very close to the holotype (V 412) in structure. The m1 of V 16286.1 is oval in outline, longer than wide, with narrower anterior side. The protosinusid is worn away, and the sinusid and mesosinusid have become closed basins. The mesosinusid is long and strongly bends, with its buccal part extending forward. The sinusid extends slightly posterolingually behind the mesosinusid. The m2 is trapezoid in outline, with a longer buccal side. The protosinusid and sinusid have become transversely closed basins. The former is shorter than the latter. The mesosinusid slightly bends anteriorly towards the protosinusid and opens lingually. The m3 is subquadrate in outline, with slightly narrower and convex posterior side. It has only one buccal and one lingual reentrant (sinusid and mesosinusid). The mesosinusid is slightly longer than sinusid, with its buccal part slightly bending forward. The sinusid is transverse, almost directly opposite the mesosinusid.

The robust i2 (Fig. 9 B, C) has a long exposed portion out of the alveolus, which is more or less straight, extending anterodorsally. It originates in the ascending ramus of the mandible. The cross section of the i2 is a narrow triangle, with flat labial and lateral sides, but a narrow and rounded lingual angle. The enamel covering is like that of I2. There is only one labial longitudinal ridge on the left i2, but two on the right i2 (one distinct and one weak). It means that the reduction of the number of labial ridges in i2 appeared in Late Miocene in *P. hipparionum*.

(4) Enamel microstructure of lower incisor (Fig. 11)

The specimen under study is the anterior end of a left lower incisor cut from the mandible of V 16286.1. It is incomplete, preserving only labial part and a small lingual part (Fig. 11 A). The cross section shows that the thickness of the enamel varies transversely: the lateral 1/3 is thicker than the medial 2/3 (Fig. 11 A). At present, the reason of the variation in thickness of the enamel is not clear. Probably the medial part is worn more heavily than the lateral part because the observed part represents only the anterior end of i2. Thus, the thickness of the enamel of i2 is measured near the lateral part in this monograph.

The section of i2 shows that the microstructure of enamel is uniserial. The total enamel thickness (T) is 168 μm. The incisor enamel consists of two layers: inner portion (PI) and outer portion (PE). No prismless external layer is seen. The PE of V 16286.1 is very thin, about 38 μm in thickness. The ratio of PE/T is about 23%. The PI is thick, about 130 μm in thickness. The ratio of PE/PI is about 29%.

The cross section of i2 shows that the Hunter-Schreger bands (HSB) are nearly vertical to the labial surface in the PE, but slightly inclined from the enamel-dentine junction (EDJ) to the medial longitudinal ridge (mr) in the PI. The inclinations of the HSB of the PI are variable: they are slightly decreased from the EDJ to the boundary between the PI and PE; and the differences of the inclinations between the lingual and labial parts are also decreased from near the longitudinal axis of i2 to both medial and lateral sides.

The left i2 of V 16286.1 has one longitudinal ridge (medial ridge, mr). The cross section shows that the enamel increases in thickness along the ridge and the HSB of the medial and lateral parts of the i2 are convergent toward the medial ridge (Fig. 11 A, C). In the medial ridge, the inclinations of the HSB, prismas (P) and interprismatic matrix (IPM) of the medial and lateral sides turned from meeting at right angles with each other in lingual part near the EDJ into paralleling each other in labial side, then gradually into the PE.

(5) Postcranial bones

Of the long limb bones of *Pa. hipparionum* collected from the Linxia Basin, only a partial left radius (V 16286.2) and a right femur (V 16286.3) are preserved.

Radius (Fig. 12; Tab. 15)   Of the radius (V 16286.2) only the distal part is preserved. The length of the preserved part is 22 mm. The shaft bends slightly posteriorly and gradually becomes more robust downwards. On the anterior surface of the middle part of the shaft there is a groove extending in superolateral to inferomedial direction, which may serve for passing of the tendon of some muscle. On the lower part of the lateral side of the shaft there are some longitudinal grooves and ridges for attachments of radio-ulnal ligaments.

The maximum width of the distal end (Tab. 15: Max W of DE) is 4.7 mm and its maximum anteroposterior distance (Tab. 15: Max APD of DE) is 4.5 mm. The distal articular surface is trapezoid in outline, 3.8 mm wide. The articular surface is smooth and concave in major part, with a transverse convexity on the medial half of the posterior border. No vestige of line separating the articular facets for the scaphoid and lunar can be observed. Probably the scaphoid and lunar are fused into a scapho-lunar bone. The styloid process of radius (stypr) is distinct. On the anterior side of the distal end there is a distinct longitudinal groove serving for the passing of the tendon of extensor. The lateral side of the distal end is roughly concave, probably for attachment of the interosseous ligaments between radius and ulna. On the medial side of the distal end there is a longitudinal groove, where tendons of *m. extensor* (*m. extensor pollicis* and *m. abductor pollicis*) may pass.

Femur   Of the femur (V 16286.3) only the middle part of shaft with third trochanter and the lower part of lesser trochanter are preserved. The third trochanter is distinct. The anterior side of the middle part of the femur is convex transversely, and the posterior side is transversely concave near the third trochanter and flat below. The width of the middle part of femur below the third trochanterthe is 5 mm and the anteroposterior distance is 4 mm.

Li Qiang (2010) reported several postcranial bones of *Pa. hipparionum* from Baogeda Ula, Nei Mongol. He listed one left talus (V 16306.15) and three metatarsals (V 16306.16–18), but did not describe them. Having observed these bones, we found that the three metatarsals are in fact three metacarpals. Among them V 16306.16 is a right McIII and V 16303.17 is a left McIV. The proximal part of the third bone (V 16303.18) is broken. However, based on the preserved part it is similar to V 16306.16 and V 16303.17 in form and size, especially to the former. Thus, V 16303.18 may be a left Mc III. In order to know some features of the postcranial bones of *Pa. hipparionum*, the above specimens from Baogeda Ula are described in the following lines.

Third metacarpal bone (McIII; Fig. 13; Tab. 16)   The right McIII (V 16306.16) is preserved very well and the left McIII (V 16306.18) preserved only the lower part of shaft and distal end. From the proximal view, the proximal end is triangular in outline. There are three facets. The largest middle one is a facet articulating with the magnum (fmg). It is trapezoid in outline. Its surface is concave transversely and convex dorsovolarly. The lateral one is a facet articulating with the uniform (fun), which is approximately triangular in shape, with a narrow volar end. Its surface is flat and faces laterosuperiorly, and forms a right angle with that articulating with magnum. These two facets are separated by a longitudinal ridge. The medial facet is the smallest, articulating with McII (fmcII). It is triangular in shape and located at the dorsomedial corner. Its surface is slightly convex and faces mediosuperiorly, and forms an obtuse angle with fmg. The ridge between these two facets is short and low. On the lateral side of the proximal end of McIII the facet articulating with McIV (fmcIV) is about lunar-formed, facing inferovolarly and separated from the facet for the uniform by a distinct dorsovolar ridge. On the lateral margin of the dorsal side there is a distinct rough crest, which may be the attachment area for proximal intermetacarpal ligament, and is here called crest for ligament (cl).

The shaft is straight. Its upper part is narrow and has a triangular cross section. Its lower part gradually becomes larger, transforming into an oblate cylinder. The dorsal surface of the shaft is flat vertically and convex transversely. The metacarpal tubercle (tmc) is a prominent and rough concavity, located on the medial margin of the dorsal side of the proximal end, about 1 mm below the facet articulating with the magnum. It may be the attachment area for *m. extensor carpi radialis brevior*. There are several distinct nutrient foramina near the upper part of the medial margin on posterior side.

On the distal end of McIII the facet articulating with the PhI of the third digit is smooth and convex on the dorsal part and has a sagittal ridge on the volar part, accompanied by symmetric grooves bilaterally. There is a deep fossa on both medial and lateral sides of the distal end, which may be the attachment areas for the collateral ligaments.

Forth metacarpal bone (McIV; Fig. 14; Tab. 16) V 16306.17 is a left McIV, with its proximal part damaged. It is slightly shorter, but stouter, than McIII. The facet articulating with McIII (fmcIII) is partially preserved. It is somewhat triangular in outline, with a flat surface facing mediosuperiorly. Along the medial margin of the dorsal side of the proximal end a well developed crest extends to the middle of the shaft. The crest may be the attachment area for the ligaments between McIII and McIV, here also called crest for ligament (cl).

The shaft is robust and straight. The cross section of its upper part is trapezoid in outline. The shaft is thin in the middle part, becomes wider in the lower part, cross section of which is oblong in shape. The distal part of McIV is similar to that of McIII in general form. But, unlike McIII, its medial part is slightly thicker than the lateral one, the distal articular facet has a slightly longer and narrower dorsal part than that on McIII and asymmetric grooves on the volar part, with lateral groove wider than the medial one and facing distolaterally.

Astragalus (As; Fig. 15) The left astragalus (V 16306.15) is well preserved. It is narrow and high, with its trochlea (tro) occupying the proximal and dorsal surfaces of the proximal end, extending in anterolateral and distal direction. The volar portion of the trochlea becomes narrower. The intermediate groove between the medial and lateral ridges is very weak, situated closer to the medial ridge. The medial and lateral ridges of the trochlea are asymmetrical. The medial ridge (mr) has a narrow and slightly steep lateral slope and a steeper and transversely convex medial slope. The lateral slope of the lateral ridge (lr) is steep and flat, and articulates with the distal end of the fibula. The medial slope of the lateral ridge is wide and gentle on the top. The anterior part of the trochlear splits into two branches distally. Its medial branch is larger than the lateral one and extends downwards to the dorsal side of the neck of astragalus, with its narrow distal end turning in dorsomedial direction.

The neck of the astragalus is tall and slender. On the plantar side of the astragalus there are two facets articulating with the calcaneus. The larger one is called proximal calcaneal facet (pcf) in this monograph (= ectal facet, posterior calcaneal facet, calcaneo-astragalar facet, or proximal calcaneo-astragalar facet, etc.) and located at the laterosuperior corner of the plantar side. It is trapezoid in form, extending in medioproximal-laterodistal direction. Its surface is concave in major axis, but straight transversely. Below the pcf there is a small tongue-shaped facet (tosf), extending transversely and facing plantodistally, forming an angle of about 90° with the pcf. The sustentacular facet (suf) is located near the medial margin and separated from the pcf by a rough groove of astragalus (gras). The suf is oval in outline, higher than wide, extending in medioproximal-laterodistal direction, with its medioproximal end extending to the lower border of the knob on the plantomedial corner of the trochlear and its distal corner closer to the distal articular facet. Its surface is slightly convex. On the medioproximal corner of the plantar side there is a distinct knob for attachment of medial collateral ligament.

On the distal end the articular facet with the navicular (fna) is oval-shaped, wider than long, with a spherical surface, and faces mainly downwards. On the medial side there is another articular facet. It is also oval in form, but slightly narrower transversely than the former, facing mainly medially and its proximal end almost meets the suf. It may be the articular facet for the tibial sesamoid (fts). Probably a tibial sesamoid was present on the medial side of the astragalus in *Pa. hipparionum*. The two facets on distal side are continuous and no distinct boundary can be seen between them.

Max H: 77 mm; Max W of PE: 55 mm; Max D of PE: 36 mm; Min W of neck: 26 mm, APD of neck: 22 mm; L of pcf: 3.2 mm; W of pcf: 2.7 mm; L of suf: 3.2 mm; W of suf: 2 mm; W of DE: 39 mm; APD of DE: 26 mm.

2) Comparison

The holotype of *Pararhizomys hipparionum* is a partial left horizontal ramus. Compared with the holotype, the Linxia mandible (V 16286.1) possesses almost the same diagnostic features as the holotype. The mandible is short and thick, with its mental foramen located below the anterior side of m1 and lower than the anterior end of the masseteric fossa. The lower molars are rooted and distally hypsodont. The buccal and lingual reentrants are subequal in depth on m1–3. The m1 and m2 have two buccal reentrants and one lingual one. The mesosinusid of m1 strongly bends anteriorly. The m3 is subquadrate in outline and has nearly opposite sinusid and mesosinusid. The i2 has one distinct labial longitudinal ridge. In size the molars are also subequal to those of the holotype of *Pa. hipparionum* (see Tabs. 4, 5).

Kowalski (1968) referred a skull from Altan Teli, Mongolia, to *Pa. hipparionum*. The Linxia specimens are also close to the specimens from Mongolia in the following features: the skull is low; the rostrum is narrow, with parallel lateral sides; the anterior ends of the premaxilla and nasal are nearly at the same vertical line; the prominent dorsolateral crest of premaxillary is parallel to anterior part of the naso-premaxillary suture and the dorsal part of the premaxilla forms a long band; the nuchal surface slightly slants forward; I2 bends strongly and has a prominent labial longitudinal ridge; the molars are mesiodistally hypsodont and buccal part of mesosinusid of m1 strongly bends anteriorly, etc. In addition, their skulls and teeth are also subequal to each other in size (see Tabs. 3–5). Therefore, there is no doubt that the Linxia specimens should be assigned to *Pa. hipparionum*.

3) Discussion

Li Qiang (2010: 55–57, Fig. 8) made an arrangement of all the known specimens of *Pa. hipparionum* in stratigraphic sequence. He stated that the evolutionary tendencies of molars in *Pararhizomys* are: "cheek teeth increasing in size, the 'index H' gradually becoming small and the crown gradually becoming higher" (Li, 2010: 48).

Having compared all known specimens of *Pa. hipparionum*, those of Linxia included, the authors found that the hypsodonty proxy as redefined in this monograph (Hp) reflects the variation of the crown height better than Li's "index H". For example, the hypsodonty indices of lingual and buccal reentrants (Hpl and Hpb2) of M2 are 3 mm and 2.6 mm in V 14178 (about 8 Ma) and 3.5 mm and 3.2 mm in V 16306.8 (about 6 Ma), and the Hpb2 of 16286.1 (about 9.5 Ma) is 2.5 mm in *Pa. hipparionum* (see Tabs. 4, 5). This means that in *Pa. hipparionum* the larger is the hypsodonty index as defined in this monograph, the higher is the molar crown.

Based on the evolutionary tendencies suggested by him, Li (2010: 56) inferred that the holotype (V 412) from Loc. 10 might be between those of Qin'an (7 Ma) and Baogeda Ula (6 Ma) in age. However, in fact, his Tab. 4 indicated that LNH of m1 and m2 of the holotype are smaller than both those of Baogeda Ula and Qin'an rather than between them. Using the method of measurement proposed in this monograph, the holotype is subequal to V 16286.1 in both size and hypsodonty proxy (see Tabs. 4, 5). More interesting is that the hypsodonty proxies of the two specimens are smaller than that of V 14179 from Qin'an. Probably the deposits yielding the holotype are close to that of V 16286 and slightly older in age than that of V 14179.

Li (2010: 55) doubted whether the mandible (V 14179) from Qin'an should be referred to *Pa. hipparionum* "based on its small size and larger dentine tract height of teeth." The method of measurement proposed in this monograph shows that the sizes of the molars of V 14179 are within the variation range of *Pa. hipparionum*. The hypsodonty proxies of reentrants (Hpl and Hpb2) of the m2 and m3 of V 14179 are subequal to each other as well (see Tab. 5). The latter is one of the most diagnostic features of the holotype of *Pa. hipparionum*. Thus, V 14179 is better retained in *Pa. hipparionum*.

Among the isolated teeth of *Pa. hipparionum* from Baogeda Ula, two of them [V 16306.13 (m1) and V 16306.14 (m2)] are similar to other specimens of *Pa. hipparionum* in having strongly bending mesosinusid. However, their lingual reentrants are much deeper than the buccal ones [Hpl/Hpb2: 66% at m1 (V 16306.13) and 77% at m2 (V 16306.14)], which are different from the other specimens referred to *Pa. hipparionum* (Hpl/Hpb2: 105%–107% on m2; see Tab. 5). Similar case also holds true for the M2 of V 16307. Their assignment to *Pa. hipparionum* is thus doubtful. They are listed as *Pararhizomys hipparionum*? in Tab. 5.

In summary, *Pa. hipparionum* is so far known to be represented by four skulls, two of them with mandibles (V 16286.1 and Z. Pal. No. MgM-V/65), three partial mandibles (holotype, V 14179 and V 16306.1) and some isolated teeth. Among them, V 16287 differs from the others in having slender rostrum and thinner incisors. It may represent a female and the others, males.

Pararhizomys huaxiaensis sp. nov.
(Figs. 16–21, 75; Tabs. 1–4, 6–8, 14)

Holotype HMV 1413, complete skull with mandible, atlas, axis, third cervical, proximal part of humerus and glenoid angle of the scapula of the same individual, from LX 200041 (Shanchengcun); upper part of middle Liushu Fm. (for details see Tabs. 1, 2).

Paratypes HMV 1924 and V 16291, two skulls: HMV 1924 from LX 200503 (Niuzhawan), V 16291 from LX 200030 (Songshugoucun); both from upper part of middle Liushu Fm. (for details see Tabs. 1, 2).

Locality and horizon Gansu: Linxia Basin: Shanchengcun and Niuzhawan of Hezheng County and Songshugoucun of Guanghe County, upper part of middle Liushu Fm. (for details see Tabs. 1, 2); Qingyang County (?), Baodean ALMA (*vide infra*).

Diagnosis Premaxilla has a shallow upper groove and a gentle longitudinal bulge under it on the lateral side. The arched posterior border of the zygomatic plate lies at about the same level as the anterior margin of M1, with its anteriormost point slightly anterior to M1. Mandibular diastema is markedly longer than lower cheek tooth row. Chin process and masseteric ridge are more prominently developed. Mandibular foramen is more anteriorly located, below the coronoid process. Molars are higher crowned, mesiodistally and lingually hypsodont. Lingual reentrants extend deeper than buccal ones on M1–2 and m1–3, but shallower than the latter on M3. Sinus and mesosinus are nearly transverse and directly opposite each other on occlusal surface of M3. Mesosinusid of m1 slightly bends anteriorly, with its buccal end extending towards the protosinusid.

Differential diagnosis The new species differs from *Pa. hipparionum* and *Pa. qinensis* in molars being higher crowned and distinctly lingually hypsodont. It differs from *Pa. hipparionum* in having longer mandibular diastema, much longer than lower molar row, and M3 with sinus and mesosinus opposite to each other, and m1 having shorter, less bending mesosinusid toward prosinusid; from *Pa. qinensis* in having longer rostrum and maxillary diastema, and longer M2.

Etymology Huaxia is the ancient name of China, and -ensis, a Latin masculine suffix denoting the provenance of the material.

1) Description

(1) Skull (Figs. 16, 17 A1–A5, B; Tab. 3)

The skull and mandible of HMV 1413 were originally tightly occluded with each other (Fig. 16), which was then disassembled by manual preparation (Fig. 17 A).

Like *Pa. hipparionum*, the skull is myomorphous, relatively narrow, long and low. It has a narrow rostrum, with lateral sides parallel with each other. The anterior part of the skull anterior to M1 is subequal in length to the posterior part of the skull (including M1). The nasal is wedge-shaped, with its anterior end aligned with that of the premaxilla. The narrow posterior end of the nasal is at about the same level as the median end of the posterior border of the premaxilla and the anterior border of the orbit. The dorsal surface of the nasal is convex transversely and longitudinally. The premaxilla bears a remarkable dorsolateral crest, which is subparallel to the anterior part of the naso-premaxillary suture. The dorsal surface of the premaxilla forms a long band. On the lateral side of the premaxilla the longitudinal bulge formed by I2 is gentle and the longitudinal groove above it is shallow. The dorsal part of the premaxillo-maxillary suture is a straight line extending longitudinally to meet the maxillo-frontal suture at an obtuse angle. The infraorbital foramen is oval in form and has a ventral slit. The infraorbital canal is large, with deeply concave inner wall, and bordered by distinct medial, upper and lateral crests. Within the infraorbital canal there are distinct septum and nasolacrimal foramen. The attachment area of *m. masseter medialis* slightly extends to the dorsal surface of the maxillary over the lateral border. The attachment area of *m. masseter lateralis* is restricted within the maxilla. Its anterior border formed by a curved crest is located behind the ventral slit of the infraorbital foramen and the anterior end of the incisive foramen. The emarginated posterior border of the zygomatic plate is nearly at the same level as the anterior margin of M1, with its anteriormost point situated slightly anterior to the M1. The zygomatic arch is rounded, with its middle part thick, cross section of which is triangular in form. The jugal bone is short and its anterior end does not extend to the anterior border of the orbit. The lacrimal is mainly located within the orbit and appears tiny on the dorsal side. The lacrimal process is weak. The postorbital constriction is pronounced. The frontal lines are faintly developed, converging posteriorly to form a prominent sagittal crest. The parietal is roughly hexagonal in outline, longer than wide. The glenoid fossa is very long, extending posteriorly to about the nuchal crest, pushing the external auditory meatus downward. The postglenoid foramen is long and large, situated on the posterior part of the glenoid fossa, along the petro-squamosal suture. The posterior border of the hard palate is located behind the M3. The parapterygoid fossa is also distinct on the posterior part of the hard palate. The mesopterygoid fossa is wider than the pterygoid fossa. The masticatory foramen is separated from the buccinator one. The deep temporal foramen is confluent with the masticatory one. The posterior apertures of the canals of deep temporal nerve, masseteric nerve and buccinator nerve are confluent. The posterior foramen of alisphenoid canal is located medial to the confluent posterior apertures mentioned above. The posterior foramen of pterygoid canal (pcpf) is located at posterior base of the internal pterygoid process near the medial side of the pterygoid fossa. The *foramen ovale* is confluent with the middle lacerate foramen. The basicranial part is similar to those of *Pa. hipparionum* in shape and structures. But the auditory bulla is more convex than in *Pa. hipparionum*. The internal carotid foramen is located posterior to the spine of the bulla. The jugular foramen is slit-like. Along the medial side of the auditory bulla neither distinct longitudinal groove connecting the internal carotid foramen with the jugular foramen, nor foramen for stapedial artery is seen. The hypoglossal foramen is separated by bony septa into two or three apertures. The ventral side of the occipital condyle is short and wide. The left and right occipital condyles are located closely, separated from each other only by a narrow groove.

The nuchal surface is similar to that of *Pa. hipparionum* in form and structure. It is about semicircular in outline, with a roughly flat surface slanting slightly forwards toward the top. A weak external occipital crest is visible, extending vertically from the nuchal crest to the upper border of the foramen magnum on HMV 1413, but no such a crest is seen on HMV 1924. The petromastoid portion (Pms) is roughly lozenge in shape, with its mediosuperior corner protruding medially. The *foramen magnum* is large and roughly circular in shape. The paramastoid processes (pmp) are preserved completely in HMV 1413 (Fig. 17 A). They are small and pyramidal in shape, extending ventrolaterally and slightly backwards.

(2) Mandible (Figs. 16, 17 A6–A7; Tab. 3)

As in *Pa. hipparionum*, the mandible is sciurognathous and has a short and thick horizontal ramus. The posterior part of the mandibular diastema is deeply concave. The chin process is more developed than in *Pa. hipparionum*. Its lower border protrudes more downwards and a large digastric fovea is present on its medial side. The symphysis extends anterodorsally and forms an angle of about 50° with the lower border of the horizontal ramus. The masseteric fossa extends anteriorly to below m1, with a strong and flaring masseteric ridge, which is more developed than in *Pa. hipparionum*, extending backward to the lower border of the angular process. The mental foramen is located below the anterior part of m1, lower than the anterior end of the masseteric fossa, at the lower 1/3 of the horizontal ramus. The coronoid process is similar to that of *Pa. hipparionum* in form, and also has a distinct temporal fovea on the lower part of the medial side, but with less bending posterior border. The condyloid process is situated lower than the coronoid process and extends mainly backwards, only slightly upwards, with its top bending medially. The anterosuperior border of the condyloid process is long and straight, extending antero-inferiorly. The distance from the posterior border of the coronoid process to the anterior border of the condyle is about 10 mm in length. On the medial surface of the condyloid process there is a shallow concavity (external pterygoid fovea, eptf), which may be the attachment area for external pterygoid muscle. The articular surface of the condyle forms a long ellipse, with distinct medial and lateral borders. The mandibular notch is shallow and about 1/3 the mandibular height in depth. The mandibular foramen is located below the middle of coronoid process, about 3 mm or 3.5 mm posterior to m3. The bulge formed by the posterior end of the i2 alveolus is prominent on the buccal side. The internal pterygoid fovea (iptf) is a large and deep triangular concavity, with its anterior end being below the coronoid process and an arched lower margin extending posterosuperiorly. The angular processes of HMV 1413 are preserved completely. They form long triangular pyramids, extending posteriorly, with hook-like ends. The posterior ends of the condyle of mandible and the angular process are almost at the same vertical line. The posterior border of the ascending ramus between the two processes is deeply concave.

(3) Teeth (Fig. 18; Tab. 4)

The teeth are also similar to those of *Pa. hipparionum*. The dental formula is 1·0·0·3/1·0·0·3. The molars are mesiodistally hypsodont. But the crowns of molars are higher than those of *Pa. hipparionum*, and their lingual hypsodonty is prominent. M1 and m1–2 have two buccal reentrants and one lingual one, and M2–3 and m3 have one buccal and one lingual reentrant. In M1 the sinus extends towards the paracone and slightly inserts between the anterosinus and mesosinus. In M3 the sinus is opposite the mesosinus, not overlapping the latter. The sinus is longer than the mesosinus on occlusal surface, but shallower than the latter on lateral walls. The m3 is subquadrate in outline, with nearly transverse and directly opposite mesosinusid and sinusid. The orthodont I2, originating from anterior margin of maxilla near incisive foramen, curves strongly with its anterior end turning slightly backwards, and has one prominent longitudinal ridge on its labial surface. The anterior part of the i2 out of the alveolus is long. Both the left and right i2 have only one distinct labial longitudinal ridge.

(4) Postcranial bones (Figs. 16, 19–21)

The preserved postcranial bones of HMV 1413 include atlas, axis, 3rd cervical vertebra, and proximal ends of scapula and humerus.

Atlas (At; Figs. 16, 19; Tab. 6)　The atlas of HMV 1413 is well preserved, with slightly damaged border of the wing and caudal articular facets. In frontal view, the atlas is roughly oval in general outline, wider than high. The dorsal arch (da) is anteroposteriorly very short, but very wide and bow-shaped. The robust dorsal tubercle (dt) is located posterior to the middle of the dorsal arch, sloping gently anteriorly, but steeper posteriorly. On its anterior slope no crest or pit can be seen. The ventral surface of the dorsal arch is smooth and straight longitudinally. The lateral vertebral foramen (lvf, see Fig. 31) and *foramen alare* (fa) are fused into a small foramen on the dorsal side, located near the middle of the lateral end of the dorsal arch. The fused foramen is here called lateral vertebro-alar foramen (lvaf). The lateral vertebral canal transversely extends from the internal lateral vertebral foramen (ilvf) to the *foramen alare* (fa) via the lateral vertebro-alar foramen. The internal lateral vertebral foramen (ilvf) is located posterior to the posterosuperior corner of the anterior articulating cavity on lateral wall of the vertebral canal. The ventral arch (va) is thicker but shorter than the dorsal arch. The ventral tubercle (vt) protrudes posteriorly. No bilateral concavities can clearly be seen along the ventral tubercle. The vertebral foramen (vf) is roughly egg-shaped, narrowing inferiorly.

The cranial articular fovea (craf) is kidney-shaped, with distinctly concave surface. The upper part is wider than the lower part. The upper parts of the two foveae are widely separated, becoming closer to each other in lower parts, where the separation is represented by a notch. The caudal articular facets (caf) are about of the same shape and position as the cranial articular fovea, but their surfaces are less concave transversely. The wings (w) form bending plates, stretching in anterosuperior-posteroventral direction. The lateral border is slightly curved. Its anterior end is about 1 mm posterior to the anterior border of the cranial articular fovea, and its posterior end reaches to the caudal articular facet. On the posterior side of the wing there is thick ridge. The transverse foramen (*foramen transversarium*, ftr) is distinct and located between the thick ridge and the upper margin of the caudal articular facet. The *fossa atlantis* (fat) is large and shallow. In the fossa there is a short groove joining the small *foramen alare* (fa) and the larger *foramen alare inferior* (fain). The distance between the above two foramina is short (only about 1 mm). The broken right *fossa atlantis* shows that the *canalis transversarius* (ctr) is long.

Axis (Ax; Figs. 16, 20; Tab. 7)　The axis of HMV 1413 is well preserved, with partially broken spinous process (sp) and right transverse process. The axis is longer than the atlas. The vertebral body (vb) is wide, with a flat dorsal surface. The odontoid process (odpr) is oblate-conic, with an obtuse anterior end and a longitudinally straight and transversely convex articular facet for the odontoid fossa of the atlas (fodf) on the ventral side. The anterior articular facets (aaf) situated bilateral to the odontoid process are triangular-formed with rounded angles and slightly convex upper and lower borders, and joining the odontoid process at their medioinferior corners. Their articular surfaces are slightly convex, facing anterolaterally. The posterior border of the odontoid process and the ventral borders of the anterior articular facets form an arched crest concave posteriorly. The ventral side of the vertebral body is in the shape of a strongly transversely widened ellipse. Its middle part is slightly bulging, which becomes wider backwards, but without forming a distinct ventral spine. Its lateral surfaces are broadly concave. The vertebral foramen is circular in form, wider than high, with a flat lower side. The pedicles of the vertebral arches are thick.

The basal part of the spinous process (sp) is preserved. Its cross section is lenticular in shape, anteroposteriorly longer than wide. The anterior and posterior sides of the spinous process form narrow crests. The anterior crest extends in superoposterior direction and the posterior one is vertical, without central groove. The postzygapophysis (pzy) extends posteriorly from the pedicle of the vertebral arch, and its articular facet for the 3rd cervical vertebra faces inferoposteriorly, forming an angle of about 20° with the horizontal plane.

The transverse process (trpr) forms a narrow triangle, with two roots, the dorsal root being larger than the ventral one. The dorsal side of the transverse process is wide, facing anterolaterally. The transverse foramen (*foramen transversarius*, ftr) is large and the transverse canal (*canalis transversarius*) is short.

Third cervical vertebra (C3; Figs. 16, 20; Tab. 8)   The 3rd cervical vertebra of HMV 1413 is preserved well, with its right transverse process and posterior border of arch slightly damaged. The ventral side of the body is longer than that of the atlas, but its dorsal side is shorter than that of the latter. The vertebral body (vb) is cylindrical in form, with a flat dorsal surface. The ventral side of the body is trapezoid in outline, with anterior border narrower than posterior one, and a slightly convex middle part becoming wider and more distinct backwards. No distinct middle keel is present. The lateral surfaces of the ventral side are concave. The vertebral fossa (vfs) is concave and transversely wide, oval in outline. The vertebral foramen is semicircular in outline, wider than high.

The dorsal part of the vertebral arch is short and wide, and bow-shaped. The spinous process is low and narrow. On dorsal side of the vertebral arch a pair of foramina are present bilateral to the spinous process. The pedicle of the arch is thick. The prezygapophysis (przy) extends forwards from the pedicles of the arch. The postzygapophysis (pzy) is situated higher than the prezygapophysis and extends posteriorly in lesser degree than the prezygapophysis extending anteriorly. The articular facet of postzygapophysis (fpzy) is long and narrow, oval in outline. The surface is flat, facing postero-inferolaterally. It is roughly parallel to the anterior articular facet. The distance between the anterior and posterior articular facets is about 1.1 mm.

The transverse process has two roots: the small dorsal root rising from the pedicle of the arch and the larger ventral one from lateral part of the vertebral body. It is also triangular in form, but is thicker than that of axis and extends laterally below the latter. The transverse foramen (*foramen transversarius*, ftr) is located between the dorsal and ventral roots at the base of the transverse process and the transverse canal (*canalis transversarius*) is longer than that of the axis.

Scapula (Sc; Figs. 16, 21 B)   The glenoid angle of the right scapula of HV 1413 is preserved. The glenoid cavity (glc) is an oval concave surface. Its major diameter is 7 mm, longer than the width of the neck of scapula (nsc), which is 5.5 mm wide. Only the lower part of the scapular spine (spsc) is preserved. It rises from the neck of scapula, about 4 mm away from the lateral border of the glenoid cavity. The scapular tuber (sctu) is low, but the coracoid process (cpr) is well developed and long, extending inferomedially.

Humerus (Hu; Figs. 16, 21 A; Tab.14)   Only the proximal part of the right humerus is preserved in HMV 1413. The maximum width of the proximal end is larger than the anteriorposterior distance (Tab. 14). The head of humerus (head) extends slightly backwards and overhangs the shaft. The articular surface has a form of oval hemispheroid. The greater tuberosity (gtu) is robust and extends laterally with its top situated lower than the highest point of the head. It is separated by a pit into two parts: the medial part is smaller, but higher than the lateral one. The lesser tuberosity (ltu) is low and small, and extends mainly downwards. It is separated by a vertical shallow groove into anterior and posterior parts. The upper part of the shaft is triangular in cross section, wider than anteroposteriorly long. The intertubercular sulcus (= bicipital groove, bg) and the crest of greater tuberosity (cgtu) are prominent.

2) Comparison

As mentioned above, HMV 1413 and the paratypes are similar to *Pa. hipparionum* in both skull morphology and tooth structure. Obviously, they are to be referred to the genus *Pararhizomys*. The genus *Pararhizomys* is so far known to include two species: *Pa. hipparionum* and *Pa. qinensis*. In comparison with these two species, HMV 1413 and the paratypes do show a number of features different from the first two species.

Their differential features from *Pa. hipparionum* can be listed as follows (those of *Pa. hipparionum* in brackets): the mandibular diastema is considerably longer than the lower cheek tooth row (only slightly longer than the latter, see Tabs. 3, 4); M1–2 and m1–3 are lingually hypsodont, with their lingual reentrants deeper than the buccal ones, Hpl/Hpb2: 85% in M2 and ≤ 71% in other molars (without distinct lingual hypsodonty, lingual reentrants subequal to the buccal ones in depth, Hpl/Hpb2: 109% and 115% in M2, and ≥ 90% in other molars, see Tabs. 4, 5); the lingual and buccal reentrants on occlusal surface of M3 are usually situated opposite each other (partially overlapping); the mesosinusid of m1 is shorter, slightly bent, with its buccal part stretching towards the protosinusid (longer, strongly bent, buccal part extending anteriorly beyond the protosinusid).

They differ from those of *Pa. qinensis* (those of the latter in brackets) in having longer rostrum, 22.1 mm long, and maxillary diastema, 27.2 mm long (shorter, maxillary diastema: 16.5 mm, after Zhang et al., 2005); sinus of M1 deeper than anterosinus and mesosinus, Hpl/Hpb2: 68% or 70% (subequally deep, Hpl/Hpb2: 95%).

Pararhizomys huaxiaensis sp. nov. is so far known to be represented by three specimens. Among them HMV 1413 and HMV 1924 are more robust in structure, with wider and higher rostra and stronger incisors than those of V 16291. Probably they represent male individuals, while V 16291 represents a female one.

3) Discussion

Li Qiang (2011, also in personal communication) stated that the specimen (RV 42023) from Qingyang, Gansu Province (= K'ingyan, Kansu), which had been considered left M1–2 and referred to *Prosiphneus licenti* by Teilhard de Chardin (1942), might be a right m1–2. We agree with him. In that specimen the longer reentrant of "M2" prominently bends anteriorly. This type of reentrant is similar to the mesosinusid of m1 in *Pararhizomys*, rather than to the sinus of M2 in *Prosiphneus*, in which the sinus is usually short and does not bend anteriorly. In addition, based on Figure 32 of Teilhard de Chardin (1942) the "M1" has two reentrants on both lingual and buccal sides. Having observed the original specimen, we found that the "M1" has two reentrants on one side, but only one reentrant on the other side. In fact, the area anterior to the reentrant on the other side there is a crevice rather than a reentrant. If it is so, this tooth is also similar to m2 of *Pararhizomys* rather than M1 of *Prosiphneus*. It is interesting to note that the two teeth are rather high crowned and are mesio-distally and lingually hypsodont as well. The lingual reentrant is deeper than the buccal ones. All these features are closer to those of *Pa. huaxiaensis*. If it is reasonable to assign the above specimen (right m1–2) to *Pa. huaxiaensis*, then *Pa. huaxiaensis* should also have appeared also in Qingyang district in late Late Miocene Baodean ALMA.

Pararhizomys longensis sp. nov.
(Figs. 22, 23, 75; Tabs. 1–4)

Holotype   V 16292.1, one incomplete skull with mandible, from LX 200042 (Guonigou), lower part of Liushu Fm. (for details see Tabs. 1, 2).

Paratypes   V 16292.2 and V 16292.3, two partial skulls, from the same locality and horizon as the holotype.

Diagnosis   On the lateral side of the premaxilla the longitudinal bulge is prominent, and the longitudinal groove above it is wide and deep. Zygomatic plate located more posteriorly: anteriormost point of its emarginated posterior border situated in line with the middle of M1. Mandibular diastema is subequal to lower cheek tooth row in length. Molars lower crowned, mesiodistal hypsodonty distinct in M1–2 and m3, but weak in m1–2 and indistinct in M3. Anterobuccally extending sinus and anterolingually oblique mesosinus partially overlap in M3. Buccal part of mesosinusid slightly bends anteriorly in m1. The i2 is short and strongly curved.

Differential diagnosis   The new species differs from *Pa. hipparionum*, *Pa. huaxiaensis* sp. nov. and *Pa. qinensis* in having more prominent longitudinal bulge, wider and deeper groove above it on lateral side of premaxilla, more posterior position of zygomatic plate, with its anteriormost point of the emarginated posterior border in line with the middle of M1; molars being lower crowned.

It further differs from *Pa. hipparionum* and *Pa. huaxiaensis* sp. nov. in having shorter and more strongly curved lower incisor; from *Pa. hipparionum* in having less bending mesosinusid in m1; from *Pa. huaxiaensis* sp. nov. in having mandibular diastema subequal to lower cheek tooth row in length and obliquely extending and partially overlapping sinus and mesosinus in M3; and further from *Pa. qinensis* in having longer maxillary diastema.

Etymology   Long is the abbreviated Chinese for Gansu Province, where the fossils were collected.

1) Description

(1) Skull (Fig. 22 A, B1–B4; Tab. 3)

Of the skull of V 16292.1 only its anterior 2/3 is preserved, containing I2's and cheek tooth rows of both sides, partial left zygomatic arch, anterior root of right zygomatic arch, and the anterior part of cranium. It is similar to that of *Pararhizomys* in general morphology and structure. The skull is myomorphous. The lateral sides of the rostrum are parallel to each other, but its ventral side seems to be more strongly curved. The nasals form an elongated wedge, pointed posteriorly. The anterior ends of nasals and premaxillae align in the same vertical line, and the posterior ends of the nasals are at about the same transverse line as the anterior border of the orbit. The premaxillary laterodorsal crest is remarkable, subparallel to the naso-premaxillary suture, so that the dorsal part of the premaxilla forms a long band. There are prominent longitudinal bulge and groove on the lateral side. However, the longitudinal groove on the lateral surface of the premaxilla is deep and wide. The infraorbital foramen also forms a wide oval with a ventral slit. The infraorbital canal has deeply concave inner wall and distinctly crested anterior border. The zygomatic arch is laterally curved. The jugal bone is short and its anterior end does not extend to the anterior border of the orbit. The zygomatic plate is restricted within the maxilla. The attachment area of the lateral masseter muscle is concave and bounded by an anterior ridge, which is in line with the middle of the incisive foramen. The emarginated posterior border of the zygomatic plate is located more posteriorly, with its anteriormost point in line with the middle of M1. The lacrimal process is weak. The orbit is small. The frontal crests are weak, converging posteriorly to form a distinct sagittal crest near the postorbital constriction area. The premaxillo-maxillary suture intersects the incisive foramen at its middle part. The deep temporal foramen and masticatory foramen are confluent with each other, but separated from the buccinator foramen. The left and right molar rows are nearly parallel to each other.

(2) Mandible (Fig. 22 B5, B6; Tab. 3)

The mandible of V 16292.1 is preserved well, but the ascending ramus is partly broken and the angular process, left coronoid process and condyloid process are missing. The mandible is also similar to that of *Pararhizomys* in basic structure: sciurognathous, with short and thick horizontal ramus. The symphysis extends steeply anterosuperiorly and forms an angle of about 45° with the lower border of the horizontal ramus. The chin process and digastric foveae are prominent. The posterior part of the mandibular diastema is deeply concave. The mental foramen is located antero-inferiorly to m1 at about the lower 1/3 height of the horizontal ramus, lower than the anterior end of masseteric fossa. The masseteric fossa extends anteriorly to below m1, with a strong and flaring masseteric ridge. The mandibular notch is shallow and about 1/4 of the mandibular height in depth. The bulge formed by posterior end of i2 alveolus is prominent on the buccal side of the ascending ramus.

(3) Teeth (Fig. 23; Tab. 4)

Dental formula is 1·0·0·3/1·0·0·3. The molars are mainly mesiodistally hypsodont. V 16292.1 is of an old individual, with all its molars heavily worn. All the preserved reentrants on the upper and lower molars have become closed basins on occlusal surface. The protosinusid of left m2 has become a circular basin. The anterobuccal areas of the occlusal surfaces of both right and left m1's and right m2 form broad concave surface, where no protosinusid can be seen. The protosinusids may have been worn away. It is possible that both m1's and right m2 may all have two buccal reentrants (protosinusid and sinusid) when less worn. If this proves true, V 16292.1 is also similar to *Pararhizomys* in occlusal structure of molars: M1 and m1–2 all have two buccal reentrants and one lingual one, and M2–3 and m3 have only one buccal and one lingual reentrant respectively. As in *Pa. hipparionum* the sinus and mesosinus extend obliquely and partially overlapping each other in M3. The m3 is subquadrate, with its posterior side slightly narrower than its anterior one, and has opposite sinusid and mesosinusid.

The I2 curves strongly and has one strong labial longitudinal ridge. The thick i2 bends strongly upwards and has a shorter exposed part out of the alveolus. As in *Pa. hipparionum*, there is only one labial ridge on the left i2, but two ridges on the right. It seems that the reduction of labial ridges from two to one is the case in V 16292.1 as well.

2) Comparison

As described above, the Guonigou specimens are similar to *Pararhizomys* in both skull morphology and tooth structures. There is no problem to refer them to the genus *Pararhizomys*. However, in comparison with the other three known species of *Pararhizomys*, they have some special features. The premaxilla has more prominent longitudinal bulge and deeper and wider groove on its lateral side (see Fig. 22 A, B3). The anteriormost point of the posterior border of the zygomatic plate is more posteriorly situated, nearly at level of the middle of M1. The crowns of the molars are relatively lower. The mesiodistal hypsodonty is clearly expressed only in M1–2 and m3, weaker in m1–2 and indistinctly in M3 (see Fig. 23).

Furthermore, they differ from those of *Pa. hipparionum* and *Pa. huaxiaensis* sp. nov. in having shorter but stronger bending i2; from *Pa. hipparionum* in having less curved mesosinusid in m1; from *Pa. huaxiaensis* sp. nov. in M3 having oblique and partially overlapping sinus and mesosinus and having mandibular diastema being subequal to the lower cheek tooth row in length (see Tabs. 3, 4); and from *Pa. qinensis* in having longer maxillary diastema [19 mm versus 16 mm in *Pa. qinensis* (after Zhang et al., 2005)].

The Guonigou specimens apparently represent a species distinct from all known species of *Pararhizomys*, here named *Pa. longensis*.

Pseudorhizomys gen. nov.

Type species   *Pseudorhizomys indigenus* gen. et sp. nov.

Other included species   *Pseudorhizomys gansuensis* gen. et sp. nov., *Pseudorhizomys planus* gen. et sp. nov., *Pseudorhizomys pristinus* gen. et sp. nov., and *Pseudorhizomys*? *hehoensis* (Zheng, 1980).

Geographic distribution and geologic age   Gansu and probably Xizang, China; Late Miocene, medial–late Bahean ALMA (~10–7 Ma).

Diagnosis   Anterior ends of premaxillae extend anteriorly beyond those of nasals. Dorsal surface of nasal is straight longitudinally. Dorsolateral crest of premaxilla is short, extending anteromedially to meet the naso-premaxillary suture, then disappearing anteriorly so that the dorsal part of premaxilla becomes narrow and triangular in outline. The juncture of the curved premaxillo-maxillary and the maxillo-frontal sutures is curvilineal, instead of being angled. Infraorbital foramen is circular in outline. Anterior root of zygomatic arch and zygomatic plate are located more anteriorly. The anterior border of the orbit is situated anterior to the posterior ends of nasal and premaxilla. The anteriormost point of posterior border of zygomatic plate is at the same transverse line as the posterior end of incisive foramen. Anterior border of attachment area of *m. masseter lateralis* is curved or S-shaped. The jugal bone is long, extending anteriorly to the anterior border of the orbit. Parietal bones roughly form a hexagon, wider than long. Parapterygoid fossa is small and shallow. Posterior border of hard palate is opposite to posterior part of M3. Mesopterygoid fossa is subequal to the pterygoid fossa in width, or only slightly wider. Internal pterygoid processes stretch directly backward, parallel to each other, setting vertical to the ventral side of the fossa. Deep temporal foramen, masticatory foramen and buccinator foramen are confluent. Posterior aperture of deep temporal canal is separated from those of the two latter canals. Lower part of the nuchal surface is relatively wide, but the petromastoid portion relatively small. Posterior concavity of the mandibular diastema is shallower. Lower border of horizontal ramus is convex downwards. Mental foramen is located at the middle part of the height of horizontal ramus, nearly in the same horizontal line as the anterior end of masseteric fossa. Mandibular notch is deeper, about 2/5 the height of mandible.

Molars are lower crowned than those of *Pararhizomys*. Mesiodistal hypsodonty is distinct only in m3, scarcely present or absent on M1, and absent on M2–3 and m1–2. M2–3 with one or two buccal reentrants (with or without anterosinus) and one lingual reentrant, lower molars with one or two lingual reentrants (posterosinusid may present or absent). The m3 is oval in outline, longer than wide, with mesosinusid extending anterobuccally to overlap sinusid partially, and usually with protosinusid. I2 originates from the middle part of maxilla near M1. It is proodont, but less curved, without distinct longitudinal ridge on labial surface. The i2 usually has two longitudinal ridges on labial surface.

Differential diagnosis   The differences between *Pseudorhizomys* and *Pararhizomys* are listed in Tab. 9.

Etymology   Pseudo, false in Greek; rhizomys, bamboo mice.

Remarks   *Brachyrhizomys hehoensis* described by Zheng (1980) may be a member of *Pseudorhizomys*, thus to be named as *Pseudorhizomys*? *hehoensis* (*vide infra*).

Pseudorhizomys indigenus gen. et sp. nov.

(Figs. 24–55, 76; Tabs. 1, 6–8, 10–17, 21)

Holotype   V 16293, a skull associated with mandible and partial postcranial skeleton of the same young individual, from LX 200004 (Yangjiashan); upper part of middle Liushu Fm. (for details see Tabs. 1, 10).

Paratype   V 16294, a skull associated with mandible of a middle-aged individual, from LX 200020 (Nanmiangoucun); upper part of middle Liushu Fm. (for details see Tabs. 1, 10).

Diagnosis   Rostrum and maxillary diastema are long. Anterior border of attachment area of lateral masseter muscle is S-shaped, with its lower part convex anteriorly beyond the ventral slit of infraorbital foramen. Temporal crests become weaker posteriorly. Anteromedial part of auditory bulla is deeply concave. Posterior concavity of mandibular diastema is relatively shorter. Coronoid process is long and inclines posteriorly, with its posterior border deeply concave. Articular facet of condyle is spindle-shaped.

Mesiodistal hypsodonty is distinct on M1 and m3. M1 has distinct undulating dentine tract on buccal side. All upper molars have two buccal reentrants (anterosinus always present) and short sinus not overlapping anterosinus. In M3 transverse mesosinus and sinus are opposite to each other or join together on occlusal surface, and mesosinus is much deeper than sinus. Anterosinus of M2–3 and protosinusid of m3 are reduced into closed basins. Lower molars have unforked sinus and lack posterosinusid. Sinusid of m3 is located more anteriorly and almost at the same transverse line as the lingual part of mesosinusid. Labial surface of I2 is convex transversely, but without distinct longitudinal ridge.

Tuberosity of metacarpal (tmc) of McII is distinct, located on the middle part of shaft on dorsal side. Facet for unciform of McV is oval and convex. Narrower middle part of shaft of McV forms a flattened column. Distal border of ungual phalanx of first finger is thin and unserrated.

Etymology   *Indigenus*, native in Latin.

1) Description

(1) **Skull** (Figs. 24 A, 25; Tab. 11)

Both skulls of holotype (V 16293) and paratype (V 16294) are incomplete, lacking one side of zygomatic arches and some other parts. The latter (V 16294) is better preserved than the former (V 16293) in general, and, thus, served the chief object of the following description. The less well preserved former skull, however, owing to its association with a large number of postcranial bones, has to be chosen as holotype of the new species.

Dorsal view (Figs. 24 A2, 25 D)   The rostrum is long and narrow, with two lateral sides slightly convergent forwards. The nasals also form a wedge, with a very narrow posterior end, which is nearly in the same transverse line as the posterior end of the serrated and obliquely extending premaxillo-frontal suture. The dorsal surface of the nasal is straight longitudinally rather than convex as in *Pararhizomys*. The anterior part of the nasal is also convex transversely, but not as strongly as in *Pararhizomys*. The flat posterior parts of the nasals bear a V-shaped median longitudinal groove. Except the slightly curved anterior part, the major parts of the naso-premaxillary sutures are straight and convergent posteriorly. There is also a dorsolateral crest on the premaxilla. But the crest is rather short and extends slightly anteromedially to meet the naso-premaxillary suture, and then disappears. Thus, the dorsal part of the premaxilla forms a narrow triangle, with the narrowest angle pointing forwards. The serrated premaxillo-frontal suture does not extend directly anteroexternally, but turns transversely in its distal half. The premaxillo-maxillary suture goes slightly posterolaterally rather than longitudinally as in *Pararhizomys*. It joins the medial end of the maxillo-frontal suture, forming a curved line. The infraorbital foramen is also limited within the maxilla, but it is more circular in form. The infraorbital foramen also has a ventral slit, a bony septum and a nasolacrimal foramen (nlf), which can clearly be seen from the anterior and dorsal views. The lacrimal bone cannot be distinguished on dorsal side and no distinct lacrimal process can be seen. Only the left parietal and squamosal are preserved. The two parietals would form a hexagon, wider than long, and with a middle spine extending forwards along the interfrontal suture. The anteromedial corner of the squamosal extends anteriorly to the posterior part of the postorbital constriction.

Only partial left jugal (V 16293) and anterior part of right jugal (V 16294) are preserved. The preserved parts tend to show that the jugal bone is long and extends to the anterior border of the orbit. The middle part of the zygomatic arch is robust. The upper posterior border of the anterior root of the zygomatic arch is deeply concave. The anteriormost point of the upper posterior border (= anterior border of orbit) is situated anterior to the posterior ends of both nasal and premaxilla. The postorbital constriction is also prominent. The two frontal crests converge posteriorly to meet each other near the postorbital constriction to form the distinct sagittal crest. The temporal fossa is large. The temporal crest is short and concave laterally, becoming gentle posteriorly.

Lateral view (Figs. 24 A3, 25 B)   The anterior ends of the premaxillae extend beyond those of the nasals. The maxillary diastema is much longer than upper cheek tooth row (see Tabs. 11, 12). The anterior part of the skull (before M1) is subequal to the posterior part (including M1) of the skull in length (see Tab. 11). There is also a longitudinal bulge on the lateral side of the premaxilla, which is formed by the alveolus of I2. Above the longitudinal bulge is a triangular concavity. The attachment area of lateral masseter muscle is also limited within the maxilla, with its anterior border being roughly S-shaped: its upper part being concave anteriorly and its lower part being convex anteriorly beyond the ventral slit of infraorbital foramen. The masseteric tubercle (mt) is situated closer to the incisive foramen than to the infraorbital foramen. The orbit is small.

The foramina in the orbital area are located relatively anteriorly and close to each other. The sphenopalatine foramen (spf) is large and located above the M1. The posterior superior alveolar foramen (psaf) is located antero-inferior to the spf and above the anterior border of M1. The dorsal palatine foramen (dpf) is located inferoposterior and lateral to the spf, above the M1/2. A tiny furrow can be seen to extend from the dpf below the anterior alar fissure (aafi). The ethmoid foramen (etf) is small and located above the spf and M1. The optic foramen (opf) is subequal to the etf in size and located below and slightly posterior to the etf, above the posterior part of M1 and the dpf. The anterior alar fissure (aafi) is large and deep, without separation into upper and lower portions, situated directly posterior to the spf. A small *foramen alare parvum* (fap) is visible at the lower part of the external wall of the aafi. Unlike in *Pararhizomys*, the deep temporal foramen (dtf), masticatory foramen (mf) and buccinator foramen (bf) are confluent, forming a common slit. The glenoid fossa (glf) is long, extending to the nuchal crest, and the lateral part of the auditory bulla takes part in forming the glenoid fossa. The external auditory meatus is pushed downwards below the posterior part of the glenoid fossa. The postglenoid foramen (pgf) may be confluent with the petrotympanic fissure. Its posterior border is slightly anterior to the external auditory meatus and about 4 mm anterior to the nuchal crest. The mastoid process (msp) is distinct. The stylomastoid foramen (styf) is located in the pit between the external auditory meatus and the mastoid process. The nuchal surface slightly slants forwards, facing upward and posteriorly. The lower border of the auditory bulla (ab) is much lower than that of the occipital condyle.

Ventral view (Figs. 24 A1, 25 A)   The incisive foramen is located at the middle of the maxillary diastema and is about 1/4 the length of the latter. The medial end of the premaxillo-maxillary suture intersects the incisive foramen at the middle, but its main (lateral) part stand near or slightly anterior to the anterior end of the incisive foramen, and serrated. The broad zygomatic plate mainly extends anterosuperiorly, with its lower surface slightly concave. It is nearly opposite to the incisive foramen in position: the anteriormost point of its arched anterior border is nearly in the same transverse line as the anterior end of the incisive foramen, and the anteriormost point of its emarginated posterior border is nearly opposite to the posterior end of the latter. The masseteric tubercle (mt) is a pit located slightly anterolateral to the posterior end of the incisive foramen. I2 originates from the middle of the maxilla near M1 and forms a bulge anterior to M1. The left and right upper tooth rows are slightly divergent posteriorly. The palatine sulci are deep. In the sulci there are two pairs of posterior palatal foramina (ppf) situated opposite to M1/2. The anterior pair is larger than the posterior one. The posterior maxillary foramen (pmf) is located posteromedial to the M3. The parapterygoid fossa is shallow, with multiple small foramina. The mesopterygoid fossa (mptf) is relatively narrow and subequal to the pterygoid fossa (ptf) in width. The left and right internal pterygoid processes (ipp) do not slant, nearly vertical to the ventral side, and extend parallel with each other. The left and right external pterygoid processes (epp) are low and short, extending posterolaterally. The posterior border of the hard palate is located at about the same transverse line as the posterior border of M3. The *foramen ovale* (fo) is confluent with the middle lacerate foramen (mlf), forming an unified large foramen (fo-mlf). The posterior foramen of deep temporal nerve canal (dtfp) joins with the confluent fo-mlf, while the posterior foramina of the masseteric nerve canal and buccinator nerve canal are open at the posterior end of the alisphenoid canal (asc). The posterior aperture of the alisphenoid canal (ascp) is situated anteromedial and inferior to the confluent fo-mlf. The posterior foramen of pterygoid canal (pcpf) is located at the medial side of the posterior part of the pterygoid fossa, near the base of the internal pterygoid process, and nearly in the same transverse line with the posterior aperture of the alisphenoid canal (ascp). The left pcpf is composed of a single aperture, but the right pcpf is separated into two apertures.

The basisphenoids and basioccipitals of both V 16293 and V 16294 are well preserved, but the ventral surfaces of the two bones of V 16293 are slightly damaged. The ventral sides of the combined basisphenoid and basioccipital form a narrow trapezium. A pair of well-developed basilar tubercles (bt) are separated by a distinct sagittal groove. On the ventral surface of the basioccipital a sagittal crest and a pair of broad depressions situated bilateral to the crest are present. A single and large hypoglossal foramen (hyf) is situated anterolateral to the occipital condyle. The two occipital condyles are short longitudinally and wide transversely. The anterior ends of the occipital condyles are separated from each other by a distinct notch. The auditory bulla (ab) is well inflated, oval in shape, with its major axis extending in anteromedial-posterolateral direction. A spine is well developed, extending in the same direction as the major axis of the bulla. A weak crest extending from the spine posterolaterally to the highest point of the auditory bulla is present on the surface of the auditory bulla. Lateral to the crest there is a weakly expressed concavity facing the pterygoid fossa. The foramen for the canal of Huguier (chuf) opens anteromedial to the external auditory meatus. The foramen for the Eustachian canal (eucf) opens anteromedially to the spine. The large internal carotid foramen (icf) is located behind the spine. There is a distinct groove extending from the internal carotid foramen inferoposteriorly on the medial side of the auditory bulla. The jugular foramen (juf) is large and situated posteromedial to the auditory bulla. There is also a groove extending from the jugular foramen antero-inferiorly on the medial side of the auditory bulla. But the two grooves mentioned above do not meet with each other. On the medial wall of the bulla near the juf no distinct foramen for stapedial arterial is seen.

Posterior view (Fig. 25 E)   The nuchal surface of the skull is similar to that of *Pararhizomys*. But, unlike in *Pararhizomys*, the external occipital crest (eoc) is more distinctly expressed, extending vertically. The occipital is wider in lower part. The petromastoid portion (Pms) is relatively small, with its upper part being slightly wider than the lateral part of the occipital beside the *foramen magnum* in width. The mastoid foramen (msf) is situated at the medial end of the upper part of the petromastoid portion-occipital suture. The paramastoid process (pmp) is small, extending inferolaterally. The *foramen magnum* (fm) is large and oval in outline, wider than high. The left and right occipital condyles (occ) are widely separated.

(2) Mandible (Figs. 24 B, 26; Tab. 11)

The mandible of V 16293 is poorly preserved, with its ascending rami mostly broken: coronoid and angular processes are mostly broken and the right condyloid process is missing. The mandible of V 16294 is well preserved, but the left and right angular processes and left coronoid process are broken.

The mandible is sciurognathous. The horizontal ramus is short and thick. Its lower border below the molars is convex. The symphysis extends anterosuperiorly, forming an angle of 40° with the lower border of the horizontal ramus. The chin process and digastric fovea are distinct. The mandibular diastema is longer than the lower molar row. From lateral view the diastema is also S-shaped, but its posterior concavity is shallower than that in *Pararhizomys*. The masseteric fossa extends anteriorly to and below the m1. The masseteric ridge is well-developed and flaring laterally. The single mental foramen is located below the anterior border of m1 and at the middle of the horizontal ramus, nearly in the same horizontal level as the anterior end of the masseteric fossa.

The ascending ramus of the mandible is long. Although its anterior border originates lateral to m2, the coronoid process (crp) inclines more posteriorly than in *Pararhizomys*, with its anterior border forming an angle of 120° with the alveolar border of lower molars. Its upper part bends slightly posteriorly and its posterior border is slightly concave. The fovea for attachment of the temporal muscle (temporal fovea, tf) is a large depression on lower part of the medial side of the coronoid process. The mandibular notch is deep, about 2/5 the mandibular height. The mandibular foramen (mdf) is large and located below the posterior border of the coronoid process, at about the same level as the occlusal surface of the lower molars. Its distance to the posterior side of m3 (7–7.5 mm) is longer than in *Pararhizomys*. The condyloid process (cdp) seems to be longer than the coronoid process. It extends posterosuperiorly, with its upper part bending medially. On the upper part of the medial side of the condyloid process there is a distinct depression, which may be the attachment area of external pterygoid muscle and is here called as external pterygoid fovea (eptf). The articular facet for the glenoid fossa is hemi-spindle in shape, with a sharp anterior end and rounded posterior border, and the convex surface facing posterolaterally. The internal pterygoid fovea (iptf) is large and deep, triangular in outline, with its anterior end extending to below the coronoid process. The posterior end of the i2 alveolus forms a bulge on the buccal side of the condyloid process. The posterior border of the ascending ramus is only slightly concave.

(3) Teeth (Fig. 27; Tab. 12)

Dental formula is 1·0·0·3/1·0·0·3. The molars are similar to those of *Pararhizomys* in basic morphology. They are lophodont, moderately high-crowned and rooted. The molars decrease posteriorly in size. The crowns of the molars are relatively lower than those of *Pararhizomys*. The mesiodistal hypsodonty is distinct on M1 and m3 only. The reentrants on the buccal side are deeper (shallower) than those on the lingual side in the upper (lower) molars.

The upper molars of both V 16293 and V 16294 are well preserved. The M1 is trapezoid in occlusal view, longer than wide, anterior border wider than posterior one, and buccal side longer than lingual one. It is mesially hypsodont, with undulating dental tract on buccal side. The occlusal surface has two buccal reentrants (anterosinus and mesosinus) and one lingual (sinus). The anterosinus and mesosinus are mainly transverse and open buccally on both V 16293 and V 16294. The mesosinus is longer than anterosinus transversely, with posteriorly bending lingual part. The sinus is open in V 16293, but closed in V 16294. The sinus is shorter than the two buccal reentrants transversely and extends transversely towards the paracone between the anterosinus and mesosinus. Its buccal end reaches only the level of the lingual end of the anterosinus, and only overlaps the mesosinus partially. The prosinus is distinct in V 16293, but weak in V 16294.

The M2 is oval in occlusal view, slightly longer than wide. No distinct mesial hypsodonty is seen on

M2. The undulating dental tract on buccal side is weak or indistinct. The occlusal surface also has two buccal reentrants and one lingual one. In the young V 16293 the anterosinus is present in the form of a closed basin, while in V 16294 the anterosinus becomes vestigial on M2. It seems that the anterosinus of M2 is being reduced in this species. The mesosinus is long transversely and bends posteriorly. The sinus extends transversely or slightly anterobuccally and overlaps the mesosinus partially. The sinus of M2 is longer than that of M1, but only extends to the level of the lingual end of anterosinus, without going beyond or overlapping the latter.

The M3 is oval in occlusal view, with slightly narrower posterior side. No mesial hypsodonty and undulating dental tract are seen. The occlusal surface also has two buccal reentrants and one lingual one. The anterosinus becomes a closed basin on the two third upper molars of V 16293 and the left M3 of V 16294, but disappears on the right M3 of V 16294. Probably, as in M2, the anterosinus is reduced on M3 as well. The mesosinus and sinus are transverse and opposite each other. The mesosinus is shorter than the sinus transversely but deeper than the latter. Some characters of M3 are variable. In V 16294 the mesosinus is situated opposite the sinus, but remains separate from the latter; while in V 16293, the mesosinus joins the sinus to form a transverse groove separating the M3 into anterior and posterior parts. In addition, on the right M3 of V 16293 there is a short longitudinal groove extending from the middle part of the confluent transverse groove posteriorly (see Fig. 27 A1). On the left M3 of V 16293 no such a longitudinal groove can be seen (see Fig. 27 A2).

I2 bends antero-inferiorly, but the curvature is less than in *Pararhizomys*. It is proodont, with its anterior end extending antero-inferiorly rather than backwards. The cross section of I2 is triangular in form, with a slightly convex labial side and a rounded lingual angle, on which there is a shallow longitudinal groove. The enamel covers the labial side completely, but only small parts of the medial and lateral sides. The boundary between the labial and lateral sides is rounded, but that between the labial and medial sides is angular, forming a longitudinal crest. The surface of the labial side is smooth, without distinct longitudinal ridge on it.

The m1 is oval in occlusal view, longer than wide, with a wider posterior side. No distal hypsodonty is seen. The occlusal surface has two buccal reentrants (protosinusid and sinusid) and one lingual one (mesosinusid). The m1 of V 16293 is preserved well. The two buccal reentrants are transverse and open buccally. The sinusid is longer than the prosinusid transversely. Among the reentrants the mesosinusid is the longest and bends forwards. Its buccal part extends anterobuccally towards the protosinusid, but does not reach the latter. The anterior parts of the two first lower molars of V 16294 are broken, thus the prosinusid is missing. The mesosinusid is long and still opens buccally. The mesosinusid bends forwards on the right m1, but on the left m1 it is separated into two parts: a transverse open lingual part and a closed buccal basin extending anterobuccally.

The m2 is oblate in occlusal view, subequally long and wide, with buccal side longer than lingual one. The m2 is shorter and wider than m1, but is similar to the latter in structure. No distal hypsodonty is observed. The occlusal surface has also two buccal reentrants and one lingual one. The two buccal reentrants are nearly transverse and the protosinusid is shorter than the sinusid. The mesosinusid is longer than the two buccal ones and bends forwards. Its buccal part extends towards the protosinusid, without reaching the latter. On the less worn V 16293 three reentrants of the m2 are open, while on the m2 of V 16294 the protosinusid and sinusid are closed and only the mesosinusid is open.

The m3 is oval in occlusal view, longer than wide, with a narrower posterior side. The crown is distally hypsodont. The occlusal surface has two buccal reentrants and one lingual one as well. The protosinusid is reduced into a closed basin on both third lower molars of V 16293, but disappears on those of V 16294. The mesosinusid is longer than the sinusid, with the lingual part bending anteriorly on m3 of V 16294. The mesosinusid is short in V 16293. Its lingual part bends towards the protosinusid, without reaching the latter. The sinusid extends transversely to the mesosinusid, nearly in the same transverse line with the lingual part of the mesosinusid.

The i2 is similar to the I2 in curvature. It is procumbent anterodorsally. The cross section forms an equilateral triangle, longer than wide, with a slightly convex labial side and rounded lingual angle. It is narrower than the I2 proportionally. The labial enamel layer slightly expands to both lateral sides. The boundary between the labial and lateral sides is rounded and that between the labial and medial sides is crested. On its labial side two fine longitudinal ridges can be seen on the V 16294, but only one distinct longitudinal ridge is present on that of V 16293.

(4) Postcranial bones (Figs. 28–55)

Most of the anterior part of the postcranial skeleton of V 16293 is preserved, including 7 cervical vertebrae, 7 thoracic vertebrae, several ribs, and partial forelimbs. Of the posterior part of the skeleton only one posibble phalanx of pes is found. The vertebral column is almost preserved in original articulated state, but the limb bones are mostly displaced from their original articulated positions (Figs. 28–30). In order to know their morphologies in more details, some of the limb bones were separated from the skeleton so that the buried parts or bones can be observed as well (Fig. 29).

Cervical vertebrae (C's)

The neck of *Pseudorhizomys indigenus* is short, with 7 cervical vertebrae closely arranged. The total curved length of the whole neck is 30 mm measured on dorsal side, 21 mm measured on ventral side.

Atlas (At; Figs. 28–31; Tab. 6)   The atlas of V 16293 is well preserved, but the wings are damaged. The atlas forms an oval ring, wider than high, but very short in length.

The dorsal arch (da) is bow-shaped, much wider than long. The dorsal tubercle (dt) is stout, located at the middle of the dorsal side, with its top surface roughened. A distinct pit is present on its anterior slope, which may be the attachment area of the *membrana atlanto-occipitalis dorsalis*. The lateral vertebral foramen (lvf) and the *foramen alare* (fa) are fused into a large lateral vertebro-alar foramen (lvaf), which is located on the anterior 1/3 of the lateral part of the dorsal side. The ventral side of the dorsal arch is smooth and straight longitudinally. The internal lateral vertebral foramen (ilvf) is located on the lateral parts of the ventral side, posterior to the mediosuperior end of the anterior articular fossa. The distinct groove for vertebral artery (gva) is located between the lateral mass of the arch and the caudal articular facet. On the floor of the groove there are two foramina, which may be the nutrient foramina.

The ventral arch is shorter but thicker than the dorsal arch. The odontoid fossa (odf) on the dorsal side of the ventral arch is long and smooth. On the ventral side of the ventral arch the middle part is convex, and the lateral parts are concave. The ventral tubercle (vt) is large, projecting posteriorly. On the bases of the left and right lateral sides of the vt a pair of tiny foramina is present on each side (see Fig. 31 B). Bilateral to the vt a pair of concavities can be seen.

The vertebral foramen is roughly oval in form, but separated into two parts: the wider upper part is passed by the spinal marrow; the narrower lower part is mainly for holding the odontoid process. Between the upper and lower parts there are rough areas for attachments of the transverse ligaments of atlas.

The cranial articular fovea (craf) is kidney-shaped, with a wider and concave upper part and a narrower and flat lower part. The left and right cranial articular foveae are widely separated on their upper parts, but closer to each other in lower parts, separated by a notch. The caudal articular facet (caf) is also kidney-shaped, with slightly concave surface and thin borders. The wider upper parts are also separated widely, and the narrower lower parts are closer to each other, both joining the odontoid fossa.

The wings of V 16293 are poorly preserved, with their lateral and lower parts mostly broken. The preserved parts tend to show that the wing takes the form of a bending plate extending in anterosuperior-posteroinferior direction. The anterosuperior part of the lateral border is arched, with its anterior end 1 mm posterior to the anterior border of the cranial articular fovea. On the posterior side of the wing a crest extends transversely and becomes thicker laterally. The fossa atlantis (fat) is large and deeply concave. In this fossa the large and widely separated *foramen alare* (fa) and the *foramen alare inferior* (fain) are connected by a deep groove. The transverse canal is short, but the transverse foramen (trf) itself is large.

Axis (Ax; Figs. 28–30, 32; Tab. 7)   Of the axis of V 16293 only the body, pedicle of vertebral arch and the transverse process are preserved.

The axis is longer than atlas. Its body takes the form of a flattened cylinder, cross section of which is wider than high. The odontoid process (odpr) is conical with a rounded top. The anterior end of the odpr is covered by rough areas, which may serve as the attachment area for the apical dental ligament. On the dorsal part of the odpr there is a shallow transverse groove for passing of the transverse ligament of atlas. The articular facet on ventral side of the odontoid process is convex transversely and straight longitudinally. The dorsal side of the body of the axis is flat at the middle, accompanied by two longitudinal shallow grooves on lateral sides. The surface of the anterior part of the longitudinal groove is rough, which may serve as the attachment area for the medial ligament of atlanto-axial joint. The ventral side of the body is transversely convex, anteriorly limited by an arched crest. The two lateral parts of the ventral surface are weakly concave. No distinct ventral spine is present. Along the posterior border of the ventral side there are two symmetrical tubercles. The vertebral fossa (vfs) is oval in shape, slightly concave. The epiphysial line is distinct on the dorsal side, but weak on the other sides.

The cranial articular facet of axis (aaf) is oval in outline, higher than wide, with a slightly convex surface facing anterolaterally. The lateral and ventral borders of the cranial articular facet form a semicircle. The lower end of the medial border of the articular facet tends to connect with the odontoid process, but they are separated by a small notch where they connect with each other. The dorsal border of the articular facet reaches to the pedicle of the vertebral arch.

The pedicle of vertebral arch is short. The postzygapophysis (pzy) extends from the posterior border of the pedicle. The facet articulating with the facet of the prezygapophysis of C3 (fpzy) faces inferoposteriorly and laterally, forming an angle of about 25° with the dorsal side of the body. The facet on the postzygapophysis is oval-shaped, longer than wide, with a flat surface. The caudal vertebral notch is distinct.

Of the axis of V 16293 the transverse process is only partially preserved. The transverse process extends from the junction of the pedicle and the body, and has two roots. The dorsal root of the transverse process is a narrow, triangular plate extending posterolaterally. The ventral root of the transverse process is broken. The transversal foramen is large.

C3–C7 (Figs. 28–30, 33; Tab. 8)   C3–C7 of V 16293 are in their original quasi-articulated states and well preserved. They are similar to each other in general morphology: the body is of a flattened cylinder, the dorsal part of the vertebral arch is wide and short bow-shaped, the spinous process (sp) is low, the pedicle of the vertebral arch (pva) is short, the pre- and postzygapophysis originate from the pva, the anterior and posterior vertebral notches (vn) on the anterior and posterior borders of pva form intervertebral foramina (invf), and the transverse process (trpr) originates from the lateral side of the pedicle and the body, etc. Some general rules of the variation from C3 to C7 can be observed as follows: The spinous process becomes more distinct from C3 to C7 and more posteriorly slanting from C6 and C7. In C3–C6 the transverse process has two roots and a large transverse foramen respectively, while in C7 the transverse process has only one large root and lacks transverse foramen. The transverse process extends posterolaterally in C3–C5; but in C6 it extends transversely with its lateral part enlarged and being separated into two laminae: the inferior lamella of transverse process (lvvc, *lamina ventralis vertebrae cervicalis*) is well-developed and forms an anteroposteriorly long plate bending ventrally, the dorsal lamella of transverse process (*lamina dorsalis vertebrae cervicalis*, ldvc) is thin rod-like, extending anterolaterally; then it extends slightly anterolaterally in C7. The vertical distances from the transverse process to the pre- and postzygapophysis gradually enlarge from C3 to C6, but it shortens again in C7. Finally, the facet for the head of the first rib is present on the posterior border of the ventral side in C7 only.

Thoracic vertebrae (T's)

In V 16293 seven thoracic vertebrae, including first to sixth thoracic vertebrae and a posterior thoracic vertebra, are preserved. T1–T6 are in their original quasi-articulated states.

T1–T6 (Figs. 28–30, 33, 34; Tab. 13)   T1–T6 are similar to C3–C7 in general morphology of the body and the arch. However, unlike in C3–C7, in T1–T6 the body and vertebral arch are longer and narrower, the spinal process is more developed, the pedicle of the vertebral arch is shorter and thicker, on the anterior and posterior corner of the ventral surface of the body there is an articular facet for the head of the rib respectively, and the transverse process has an articular facet for the tubercle of rib on its lateral end. They are similar to C7 but different from C3–C6 in transverse process having only one root and lacking transverse foramen.

The differences among T1–T6 can be summed up in the following lines. From T1 to T6, the bodies gradually become longer and narrower; the vertebral arches become longer; the distances between the pre- and postzygapophysis increase; and the maximum widths of the lateral borders of the pre- and postzygapophysis become narrower. The form and inclination of articular facets of the pre- and postzygapophysis vary as follows: the facets on the prezygapophysis vary from oval, longer than wide in T1 to shorter and circular in other T's in outline; their surfaces face mediosuperiorly in T1 and T2, become facing dorsally in T3 and then laterosuperiorly in T4 and T5; the facet of postzygapophysis faces lateroventrally in T1, mainly ventrally in T2, then medioventrally in T3–T5. The transverse processes become more robust from T1 to T5. Their directions are also variable: extending transversely in T1 and T2, anterolaterally in T3 and T4, and at last, bending dorsally in T5. The relative positions of the transverse processes vary as follows: lower than the zygapophysis in T1; their dorsal borders in the same level as the lateral borders of prezygapophysis in T2; higher than the lateral border of the prezygapophysis, but in the same level as the lateral border of postzygapophysis in T3; and higher than both the pre- and postzygapophysis in T4 and T5. The spinous processes gradually enlarge from T1 to T5 and become triangle-pyramid in T3–T5. Their directions vary from vertical to posterodorsally oriented, with the degree of inclination increasing from T3 to T5. The positions of the spinous processes also vary: they are located in the central parts of the vertebral arches in T1–T3, moving to posterior parts of the vertebral arches in T4–T6, etc.

Posterior thoracic vertebra (T?, Figs. 28–29, Tab. 13)   In V 16293 there is another thoracic vertebra, which is displaced from the articulated thoracic vertebrae, lying diagonally beside the T5 and T6. Of this vertebra only part of the body is preserved. The length of the body is 7.2 mm, much longer than that of T6. It may be one of posterior thoracic vertebrae.

Ribs (R; Figs. 28–30, 35)

Of V 16293 only the anterior 6 pairs of ribs and the left 7th rib are preserved. The first pair of ribs and the left 2nd–7th ribs are preserved in original quasi-articulated positions, but the right ribs are more or less displaced.

R1 (Figs. 28, 35)   The two 1st ribs are well preserved, with only the epiphyseal heads still unossified with their shafts. The two epiphyseal heads may be represented by 2 small bony pieces adhered to the skeleton bilateral to the bodies of the C7 and T1 (see Fig. 29, 33; hR1?). The 1st rib is rather short and stout, with very long neck relative to its shaft. The shaft is rather flattened in anteromedial–posterolateral direction, slightly curved and there is a weak costal groove on its lateral side. The lower part of the shaft becomes wider in anteromedial-posterolateral direction and slightly twisted posterolaterally. The angle of rib is indistinct. The sternal extremity is truncated by a rough surface of oval form, representing the costo-chondral articulation with the costal cartilage. The costal neck (cn, collum of costae) is long and straight, making the separation of the costal head (ch) from the tubercle of rib (tur) particularly wide. The costal head is larger than the tubercle of rib in size. The epiphyseal surface joining the head is irregular in form, roughly convex and transversely, and pointed anteriorly. The articular surface on the tubercle of rib is oval-spherical, serving for articulation with the transverse process of T1.

The maximum chord length of R1 is 11.3 mm (left) and 11.1 mm (right). The length from costal head to posterior border of the tubercle of rib is 6.2 mm (left) and 6.6 mm (right). The length of the shaft of R1 (distance from tubercle of rib to ventral end) is 10.5 mm (left), 10.4 mm (right). The width and thickness of the proximal end of R1 on the posterior side are 2.6 mm and 1.5 mm (left), and 2.6 mm and 1.5 mm (right). The width of the ventral end is 1.9 mm (left) and 1.8 mm (right). The anteromedial-posterolateral distance of the ventral end is 3.5 mm (left and right).

R2 (Figs. 28, 29, 34)   Only the left 2nd rib of the V 16293 is well preserved. It is longer, more slender, flatter and more curved than R1. The upper part of the shaft is flat anteroposteriorly and its lower part is oblate-cylindrical in form. The sternal extremity of the shaft is rough and oval, serving for receiving the costal cartilage. The costal neck is also long and straight, but shorter than that of R1. The costal head is slightly larger than the tubercle of rib in size. On the medial side of the tubercle of rib there is an oval facet with a saddle-shaped surface.

The maximum chord length of R2 is 16.3 mm. The length from costal head to tubercle of rib is 5.8 mm. The length of the shaft of R2 (chord length from tubercle of rib to ventral end) is 16.4 mm. The width and thickness of the proximal end of the shaft of R2 are 2.6 mm and 0.9 mm. The anteroposterior distance of the ventral end is 2.6 mm. The width of the ventral end is 1.8 mm.

R3 and R4 (Figs. 28–30, 34)   Of V 16293 the costal heads and the ventral ends of both R3 and R4 are broken. R3 and R4 are similar to R2 in morphology. But they are much longer than R2. The costal grooves on the posterior sides of the shafts are shallow on R3 and R4, but longer than those of R2, which can extend to the ventral ends. On R3 and R4 the tubercles of ribs are lower than those of R1 and R2 in height, and separated from the shafts by distinct notches. The articular facets for the transverse processes are oval and slightly convex.

The chord lengths of the preserved parts of R3 and R4 are 19 mm and 25 mm respectively. The widths and thickness of the proximal ends of the shafts are 2.6 mm and 0.7 mm in R3 and 2.5 mm and 1.1 mm in R4.

R5–R7（Figs. 28–30, 34）   The distal ends of both left and right R5 are missing. The left R5 is preserved better than the right one. The shaft of R5 is similar to those of R2–R4 in morphology, but longer than the latter. On the costal head of R5 there are two articular facets articulating with the bodies of T4 and T5 respectively. The preserved maximum chord length of left R5 is 25.5 mm.

Only ventral part of shaft of left R6 is preserved. The preserved length of the shaft of R6 is 12.3 mm. Only left R7 is preserved. The two ends of R7 are also missing. The shaft of R7 is longer and thinner than those of R1–R6. The preserved chord maximum length of R7 is 26 mm.

Forelimb

Only some parts of the forelimb bones of V 16293, including shoulder girdle (clavicle and scapula), humerus, radius, ulna and partial manus, are preserved.

Clavicle (Cla; Figs. 28, 36)   Both the left and right clavicles of V 16293 are partially preserved with their acromial ends broken. The left clavicle is preserved better than the right one, with most of its proximal part preserved. The clavicle is a slender and long bone, curved and slightly twisted, with its anterior and posterior sides slightly wider. The sternal end is robust. The articular surface for sternum is rounded and rough. The shaft gradually becomes more slender towards the acromial end.

The lengths of the preserved parts of the clavicles are: 20 mm (left) and 15.6 mm (right). The lengths and widths of the sternal ends of the clavicles are: 3.4 mm and 3.3 mm (left) and 3.4 mm and 3.1 mm (right).

Scapula (Sc; Figs. 28, 30, 37)   Only the lower parts of the left and right scapulae are preserved. The glenoid cavity (glc) is roughly pear-shaped, with its transverse diameter slightly longer than half of the major (anteroposterior) axis, which is also longer than the width of the neck of the scapula (nsc). No glenoid notch is differentiated. The line of the spine of scapula (spsc) intersects the major axis of the glenoid cavity at an acute angle (posterior to the intersection). The scapular tuber (sctu) is very prominent, extending antero-inferiorly. The coracoid process (cpr) is very long and plate-like, takes its origin from the anteromedial corner of the tuber, separated from the tuber by a shallow notch and extending anteromedially and inferiorly. The coracoid process serves for the attachment of the *m. coracobrachialis*, *m. biceps brachii* and *m. pectoralis profundus*. Of the spine of scapula (spsc) only its basal part is preserved and its lower end extends to the level of the neck of scapula (nsc), about 5 mm above the lateral border of the glenoid cavity. The acromion is broken away. The anterior border of the scapula is largely broken. The preserved part of the supraspinous fossa (sspf) forms a very narrow fan in shape, minimum width of which is only 1.5 mm. The infraspinous fossa (ispf) is preserved better than the supraspinous fossa is. It is also fan-shaped, but its lower part is wider than that of the latter. The lower part of the posterior border of the scapula is concave. It is wider and gently convex transversely near the neck of scapula, but becomes a narrow ridge upward. Near the posterior border of the infraspinous fossa there are several *lineae musculares*, which may serve for the attachment of *m. infraspinatus* and *m. teres minor*. The subscapular fossa (sscf) is also narrowly fan-shaped. A longitudinal bulge corresponding to the concavity of the infraspinous fossa is present on the medial surface of the scapula. There is also a well-developed muscular line near the posterior border of the subscapular fossa and a rough tubercle on the posterior side of the neck of scapula, which may be the attached areas of *m. triceps brachii* and *m. teres minor*.

The minimum width of the neck of the scapula is 4 mm on both left and right scapulae. The anteroposterior lengths and widths of the glenoid cavities are 7 mm and 3.6 mm (left) and 7 mm and 3.8 mm (right) respectively.

Humerus (Hu; Figs. 28, 30, 38; Tab. 14)   Both left and right humeri are well preserved, except that the head of the left humerus is slightly damaged. V 16293 is of a young individual and the epiphyseal lines can clearly be recognized on the proximal parts of the humeri.

The humerus is similar to those of *Rhizomys* and *Aplodontia* in general morphology. The proximal end is transversely wide and anteroposteriorly thick, slightly wider than thick. The head of humerus (head) extends posterosuperiorly and overhangs the posterior border of the shaft. The articular facet for the glenoid cavity of the scapula is oval-hemispheric in form, much larger than the glenoid cavity in size. The neck of humerus is distinctly below the head. The greater tuberosity (gtu) is robust, but does not surpass the upper border of the head. The depression and its surrounding rough area on the top of the greater tuberosity may serve for the attachment of the *m. supraspinatus*; the rough convexity on the lateral side of the greater tuberosity may serve as the attachment of the lateral head of *m. triceps brachii*. On the lateroposterior side of the greater tuberosity there is also a distinct depression in the upper part, which may serve as the attachment of the *m. infraspinatus*, and below the depression the small convexity may serve as the attachment of the *m. teres minor*. The lesser tuberosity (ltu) is smaller and lower than the greater tuberosity. The attachment area for *m. subscapularis* is also distinct on the top of the lesser tuberosity. The intertubercular groove (= bicipital groove, bg) is wide, without intermediate ridge in it.

The shaft of the humerus forms a twisted trihedron. The upper part of the anterior side of the shaft is wider than its lower part. The upper part faces slightly medially and its surface is flat longitudinally but slightly concave transversely, forming a longitudinal shallow groove, which is the continuation of the intertubercular groove. The lower part of the anterior side of the shaft twists medially and has a flat or slightly concave surface. The crest of the greater tuberosity (cgtu, = humeral crest) is well developed, extending from the greater tuberosity to the deltoid tuberosity (dtu). The anterior side of the crest of the greater tuberosity forms a rough area for the attachment of *m. pectoralis superficialis* and part of *m. pectoralis profundus*. The deltoid tuberosity (dtu), which serves as the attachment of *m. deltoideus*, is well developed and forms a triangular plate protruding laterally above the middle part of the shaft of humerus. On the lateral side of the shaft the upper part is narrow and slightly concave transversely, and the lower part becomes wider and twisted anteriorly to form a broad musculo-spiral groove (msg), where the *m. brachialis* passes by. The musculo-spiral groove (msg) is bounded by the well-developed lateral epicondylar crest (lepcr). The lateral epicondylar crest is about 1/3 the length of the shaft. On its dorsal side there are two depressions serving for the attachment of *m. extensor carpi radialis* and *m. extensor digitorum communis* respectively. The posterior side of the shaft is flat longitudinally, slightly curves posteriorly near the head, but does not twist. The upper part of the posterior side of the shaft is narrow and convex transversely and its lower part becomes wider and slightly concave transversely. The upper part of the medial border of the shaft is formed by the crest of lesser tuberosity (cltu), which is distinct and extends from the lesser tuberosity to the upper third of the shaft. The rough area on the anterior side of the crest of lesser tuberosity serves for the attachment of *m. latissimus dorsi*. Posterior to the crest of lesser tuberosity there is a rough concavity for attachment of *m. teres major*. Below the crest of the lesser tuberosity there is also a rough area, which may serve as the attachment of *m. coraco-brachialis*. The lower part of the medial border of the shaft is formed by the medial epicondylar crest (mepcr). There is a concavity between the lower part of medial epicondylar crest and the upper part of the medial epicondyle.

The distal end of the humerus is wider and flatter than the proximal end. Its transverse axis is parallel to that of the proximal end, and forms an acute angle with the horizontal projection of the major axis of the head of humerus. The capitulum (capm), the articular facet for the articular fovea of the radius, is semi-spindled in shape. The trochlea (tro), articulating mainly with the semilunar notch of the ulna and partially with the radius, is oblique and asymmetric. The upper part of the anterior part of the trochlea is slightly narrower than that of the capitulum, but wider downwards, then turns posteriorly and occupies the whole articular facet on the posterior side. The boundary between the capitunum and trochlea is distinct at the inferoposterior side, but indistinct on the anterior side. The coronoid fossa (cof) takes the form of a rounded triangle. It is about 1/2 the width of the articular facet. The surface of the coronoid fossa is rough, but without distinct furrows and ridges. Near the medial margin of the fossa there is a tiny nutrient (?) foramen. The olecranon fossa (olf) is wide and deep, wider than both the posterior parts of the trochlea and the coronoid fossa. The olecranon fossa and coronoid fossa are subequal in depth, but do not penetrate through. The medial epicondyle (mep) is well developed and extends laterally. There are three concavities on the medial epicondyle: the anterosuperior one is the smallest, which may be the attachment area of *m. pronator teres*; the posterosuperior one, the attachment area of *m. flexor carpi radialis*; and the lower, the largest one, for *m. flexor digitorum profundus*, *m. flexor digitorum superficialis* and *m. flexor carpi ulnaris*, etc. On the posterior side of the medial epicondyle there is a wide groove for ulnar nerve. No entepicondylar foramen is present on the medial epicondyle. The lateral epicondyle (lep) is much smaller than the medial one in size. There are three small concavities on the lateral side of the lateral epicondyle, which may be the attachment areas for the *m. extensor digitorum lateralis*, *m. extensor carpi ulnaris* and *m. supinator.* For their measurements see Table 14.

Radius (Ra; Figs. 28, 30, 39; Tab. 15)　Both the left and right radii are preserved. The right radius is better preserved. Its distal epiphysis had originally been displaced from the shaft, later relocated to its original place. Of the left radius only the upper half and the distal end are preserved.

The radius is much more slender in comparison with the humerus. Both its proximal and distal ends are wide and thick. The head of radius (head) is wider than long (anteroposterior distance, APD). The outline of the articular facet for the humerus is that of a wide oval. The sagittal crest (sacr) is gentle, slightly concave longitudinally, and located nearer to the medial margin. The articular facet lateral to the sagittal crest, articulating with the capitulum of the humerus (fcapm), is large and concave, and occupies most of the proximal articular facet. The articular facet for the lateral part of the trochlea of humerus (ftro), situated medial to the sagittal crest, is very narrow, facing mediosuperiorly. The capitular eminence (capem, = coronoid process of radius) is distinct. The articular facet for the ulna (ful) is wide and short, occupying almost the whole posterior side of the proximal end of the radius. Its surface is flat longitudinally and slightly convex transversely. The *tuberosity proximalis lateralis* is more prominent than the *tuberosity proximalis medialis*. The neck of radius (nra) is distinct.

The shaft forms an oblate cylinder, wider than long (in anteroposterior direction), bending posteriorly at the two extremities. The upper part of the shaft is narrow, becoming gradually wider from the middle of the shaft downwards. The anterior side of the shaft is slightly convex proximodistally; its upper part is flat, while the lower part is convex transversely. Along the medial border of the anterior side in the middle part of the shaft there is a rough area, which may be the attachment area of the *m. pronator teres* and *m. supinator*. Along the lateral border of the middle part of anterior side of the shaft there is a well-developed rough tuberosity extending downwards and medially, which may be the attachment area of the *m. extensor pollicis brevis* and *m. abductor pollicis*. The medial side of the shaft is narrow in its upper part, becoming wider downwards. Its surface is convex transversely and mostly flat longitudinally, except for the distal part, which is slightly concave. On the upper part of the medial side there is a smooth surface near the neck of radius, which may be the area passed by the tendon of *m. brachialis* and is called a groove for the *m. brachialis* (gbr) here. The lateral side of the shaft is straight longitudinally. Its upper part is narrow, slightly concave transversely, and its lower part becomes wider downwards and turns to face posteriorly, covered by well-marked grooves and ridges, which may be the attachment areas of *lig. interosseum antibrachii* (or interosseous membrane) of the forearm. The radial tuberosity (tura) is a well-developed rough concavity on the upper part of the lateral side of the shaft, which may be the attachment area of *m. biceps brachii*. The posterior side of the shaft is concave longitudinally and slightly twisted. Its upper part faces posteriorly and is concave transversely, forming a wide longitudinal groove. On the upper end of posterior side and just below the facet for the ulna, and lateral to the groove for *m. brachialis* there is a rough area, which may be the insertion of the *flexor digitorum profundus*. The lower part of the posterior side twists posteromedially, forming a continuous surface with the medial side.

The distal end is larger than the proximal end. From distal view it is trapezoid in outline, lateral side longer than medial side in anteroposterior direction. The articular facet for the scapho-lunar is broad: its lateral part is concave, and its medial part is flat or convex and turns to face inferolaterally. The radial styloid process (stypr) is prominent. On its medial side there is a longitudinal groove, which may be passed by the tendon of *m. flexor digitorum profundus*. The lateral side of the distal end has a rough and transversely concave facet, through which the radius may articulate with the ulna. On the dorsal side of the distal end there is a shallow longitudinal sulcus, which may be passed by the tendon of *m. extensor digitorum communis*.

Ulna (Ul; Figs. 28, 30, 40)　The right ulna of V 16293 is well preserved, but the olecranon is damaged and the distal end is preserved separate. Of the left ulna of V 16293 only the upper part is preserved, with the olecranon broken.

The ulna is more robust than the radius. Its upper part is prismatic in form, thicker in anteroposterior direction and narrower transversely. The shaft gradually becomes laterally flattened downwards. The proximal end is robust and the olecranon (ol) is much higher than that of the *processus anconaenus* (pran). The anterior border of the olecranon is very narrow, extending from the *processus anconaenus* superiorly. The medial side of the olecranon is flat, while the lateral side is concave longitudinally and convex transversely. The *processus anconaenus* is gently convex and located at the lateral part of the anterior border of the olecranon. The semilunar notch (semno) is deep. From anterior view, its upper part is narrower than the lower part. The surface is semilunar, concave longitudinally and convex transversely, without forming a distinct longitudinal crest. The medial and lateral articular facets are asymmetrical. The medial one is much larger than the lateral one, forming a long oval, facing medially. Its medial border is S-shaped, with a slightly concave upper part and a convex lower part. The lateral articular facet faces laterally and downwards. The coronoid process (copr) is prominent. The radial notch of the ulna (rano) is located below the lateral articular facet and lateral to the coronoid process. It is wide and short, and its surface is flat longitudinally and slightly concave transversely, facing anterolaterally.

The shaft is straight without twisting. The upper part is larger and its cross section forms a narrow triangle with three concave sides. The lateral side is the widest and the anteromedial side is the narrowest. The surface of the lateral side is concave transversely and presents a wide *fossa lateralis ulnae* (flul). It is subequal to that of *Rhizomys* in depth, but shallower than those in *Myospalax* and *Geomys*. The upper part of the anteromedial side of the shaft is much narrower transversely. The tuberosity of ulna (tuul) is located on the upper part of the anteromedial side and below the coronoid process. It is a rough concavity serving for the attachment of *m. brachialis*. The anterior edge of the shaft is wider transversely in upper part, becoming narrower downwards. The upper part of the medial edge of the shaft is more developed, and extends inferoposteriorly to meet the medial articular facet of the semilunar notch obliquely. Its middle and lower parts become weak downwards and merge with the posterior edge of the shaft. The posteromedial side of the shaft is wider in upper part, narrowing downwards to eventually disappear. The lower part of the anteromedial side of the shaft is wider, occupying the whole medial side. The cross section of the lower 2/3 of the shaft forms a wide oval.

Of the distal end of the right ulna only the epiphysis is preserved. From distal view, the distal end is oval in outline. Its posterior side forms a curved surface, concave longitudinally and convex transversely. The anterior side is partially broken on its lateral part. The styloid process of ulna is prominent and has a hemicylindrical facet articulating with cuneiform and pisiform on its distal surface.

Manus (Ms; Figs. 28, 29, 41)　The hand (manus) bones of V 16293 are only partially preserved in original position: only the right McI–McV are preserved in quasi-original position or only slightly dislocated. All the other bones are dispersed around. Therefore, it is difficult to identify every hand bone based purely on its relative position. A careful comparison of these hand bones with those of some extant rodents (*Rhizomys, Myospalas, Ondatra, Aplodontia*) and HMV 1942 (*vide infra* pp. 234–240) renders it possible to recognize and arrange them as follows (see Fig. 41).

Carpal bones (Carp; Fig. 41 A)　Of the carpal bones only the right trapezium and unciform are recognized.

Trapezium (Trm = Carpale I; Figs. 41 A, 42)　The dorsal side of the trapezium is trapezoid in outline. The articular facet for the scapho-lunar bone (fsc-lu) is fan-shaped, widened volarly. It is convex in dorsovolar direction and slightly concave transversely. On lateral side of the trapezium the articular facet for the trapezoid (ftrd) forms a narrow fan, widened volarly. The surface is concave in dorsovolar direction. On distal side of the trapezium the articular facet for McII (fmcII) is oblate, dorsovolarly longer than wide, facing inferolaterally. It forms an obtuse angle with that for the trapezoid (ftrd) and is separated from the latter by a weak crest. On the medial side there is an articular facet for the McI (fmcI). It is semicircular in outline. Its surface is slightly convex in dorsovolar direction and slightly concave proximodistally, facing inferomedially. It is separated from the articular facets for both the scapho-lunar and McII by crests.

Unciform (Un = Carpal IV+V; Figs. 41 A, 43)　The right unciform of V 16293 is well preserved. It is irregularly tetrahedral in form, much wider than long.

The dorsal side of the unciform (Fig. 43 B) is about triangular in form with its top upside down. The other three sides are: proximal, distomedial and distolateral. The two distal sides meet each other at about a right angle.

The proximal side (Fig. 43 A) is also triangular in outline, with a conical protuberance near the middle of the volar margin. Two articular facets are present on the proximal side. The medial facet articulates with the scapho-lunar (fsc-lu). It is strongly convex dorsovolarly and slightly concave transversely, facing mediosuperiorly. The lateral articular facet is for the cuneiform (fcu), larger in size and less steeply inclined than the medial one (fsc-lu). Its surface is strongly concave transversely and convex dorsovolarly, with the volar end almost reaching the facet for McIV on distal side. The two proximal facets are separated by a rough area behind the conical protuberance only.

The distomedial side is trapezoid in outline with a flat surface facing inferomedially. There are two articular facets: the mediosuperior articular facet for the magnus (fmg) is square in outline, while the inferolateral articular facet for McIII (fmcIII) is trapezoid in outline. No distinct boundary is seen between the two facets.

The distolateral side is triangular in outline and faces inferolaterally. There are also two articular facets on it. The medial facet for McIV (fmcIV) is a curved, concave surface, with its dorsal part wider than volar part. It is separated from the facet for McIII by an arched and concave crest. The lateral articular facet for McV (fmcV) is oval in outline, with its dorsal part wider than volar part. Its major axis extends from the dorsolateral end mediovolarly. The surface is concave in the direction of the major axis and flat in the direction of the minor axis. The crest separating the two facets is concave and arched. The distolateral side meets the proximal side at an acute angle and separated from the latter by an arched ridge stretching in dorsovolar direction.

Metacarpus (Mc)　Only right 2nd–5th and left 1st–3rd metacarpals are well preserved.

First metacarpal bone (McI; Figs. 41 B, 44; Tab. 16)　The left McI is very short and irregular in form. On proximal side there are two articular facets facing laterosuperiorly. The medial one is larger, articulating with the trapezium (ftrm). It takes the form of a rounded triangle, with its surface slightly convex dorsovolarly and slightly concave transversely. The lateral facet is smaller, articulating with the McII (fmcII). It is semicircular in outline, with a slightly concave surface. The two articular facets are separated by a weak crest.

The distal end is much narrower than the proximal one in transverse direction. The articular surface of the distal end is strongly diagonal relative to that of the proximal end so that the two form a sharp angle. The articular facet for the proximal phalanx (fphI) is extraordinarily large, forming a strongly arched surface in dorsovolar direction. Its dorsal part is semicircular in outline, convex, but without sagittal ridge. Its upper border is asymmetric, strongly bending medially. The volar part of the fphI has a weak sagittal ridge, becoming more distinct to the volar end. The sagittal ridge is situated closer to the lateral side. The bilateral grooves along the sagittal ridge are also asymmetrical: the medial one is well developed and deepens posteriorly, while the lateral one is smaller and weakly concave.

The shaft is very short. There is a deep pit on the dorsal side, but a broad and rough concavity on the volar side of the shaft. The medial side of the shaft is much longer than the lateral one. On the upper part of the medial side there is a large tubercle with a rough central concavity, which may be the attachment area for *m. abductor pollicis* (abp). Above the distal articular surface is a depression, surrounded by an eminence, which may be the attachment area of the medial collateral ligament. The lateral side is very short, and is represented only by a depression for the attachment of the lateral collateral ligament.

Second metacarpal bone (McII; Figs. 41, 45; Tab. 16)   Both the left and right McII of V 16293 are preserved. The right McII is preserved better than the left one, of which only the lower part is preserved.

The proximal end of McII is large. Its top view is trapezoid in outline, lateral border longer than medial one in dorsovolar direction. Of the three proximal articular facets the most medial facet articulates with the trapezium (ftrm). It is oval in outline, convex and facing mediosuperiorly. The articular facet for the trapezoid (ftrd) is the largest among the three facets. It is trapezoid in outline, with a notch on dorsal border. Its surface is slightly convex dorsovolarly and strongly concave transversely, facing dorsomedially. The lateral one is the articular facet for the *centrale* (fce). It is strip-shaped, very long in dorsovolar direction but narrow transversely. Its surface is convex in dorsovolar direction and separated from the facet for the trapezoid by a crest. On the medial side of the proximal end there is an articular facet for McI (fmcI). It is triangular in outline, with convex surface facing medially and separated from the facet for the trapezium by a weak line. On the lateral side of the proximal end there is an articular facet for McIII (fmcIII) below the facet for the *centrale*. It is a slightly curved strip, extending dorsovolarly, with its dorsal part wider than volar part. Its surface is concave dorsovolarly and faces laterodistally, forming a right angle with the facet for *centrale*. The area below the facet is rough and concave. On the volar side of the proximal end there is a wide facet below the facet for the trapezoid, which may articulate with the trapezoid when the carpal bends strongly. The dorsal side of the proximal end is slightly concave in its middle part. On its medial and lateral margins the crests for ligaments (cl) are distinct.

The shaft of McII is straight. Its upper part is narrower transversely and thicker dorsovolarly. Its cross section is trapezoid in outline, medial side being narrower than the lateral one. The shaft becomes wider but thinner in the middle part, becoming thicker again from there downwards. The cross section of the lower part becomes oblate. The dorsal side is straight longitudinally and slightly convex transversely. Near the medial margin of the dorsal side of the middle part of the shaft there is a distinct tuberosity, the attachment area for *m. extensor carpi radialis longus*, which is here called tuberosity of metacarpal bone (tmc). The volar side is slightly concave longitudinally and flat transversely. Along the medial margin of the upper part of the volar side there is a tuberosity, which may serve for the attachment of *m. flexor carpi radialis* and *m. adductor pollicis* (adp). On the lower end of the volar side there is middle longitudinal bulge accompanied by a pair of concavities. The medial concavity is larger than the lateral one. The medial and lateral sides of the shaft are straight and divergent downwards.

The distal end of McII is larger than the shaft in width and thickness, but slightly thinner dorsovolarly than the proximal end. The articular facet for first phalanx forms a transverse semicylinder. Its dorsal part is flat transversely and strongly convex longitudinally, but without sagittal ridge on it. On the volar side of the articular facet the sagittal ridge is well developed, accompanied by broad bilateral grooves, which are continuous with the paired concavities situated above them. The medial groove is wider and deeper than the lateral one. On both medial and lateral sides of the distal end there is a depression surrounded by eminence respectively, which is the attachment area for collateral ligament. The distal ends of both left and right McII show distinct epiphysial lines.

Third metacarpal bone (McIII; Figs. 41, 46; Tab. 16)   Both the left and right McIII of V 16293 are preserved. The proximal end of the left McIII and the distal end of right McIII are more or less damaged.

McIII is longer and thicker than McII. The proximal end is larger. From the top view it is trapezoid in outline. There are three articular facets on it. The middle is the largest one, articulating with the magnus (fmg). It is trapezoid in outline, lateral side longer than medial one in dorsovolar direction. Most part of the surface is convex dorsovolarly and concave transversely, facing mediosuperiorly. The laterovolar corner of the facet forms a triangular convexity, facing mainly volarly. The lateral one is the articular facet for unciform (fun), which is roughly semicircular in outline, narrowing volarly. Most part of the surface is flat, facing mainly laterosuperiorly and slightly volarly. Only the most volar part of the facet turns to face dorsolaterally. The boundary between the facets for the magnum and unciform is an arcuate ridge. The articular facet for McII is located on the dorsal part of the medial side. It is triangular in outline, with a wider dorsal border. The surface is slightly convex, facing mediosuperiorly. On the lateral side of the proximal end there is an articular facet for McIV (fmcIV). It is band-shaped, extending dorsovolarly and its surface is concave strongly dorsovolarly and slightly concave vertically, facing mainly inferolaterally and slightly volarly. It forms a right angle with the facet for the unciform. Along the lateral margin of the dorsal side of the proximal end the crest for ligament (cl) is well developed and long.

The shaft is straight. Its upper part is narrow, and its cross section is trapezoid, subequally wide and long. The shaft thins at the middle and then widens downwards. The cross section of the lower part of the shaft is broadly oval. The dorsal side of the shaft is flat longitudinally and slightly convex transversely. The tuberosity of metacarpal bone (tmc) is well developed and forms a large rough concavity surrounded by eminence located near the medial margin of the dorsal side of the shaft, about 1 mm below the proximal facets, which serves as the attachment of *m. extensor carpi radialis brevis*. The volar side of the shaft is concave longitudinally. Along each of the medial and lateral margins in the upper part there is a rough tuberosity, which may be the attachment areas of *m. flexor carpi radialis* and *m. adductor pollicis* respectively. The middle part of the volar side is slightly convex transversely. The lower part of the volar side is similar to that of McII.

The distal end of McIII is large, slightly wider, but thinner in dorsovolar direction than the proximal end. It is similar to that of McII in basic morphology. But it is slightly wider than that of McII. The epiphysial lines of the distal ends are visible in both McIII's.

The McIII of V 16293 is similar to that of V 16306.16 (*Pa. hipparionum*) in general morphology. It is different from the latter only in a few features: the transverse concavity of the facet for magnum and the concavity below the facet for McIV are shallower in depth, the lateral crest for ligament is weaker and the distal end is slightly more slender. Since the epiphysial lines are more clearly shown in McIII of V 16293 than in V 16306. 16, V 16293 should represent a younger individual. It seems that the above differences may represent largely the age variations. However, the less concave facets in V 16293 tend to show a weaker transverse movement between the magnus and McIII.

Fourth metacarpal bone (McIV; Figs. 41, 47; Tab. 16)   Only right McIV is well preserved. McIV is shorter than McIII, shorter but thicker than McII. The proximal end is large. The top side is triangular in outline, with its dorsal and lateral sides concave but mediovolar side slightly convex. There are two articular facets on the top side. The medial one articulates with McIII (fmcIII). It is oval in outline, dorsal part wider than volar part, and faces dorsosuperiorly and slightly medially. Its surface is convex, more so dorsovolarly than transversely. The articular facet for the unciform (fun) is larger, trapezoid in outline, lateral side wider than medial side dorsovolarly and faces slightly dorsomedially. Its surface can be subdivided into three parts: the dorsal part is concave, the middle part is slightly convex, and the triangular volar part is slightly concave. On the lateral side of the proximal end there is an articular facet for McV (fmcV). It is a dorsovolarly extending band, facing laterovolarly and forms a right angle with the facet for unciform. Its surface is concave dorsovolarly and straight vertically. Below the facet for McV there is a tuberosity, which may be the attached area of interosseous ligaments. On the dorsal side of the proximal end the medial and lateral crests for ligaments are well developed.

The shaft is straight. Its upper part is narrow and slightly thicker, trapezoid in cross section, wider than thick. The shaft becomes the thinnest at its middle and then widens downwards. The lower part of the shaft is wider than the proximal end, oblate in cross section. The dorsal side is flat longitudinally and convex transversely. The volar side is slightly concave longitudinally and slightly convex transversely. On the lower part of the volar side the middle bulge and the bilateral concavities are longer, occupying about the lower 1/3 of the shaft. The medial side is thinner and convex transversely, with its lower part bending medially. The lateral side is slightly longer than the medial side. It is also thin and convex transversely, but concave longitudinally.

The distal end is similar to that of McIII in basic morphology. But its articular facet for PhI is symmetrical: the sagittal ridge on the volar side is located at the middle and the two lateral grooves are subequal to each other in size.

Fifth metacarpal bone (McV; Figs. 41, 48; Tab. 16)   Only the right McV is preserved. McV is larger than McI in general size, shorter and proportionally thicker than the other three metacarpals (McII, McIII and McIV). On the top of the proximal end there are also two articular facets. The lateral facet for the unciform (fun) occupies almost the whole top side. It is oval in outline, convex, with its major axis extending dorsolaterally. The medial articular facet for McIV (fmIV) is small and triangular in outline, flat, facing mainly medially and slightly superiorly, and forms a right angle with the facet for unciform. On volar part of the lateral side of the proximal end there is a well-developed depression surrounded by eminence, which is the attached area for *m. extensor carpi ulnaris.*

The shaft takes the form of a straight and flattened cylinder. The upper part is narrow and thick, becoming more flattened downwards. The dorsal side is flat, with a groove extending in superomedial-inferolateral direction. The volar side is concave longitudinally, with a distinct middle longitudinal ridge extending inferomedially, which becomes more developed downwards. Distinct concavities are developed on both sides of the longitudinal ridge. The medial concavity is slightly larger than the lateral one. The medial and lateral sides of the shaft are thinner dorsovolarly and slightly diverge downwards.

The distal end is larger than the proximal end. It is similar to that of McIV in basic morphology. But the articular facet for PhI is asymmetrical: the medial part is longer in dorsovolar direction and with a more convex anterior part than the lateral one. On the volar side the middle sagittal ridge is more developed and continues with that on lower part of the shaft, and the lateral groove is slightly larger than the medial one. The lateral depression for the attachment of the lateral collateral ligament is larger than the medial one as well.

Phalanges of fingers (Ph; Fig. 41)   Of the right manus of V 16293 six phalanges are preserved, including proximal and distal phalanges of 1st finger, proximal phalanx of 2nd finger, proximal and middle phalanges of 4th finger and distal phalanx of 5th finger. Nine phalanges are preserved in the left manus of V 16293. They are three phalanges of the 2nd, 3rd and 4th fingers. The first finger of *Pseudorhizomys* is composed of two phalanges, all of the other fingers are composed of three phalanges.

First finger (FI, thumb)

Proximal phalanx of FI (FI-Ph1; Figs. 28, 41, 49; Tab. 17)   The proximal end of FI-Ph1 is large, wider than thick. The proximal articular facet for McI is semicircular in outline, with wider volar side. It is concave and faces dorsosuperiorly and forms an acute angle with the axis of the shaft. On the volar side of the proximal end there are two tuberosities separated by a middle groove. The lateral tuberosity is larger than the medial one. The medial one is the attachment area for *m. abductor pollicis*, while the lateral one, for *m. flexor pollicis brevis* and *m. adductor pollicis.*

The shaft is trapezoid in lateral view. The upper part of the shaft is wide and thick, about rectangular in cross section; the lower part is slender, being a flattened quadrilateral in cross section. The upper part of the dorsal side is convex transversely. On the upper part of the dorsal side there is a distinct tuberosity, the attachment area for *m. extensor pollicis brevis*, which is named here as extensor tuberosity (extu). The volar side is straight longitudinally and extends inferodorsally. It is slightly convex transversely on the upper part and flat on the lower part.

The distal end is smaller than the proximal end and only slightly wider than the lower part of the shaft. The distal articular facet for the distal phalanx is semicylindrical. Both its dorsal and volar parts extend superiorly. The volar border is situated higher than the dorsal border. The surface is strongly convex longitudinally. It is narrow and flat transversely on the dorsal part and becomes wider and slightly concave transversely on the volar part. The dorsosuperior border of the articular facet is slightly convexly arched, while its superovolar border is nearly straight transversely. On each of the medial and lateral sides there is a rough depression surrounded by eminence. The lateral depression is larger than the medial one, to both of which the collateral ligaments are attached.

Distal phalanx of FI (FI-Ph2; Figs. 29, 41, 50; Tab. 17)　The distal phalanx of the right first finger of V 16293 is preserved, with the top of its proximal end broken. The preserved part of FI-Ph2 is conical in form. Its proximal end is robust, thicker than wide. The surfaces of the dorsal and the two lateral sides are all smooth and transversely convex. Since the most part of the proximal end is broken, only the lower part of the extensor process (expr) is available for observation. On the volar side the tuberosity of ungual phalanx (tunph) is well developed, which affords the attachment for *m. flexor digitorum profundus*. Bilateral to the tuberosity the medial and lateral volar sulci (vos) are distinctly shown. There is a pair of volar foramina (vof) in the medial and lateral sulci respectively. The lateral vof is larger than the medial one.

The distal end of FI-Ph2 is shovel-formed, with a sharp and thin distal margin without sagittal notch. The dorsal side is stronger convex transversely than longitudinally. The volar side is concave longitudinally but convex transversely. The distal border is thin and sharp, arched and smooth, unserrated, and separated into two parts by the tip. The medial part of the border is much longer than the lateral one. There are two lateral foramina (lf) on the volar side, the medial one is situated higher than the lateral one.

Second finger (FII)

Proximal phalanx of FII (FII-Ph1; Figs. 28, 29, 41, 51; Tab. 17)　The FII-Ph1 is only about 2/3 the length of the McII, but larger than FI-Ph1. Its proximal end is large, wider than thick, and tilts dorsally. The articular facet for McII is concave on the dorsal part and has a short and wide middle groove separating the two volar tuberosities, of which the lateral one is slightly larger than the medial one.

The shaft is about semicylindrical in form, becoming slenderer and more flattened downwards. The extensor tuberosity (extu) is well developed, serving as attachment for *m. extensor digitorum communis*. The volar side of the shaft is flat, with slightly longitudinally concave lower part, which is bounded by distinct medial and lateral longitudinal crests.

The distal end is wider than thick. The articular facet for the middle phalanx is trapezoid in outline and tilts volarly. It is less convex longitudinally, and separated by a shallow sagittal groove into two slightly convex condyles. Its dorsal part slightly turns dorsally, and with a transversely straight dorsosuperior border. Its volar part extends superiorly, much more than the dorsal part, and with a concave superovolar border. On each of the medial and lateral sides of the distal end there is a roughened depression surrounded by eminences for the attachment of the collateral ligaments.

Middle phalanx of FII (FII-Ph2; Figs. 28, 29, 41; Tab. 17)　The FII-Ph2 is shorter than FII-Ph1. The proximal end is large. Its proximal articular facet is oblate in outline and tilts slightly dorsally. The surface is concave, and separated by a weak sagittal ridge into two concavities. The well-developed extensor tuberosity (extu) is pillow-shaped. On the volar side, the groove by which the tendon of *m. flexor digitorum profundus* passes is wide, with bilaterally situated prominent tuberosities. They are the attachment areas for *m. flexor digitorum superficialis* and lateral collateral ligaments.

The shaft of FII-Ph2 forms a truncated pyramid, becoming slenderer downwards. The dorsal, medial and lateral sides are concave longitudinally and flat transversely. The volar side is flat.

The distal end forms a transverse cylinder and protrudes more dorsally than the shaft. The articular facet is strongly convex dorsovolarly, semicircular in lateral view. Its dorsal part extends slightly upwards. Its volar part extends upwards much more than that of the dorsal part, and has a concave superovolar border. Like in FII-Ph1, the surface of the volar side is longitudinally convex and separated by a shallow sagittal groove into two condyles. On the medial and lateral sides of the distal end there is a pair of large rough depressions for the attachments of the collateral ligaments.

Distal phalanx of FII (FII-Ph3; Figs. 28, 29, 41; Tab. 17)   The FII-Ph3 is thinner and longer than FII-Ph2, and claw-shaped. The proximal end is large, thicker than wide. The preserved part of the articular facet forms a shallow transverse groove, concave dorsovolarly and straight transversely.

The shaft forms a trihedral pyramid, composed of volar, laterodorsal and mediodorsal sides. The two latter sides are separated by a distinct dorsal longitudinal crest extending to the distal end. They form two subequally narrow and long triangles, both straight longitudinally and convex transversely. Their surfaces are smooth on the upper part but become roughly sculptured by longitudinal grooves and ridges downwards. On the volar side the tuberosity of ungual phalanx (tunph) for the attachment of *m. flexor digitorum profundus* is well marked. Bilateral to the tuberosity are situated prominent medial and lateral sulci. The large volar foramina are located at the proximal parts of the two volar sulci respectively. There is also a longitudinal crest extending from the tuberosity of ungual phalanx to the sharp distal end. On the medial and lateral sides of the longitudinal crest there are also some longitudinal grooves and ridges extending to the distal end. No sagittal notch is seen at distal end. The distal end of FII-Ph3 is quite different from that of FI-Ph2: it is a flattened cone in form, sharply pointed distally, without lateral foramina as in FI-Ph2.

Third finger (FIII)

Proximal phalanx of FIII (FIII-Ph1; Figs. 28, 29, 41; Tab.17)   The FIII-PhI is about 3/5 the length of McIII, but longer and thicker than FII-Ph1. The proximal end is large. Its top view is semicircular in form, and tilts dorsally. The proximal articular facet for McIII can be divided into two parts: the dorsal part is oblate in outline and concave, and the volar part composed of a middle groove and bilaterally situated tuberosities. The lateral tuberosity is larger than the medial one. The medial and lateral sides of proximal end are bulging and rough. FIII-PhI is similar to FII-Ph1 in general morphology, but the extensor tuberosity is more developed and the sagittal groove on the distal articular facet is slightly shallower than in the latter.

Middle phalanx of FIII (FIII-Ph2; Fig. 41; Tab. 17)   The FIII-Ph2 is shorter and slightly thinner than FIII-Ph1. FIII-Ph2 is similar to FII-Ph2 in general morphology, but is longer and thicker. The proximal articular facet is deeply concave and has a weak sagittal ridge. The distal articular facet is semicylindrical in form and tilts volarly. Its dorsal margin slightly turns dorsally. The superovolar border is transversely straight. The surface of the distal articular facet is convex dorsovolarly, with a wide and gentle sagittal groove and two condyles. The medial condyle is slightly larger than the lateral one.

Distal phalanx of FIII (FIII-Ph3; Figs. 29, 41, 52; Tab. 17)   The FIII-Ph3 is longer and narrower than FIII-Ph2, but larger and longer than FII-Ph3. The proximal end of FIII-Ph3 is narrow and thick. The extensor process (expr) is well developed and tile-like in form. From the dorsal and volar views it is trapezoid in outline, with a narrower top. The proximal articular facet for FIII-Ph2 is semicylindrically concave. It can be divided into two parts: the larger volar part and the smaller dorsal part. The volar part is oblate in outline, separated by a medial sagittal ridge accompanied by two concavities, of which the medial one is larger than the lateral one. The dorsal part, located on the volar side of the extensor process, is straight transversely. FIII-Ph3 is similar to FII-Ph3 in general morphology. But, unlike in FII-Ph3, the dorsal longitudinal crest of FIII-Ph3 extends distomedially. Thus the dorsolateral side is wider and more convex transversely than the dorsomedial side. The longitudinal grooves and ridges on the distal part are more developed than those of FII-Ph3. The volar longitudinal crest is short and does not reach to the distal end.

Fourth finger (FIV)

Proximal phalanx of FIV (FIV-Ph1; Figs. 28, 29, 41, 53; Tab. 17)　The FIV-Ph1 is about 2/3 the length of McIV. It is much larger and longer than FI-Ph1, slightly stouter than FII-Ph1 and FIII-Ph1, but shorter than FIII-Ph1. The proximal end of FIV-Ph1 is wider than that of FII-Ph1, subequal to that of FIII-Ph1. FIV-Ph1 is similar to FII-Ph1 and FIII-Ph1 in general morphology, with a prominent extensor tuberosity as well. Unlike FII-Ph1 and FIII-Ph1, the medial volar tuberosity in FIV-Ph1 is larger than the lateral one.

Middle phalanx of FIV (FIV-PhII; Figs. 28, 29, 41, 54; Tab. 17)　The FIV-Ph2 is shorter and thinner than FIV-Ph1, but slightly thicker than FII-Ph2 and shorter than FIII-Ph2. FIV-Ph2 is similar to FII-Ph2 and FIII-Ph2 in general morphology. The articular facet on the proximal end is large and concave, with a weak sagittal ridge. The extensor tuberosity is also pillow-shaped. On the volar side of the proximal end there is also a wide groove, where the tendon of *m. flexor digitorum profundus* passes by. Bilateral to the groove there are lateral and medial tuberosities for *m. flexor digitorum superficialis* and lateral ligaments.

The articular facet of the distal end is also hemicylindrical in form, extending mainly superovolarly and with a transversely straight superovolar border. The facet has a gentle sagittal groove and convex lateral and medial condyles. The medial condyle is larger than the lateral one. The well-preserved dorsal part of the distal articular facet forms an arcuate border, which is narrower than that on the volar side. On both sides of the distal end the depressions surrounded by eminences are prominent.

Distal phalanx of FIV (FIV-Ph3; Figs. 29, 41; Tab. 17)　The preserved part of FIV-Ph3 shows great similarity with FIII-Ph3 in morphology and size.

Fifth finger (FV)　Of the phalanges of the 5th finger in V 16293 only the right distal phalanx is partially preserved, with its extensor process broken.

Distal phalanx of FV (FV-Ph3; Fig. 41 A; Tab. 17)　The FV-Ph3 is similar to FIII-Ph3 and FIV-Ph3 in general morphology. Unlike the latter two, FV-Ph3 is shorter and thinner. FV-Ph3 forms a flattened trihedral pyramid. The dorsal longitudinal crest is located nearer to the medial border, thus meeting the medial border in its distal part. The dorsolateral side is much wider than the dorsomedial side. The distal part of dorsomedial side becomes narrower and sharp, which is separated from the volar side by a prominent medial edge. The boundary between the volar side and dorsolateral side becomes distinct only in distal part. The lateral and medial borders form a sharp-pointed trowel at their distal ends. The lateral part of the distal end is longer than the medial part; both parts are smooth, unserrated. No sagittal notch is present at the distal end. The longitudinal crest extending from the tuberosity of the ungual phalanx is weak. But there are two lateral foramina: the lateral one is larger and located at the distal end of the lateral volar sulcus; the medial one is smaller and located at the proximal end of the medial border. The dorsal side of the distal end is slightly convex longitudinally and straight transversely, while the volar side is slightly concave longitudinally and convex transversely. The grooves and ridges are rough, only distinct near the medial and lateral borders.

Hindlimb

Of the hindlimb of V 16293 only one isolated primary phalanx is preserved. It is longer and thinner than all the above mentioned phalanges of the fingers. If it belongs to the same individual of V 16293, it should be a phalanx of toe rather than of finger, and probably a proximal phalanx of a middle toe (ToII, ToIII or ToIV), or more probably the proximal phalanx of right ToIII or left ToIV, because the phalanx is rather symmetrical. For convenience in description, this phalanx is tentatively inferred as a right (?) proximal phalanx of toe III (RToIII?-Ph1).

The proximal and distal ends of RToIII?-Ph1 (Fig. 55; Tab. 17) are slightly damaged, otherwise is well preserved. The proximal end is large. When viewed from above, it is a rounded equilateral triangle in outline.

The preserved part of the proximal end shows that the proximal articular facet is gently concave. The plantar side of the proximal end is composed of a wide middle sulcus, accompanied by medial and lateral tuberosities. The medial tuberosity is slightly larger than the lateral one. The extensor tuberosity on the dorsal side of the proximal end is a well-developed and rough one, which serves as the attachment of *m. extensor digitorum longus*. On medial and lateral sides of the proximal end there are rough areas near the plantar border.

The shaft is slender and straight. Its upper part is larger in size and trapezoid in cross section. Its middle part becomes slenderer. Its lower part becomes wider again, but thinner, oblate in cross section. The dorsal side of the shaft is straight longitudinal and convex transversely. The medial and lateral sides are slightly concave longitudinally and convex transversely. The plantar side is flat.

The distal end is smaller than the proximal one and its cross section is a flattened rectangle in outline. The articular facet for middle phalanx of toe is semicircular. Its dorsal part is small and turns dorsally, extending slightly upwards. The plantar part is large and extends upwards, much beyond the dorsal part, and has a deeply concave upper border. The surface of the facet is convex dorsoplantarly and separated by a wide sagittal sulcus.

For measurements see Tab. 17.

2) Comparison

From the above description it is evident that the skulls of V 16293 and V 16294 are similar to those of *Pararhizomys* in general morphology. The shared features are: skull low with myomorphous zygomasseteric structure; anterior and posterior parts of skull subequal in length, with narrow rostrum, but wide cranial part; premaxilla possessing dorsolateral crest; zygomatic plate confined to maxilla; infraorbital foramen with ventral slit; canals of *masseteric* n. and *buccinators* n. very short; *foramen ovale* confluent with middle lacerate foramen; glenoid fossa elongated; and nuchal surface slightly slanting forwards; mandible sciurognathous with short and thick horizontal ramus, well developed masseteric ridge of masseteric fossa, and sharp angular process extending backwards; posterior end of i2 alveolus forming a large bulge on buccal side of ascending ramus; and some molars being mesiodistally hypsodont, with simple occlusal structure, etc.

However, the above described specimens do have some features distinct from those of *Pararhizomys*: the dorsal surface of the nasal being more straight, with less curved anterior part and flat posterior part; the anterior end of the premaxilla extending anteriorly beyond nasal; the dorsolateral crest of the premaxilla being shorter, disappearing anteriorly; the dorsal part of premaxilla being triangular in form; dorsal part of the premaxillo-maxillary suture and maxillo-frontal suture forming a curved line; infraorbital foramen more circular in form; anterior root of the zygomatic arch and zygomatic plate extending more anteriorly; jugal very long, reaching to lacrimal; the parietal short and wide, hexagonal in form; posterior border of hard palate located at the same transverse line as the posterior border of M3; the mesopterygoid fossa narrow and subequal to pterygoid fossa in width; internal pterygoid process more vertically planted, parallel to each other; deep temporal foramen, masticatory foramen and buccinator foramen confluent but the posterior foramen of deep temporal nerve canal is separated from those of the latter two; mandibular diastema less concave in its posterior part; lower border of horizontal ramus convex; mental foramen situated relatively higher; mandibular notch deeper; molars relatively lower crowned, only M1 and m3 with distinct mesiodistal hypsodonty; I2 proodont, without longitudinal crest on labial side, taking its origin from the middle part of maxilla before M1, etc. (see Tab. 9). The wide range of differences between V 16293 and V 16294 and the genus *Pararhizomys* inclined us to establish a new genus and species, named here as *Pseudorhizomys indigenus*.

<h2 style="text-align:center">*Pseudorhizomys gansuensis* gen. et sp. nov.</h2>

(Figs. 56–70, 76; Tabs. 1, 6–8, 10–19)

Holotype   HMV 1942, a skull associated with mandible and some postcranial bones, from LX 200502 (Songjianao); lower part of middle Liushu Fm. (for details see Tabs. 1 and 10).

Paratypes   1) V 16297, a skull with mandible, from LX 200007 (Guchengcun); 2) V 16298, a complete skull, from LX 200011 (Dashengou); 3) V 16299, a partial skull with left hemimandible, from LX 200046 (Hejiazhuang); 4) V 16300, an anterior part of a skull, from LX 200037 (Panyangcun); all the above 4 specimens are from the lower part of middle Liushu Fm. (for details see Tabs. 1 and 10).

Referred specimens   1) V 16296, a skull with mandible, from LX 200019 (Xiaozhaicun), upper part of middle Liushu Fm. (for details see Tabs. 1, 10); 2) V 16301, a partial skull associated with right hemimandible; 3) V 16302, a partial skull; and 4) V 16303, an anterior part of skull. All the latter three specimens are from the Linxia Basin, without further information of exact provenance.

Locality and horizon   So far known only from the Linxia Basin, Gansu: Songjianao, Dashengou and Panyangcun of Hezheng County and Guchengcun, Hejiazhuang and Xiaozhaicun of Guanghe County; Late Miocene, medial–late Bahean ALMA (~10.5–7 Ma).

Diagnosis   Rostrum and maxillary diastema long; anterior border of attachment area of lateral masseter muscle S-shaped, with its lower part convex forwards beyond the ventral slit of infraorbital foramen; temporal crest weak, becoming gentle posteriorly; anteromedial part of auditory bulla slightly concave; posterior concavity of mandibular diastema relatively short; coronoid process longer, less inclining posteriorly, with relatively straight posterior border; articular facet of condyle of mandible elliptic in form; only M1 and m3 distinctly mesiodistally hypsodont; sinus long transversely on M1–2 and inserting between anterosinus and mesosinus on M1; M2–3 lacking anterosinus; sinus extending anterolaterally and obliquely toward mesosinus on M3; lower molars having forked sinusid, but lacking posterosinusid; sinusid located more posteriorly than mesosinusid and lacking protosinusid on m3; labial side of I2 transversely convex, without longitudinal ridge.

Tuberosity of McII (tmc) forms a rough depression, located on medial side of upper part of the dorsal side; articular facet for unciform of McV trapezoid in outline, saddle-shaped; middle part of shaft narrower and quadrilatero-prismatic. Distal border of ungual phalanx of first finger serrated.

Differential diagnosis It differs from *Ps. indigenus* gen. et sp. nov. in auditory bulla lacking deep anteromedial concavity; coronoid process less inclining posteriorly, with relatively straight posterior border; condyle of mandible having elliptic articular facet; sinus longer on M1–2, inserting between anterosinus and mesosinus on M1; M2–3 and m3 with two reentrants only; M2–3 lacking anterosinus and m3 lacking protosinusid; M3 having sinus oblique to mesosinus; lower molars having forked sinusids; sinusid of m3 more posteriorly situated than mesosinusid; tuberosity of McII (tmc) forming a rough depression, located on medial border of upper part of the dorsal side; articular facet for unciform of McV trapezoid in form, saddle-shaped, middle part of shaft of McV slender, quadrilatero-prismatic in cross section; and the distal border of ungual phalanx of FI-Ph2 serrated, etc.

Etymology Gansu, name of the province, where the fossils were collected.

1) Description

(1) **Skull** (Fig. 56; Tabs. 11, 18)

Dorsal view (Fig. 56 A2, B2)

The dorsal side of the nasal is rather straight longitudinally with its anterior part slightly convex and posterior part flat transversely. The dorsolateral crest of premaxilla is short, extending anteromedially to join the naso-premaxillary suture. The dorsal side of the premaxilla is triangular in outline, with a narrow anterior part culminating in a sharply pointed anterior tip. The premaxillo-frontal suture is strongly serrated, extending rather straight and anterolaterally to meet the premaxillo-maxillary suture at an obtuse angle. The curved premaxillo-maxillary suture on the dorsal side extends posterolaterally and meets the medial end of the maxillo-frontal suture, forming a continuous curved line with the latter. The lacrimals of V 16196, V 16298 and V 16302 are well preserved, but only a small part of the lacrimal appears on the dorsal side and a small lacrimal tubercle can be seen. The jugal is so long that its anterior end reaches the lacrimal. The zygomatic arch is strongly arched and its posterior end is separated from nuchal crest by a notch. The anterior root of the zygomatic arch is situated more anteriorly and its posterior border is deeply concave. The anteriormost point of its posterior border (= anterior border of orbit) lies anterior to the posterior ends of both the nasal and premaxilla. The postorbital constriction is distinct. The frontal crest is weaker than that of *Ps. indigenus* gen. et sp. nov., and the sagittal crest is prominent. The parietal is hexagonal in form, wider than long, with an anterior middle spine. As in *Ps. indigenus* gen. et sp. nov. the temporal fossa is large and the temporal crest is short and becomes gentle posteriorly.

Lateral view (Fig. 56 B4)

The dorsal outline of the skull is slightly convex. The anterior end of the premaxilla extends anteriorly, beyond that of the nasal. I2 curves weaker, with its anterior tip extending antero-inferiorly rather than posteriorly, and its posterior end originates from the middle of maxilla, forming a bulge in front of M1. The maxillary diastema is much longer than the upper molar row. The anterior part of the skull before M1 is subequal to the posterior part (including M1) of the skull in length (see Tab. 11). The infraorbital foramen is a large circle in outline and has a distinct ventral slit and a bony septum. The zygomatic plate has a concave inferior surface and extends anterosuperiorly. The attachment area of *m. masseteric lateralis* is limited within the maxilla and its anterior border is S-shaped, with its lower part convex anteriorly beyond the ventral slit of the infraorbital foramen. The masseteric tubercle takes the form of a concavity, situated near the incisive foramen. The orbit is small. The long glenoid fossa for the mandible extends to the nuchal crest, pushing the external auditory meatus downwards. The mastoid process is distinct and the nuchal surface slants slightly forwards towards its top.

The foramina in the orbital area are similar to those of *Ps. indigenus* gen. et sp. nov., but they are positioned more anteriorly as a whole. The dorsal palatine foramen (dpf) is tiny and indistinct, situated between M1 and M2. The deep temporal foramen, masticatory foramen and buccinator foramen are confluent to form a common slit. The posterior aperture of the deep temporal canal is separated from those of the masseteric and the buccinator canals. The former opens in *foramen ovale*, while the latter two open in the posterior end of the alisphenoid canal and may be confluent with the posterior foramen of the alisphenoid canal. There are many similarities between the other orbital foramina of *Ps. indigenus* gen. et sp. nov. and those of the specimens described above in position, form, and size. Their descriptions are omitted here (see Fig. 56 B4).

Ventral view (Fig. 56 A1, B1)

The incisive foramen is located at the middle of the maxillary diastema and is about 1/4–2/5 the length of the diastema. The serrated premaxillo-maxillary suture is anteriorly convex and intersects the incisive foramen at the latter's middle part. The zygomatic plate is nearly in the same transverse level as the incisive foramen: the anteriormost point of the arched anterior border of the zygomatic plate is roughly opposite to the anterior end of the incisive foramen, and the anteriormost point of the emarginated posterior border of the former aligns with the posterior end of the latter. The palatine sulci are deep and there is only one pair of posterior palatal foramina in the palatine sulci, situated medial to M1/M2. The parapterygoid fossa is shallow, without clearly defined boundary line. The posterior maxillary foramen (pmf) is located at the posterolateral corner of the parapterygoid fossa. There is another small foramen situated medial to the pmf. The mesopterygoid fossa is slightly wider than the pterygoid fossa. The internal pterygoid processes are planted vertical to the ventral side of the skull, extending longitudinally and parallel to each other. The posterior border of the hard palate aligns with posterior border of M3. The posterior foramen of the alisphenoid canal (ascp) is situated posterolateral to the pterygoid fossa, and posterior to the external pterygoid process. The single posterior foramen of the pterygoid canal (pcpf) is located at the medioposterior margin of the pterygoid fossa, near the posterior base of lateral side of the internal pterygoid process and is slightly posterior to the posterior foramen of the alisphenoid canal. The *foramen ovale* and middle lacerate foramen seem to be confluent with each other. The auditory bulla is less inflated than that of *Ps. indigenus* gen. et sp. nov. The concavity on the anteromedial part of the auditory bulla is shallow or indistinct. On the medial side of the auditory bulla there is a large and deep groove extending inferoposteriorly from the internal carotid foramen. But the groove extending from the jugular foramen is indistinct. On the medial wall of the bulla near the jugular foramen no foramen for stapedial artery is visible.

Posterior view (Fig. 56 B5)

The nuchal surface is similar to that of *Ps. indigenus* gen. et sp. nov. in general. It is semicircular in outline, with a rather flat surface and slightly bulged external occipital crest. The lower part of the occipital is rather wide and the petromastoid portion is relatively smaller. The *foramen magnum* is oval in outline, wider than high. The posterior parts of the left and right occipital condyles are separated widely. The paramastoid process is prismatic in form, extending inferolaterally, with a blunt inferior end and a concave medioventral side. The mastoid process is shorter and smaller than the paramastoid one.

(2) Mandible (Fig. 57; Tabs. 11, 18)

The mandible of HMV 1942 and the specimens listed above are similar to that of *Ps. indigenus* gen. et sp. nov. in general morphology. It is sciurognathous and has a slightly convex lower border of the horizontal ramus below the molars. The chin is distinct, with an oblique ridge along its inferior border, so that the lower borders of the symphysis and the horizontal ramus form an angle of about 40°. The digastric fovea is present on the medial side of the chin, in the form of a narrow crescent. The mandible diastema is longer than that of the lower cheek tooth row. Its posterior concavity is longer but shallower than in *Pararhizomys*. The masseteric fossa extends anteriorly to below m1, with a weak upper crest but a well-developed and flaring masseteric ridge. The mental foramen is composed of one or two apertures, located below the anterior margin of m1 at the same level of, or slightly lower than the anterior border of the masseteric fossa.

The ascending ramus is long. The coronoid process reclines less posteriorly, with its anterior border forming an angle of 100° with the alveolar border of lower molars and a relatively straight posterior border. The temporal fovea is distinct on the lower part of the medial side of the coronoid. The mandibular notch is deep, about 2/5 the height of the mandible. The mandibular foramen is large, located below the posterior border of the coronoid process, nearly at the same horizontal level as the occlusal surface of the lower molars. The condyloid process is lower than the coronoid process, extending posterosuperiorly, and its upper part bends lingually. The attachment area of the external pterygoid muscle is represented by a small and shallow concavity. However, the articular facet of the condyle of mandible is of a rather wide ellipse. The bulge formed by the posterior end of the i2 alveolus is prominent on the buccal side. The internal pterygoid fovea is large and deep, extending anteriorly to and below the middle of the coronoid process. The ventral border of the ascending ramus is convex. The angular process is well developed, but smaller than that of *Pa. huaxiaensis* sp. nov. It extends mainly posteriorly, only slightly superiorly, with a convex lower border. Its posterior end bends slightly medially, nearly in the same vertical line as the posterior border of the condyle. The curvature of the posterior border of the ascending ramus is shallower than in *Pa. huaxiaensis* sp. nov.

(3) Teeth (Fig. 58; Tabs. 12, 19)

Dental formula: $1 \cdot 0 \cdot 0 \cdot 3 / 1 \cdot 0 \cdot 0 \cdot 3$. The teeth of HMV 1942 and other specimens described here are similar to *Ps. indigenus* gen. et sp. nov. in general morphology. They are lophodont, mid-high-crowned and rooted. The mesiodistal hypsodonty is distinct in M1 and m3 only. The molars decrease from M1 (m1) to M3 (m3) in size.

M1 is trapezoid in occlusal view, wider than long, and buccal side longer than lingual one. The crown is mesially hypsodont, with undulating dentine tract on buccal side. M1 has three reentrants (two buccal and one lingual) on the occlusal surface. The three reentrants are open on less worn specimens (V 16296, V 16297 and V 16300). Only the lingual sinus becomes a closed basin in more worn V 16298, all three reentrants are closed on the other heavier worn specimens. The anterosinus extends transversely to the middle of the tooth (7/8) or slightly beyond (V 16300). The mesosinus is the longest one of the three reentrants, usually straight and extending transversely or slightly posteriorly (6/8), but its lingual end bends posteriorly on the less worn V 16296 and V 16300. The sinus extends transversely or slightly anterolaterally, inserting between the anterosinus and mesosinus. Thus, it overlaps both the anterosinus and mesosinus partially.

M2 is about oval in occlusal outline, wider than long. Only the M2 of V 16300 shows some degree of mesial hypsodonty. The undulation of dentine tract is indistinct. There are only two reentrants on the occlusal surface (mesosinus and sinus) in M2, but no anterosinus is seen. The two reentrants are open on less worn M2 (V 16296–V 16298 and V 16300) and closed on the others. The mesosinus is more or less curved posteriorly and longer than or subequal to the sinus in length. The sinus extends anterobuccally anterior to the mesosinus, partially overlapping the latter.

M3 is oval in occlusal outline, with narrower posterior side. No mesial hypsodonty is observed. There are also only two reentrants on the occlusal surface in M3: mesosinus and sinus. No anterosinus is visible. The mesosinus is short transversely and extends anterolingually. It is usually open buccally except in HMV 1942 and V 16301, where the mesosinuses are closed. It is also deeper than the sinus in depth. The sinus is longer than the mesosinus transversely and extends anterobuccally to overlap the latter obliquely. The sinus is open on M3 of V 16296 and V 16298 and right M3 of V 16297, but closed on all the others. There are some variations of the posterior part of M3. For example, an accessory pit is present on the posterior part of the left M3 of V 16298 and right M3 of both V 16299 and V 16300, but absent on the others. Besides, there is another accessory pit inserting the mesosinus and sinus on the left M3 of V 16298.

I2 is also similar to that of *Ps. indigenus* gen. et sp. nov. It is proodont, with its anterior end extending antero-inferiorly, rather than inferoposteriorly. The cross section is triangular, with a slightly transversely convex labial side and a rounded lingual angle, on which a longitudinal groove is present. The enamel layer covers mainly the labial side and expands only slightly on medial and lateral sides. No longitudinal ridge is seen on the smooth labial surface.

The m1 is oval in occlusal outline, longer than wide, with a wider and posteriorly convex posterior side. The distal hypsodonty is indistinct on the m1 of HMV 1942, V 16299 and V 16301, but distinct on V 16296 and V 16297. The undulation of the dentine tract is distinct on both buccal and lingual sides on the latter two specimens. There are two buccal reentrants (protosinusid and sinusid) and one lingual one (mesosinusid) on the occlusal surface of the m1. In the known first lower molars the protosinusids may become enamel pits (6/8) or completely disappeared (2/8) depending on wear. The sinusid and mesosinusid are open only on the m1 of V 16296, closed on all the others. The sinusid extends posterolingually. The sinusids are distinctly forked on the lingual part in most of the first lower molars, but are indistinct on heavily worn specimens (V 16299 and

V 16301). It seems that the fork of the sinusid disappears with wearing. The buccal part of the mesosinusid bends anteriorly toward the protosinusid. The mesosinusid is subequal to the sinusid in transverse width.

The m2 is rounded in occlusal outline, subequally long and wide. It is shorter and wider than m1. No clear distal hypsodonty is seen. The undulating dentine tract may be present on V 16296 and V 16297 or absent on the others. Like m1, there are also three reentrants on the occlusal surface. The three reentrants are open only on the m2 of V 19296, but closed on the others. The protosinusid is the shortest among the three reentrants, and mainly extends transversely or slightly anterolingually. The mesosinusid is the longest and with a bending buccal part, which extends toward the lingual end of the protosinusid, but does not reach the latter. As in m1, the sinusid also has a forked lingual part, but the anterior branch of the fork is weaker. The fork of the sinusid disappears on the heavier worn m2 of V 16301.

The m3 is oval in occlusal outline, longer than wide, with a narrower posterior side. The distal hypsodonty is clearly shown in m3. However, no undulating dentine tract is seen. There are only two reentrants (mesosinusid and sinusid) on the occlusal surface. No protosinusid is seen. The mesosinusid is longer and deeper than the sinusid, and bends anteriorly. The lingual end of the mesosinusid is closed only on the m3 of V 16301, but open on the others. The sinusid is usually transverse and also forked in most third lower molars, but weaker than that of m2. The forking of sinusid is indistinct on V 16296 and V 16301. V 16301 represents an old individual, forked sinusid of which may be worn out. The sinusid is located more posteriorly to the mesosinusid and overlaps the latter partially.

The i2 is long and robust and its curvature is greater than that of *Pa. hipparionum*, but subequal to that of *Pa. huaxiaensis* sp. nov. The anterior part of i2 is long and bends anterosuperiorly. Thus, the distance between its worn facet with that of I2 and the anterior border of the alveolus is longer. The i2 is also triangular in cross section, longer than or subequal to wide, with slightly convex labial side and rounded lingual angle. The enamel layer covers mainly the labial side and expands slightly on the medial and lateral sides. The boundary between the labial and medial sides is sharply defined, while that between the labial and lateral sides is indistinct. There are usually two longitudinal ridges on labial surface. However, only one longitudinal ridge is present on that of V 16296.

Measurements of teeth see Tab. 19

(4) **Enamel microstructure of the lower incisor** (Fig. 59)

The studied specimen is the anterior end of the lower incisor cut from the right hemimandible of V 16297. Its enamel microstructure is similar to that of *Pa. hipparionum* in basic features. It is uniserial. The total enamel thickness (T) is 140 μm. The incisor enamel consists of two layers: inner portion (PI) and outer portion (PE). No prismless external layer is seen. The PE of V 16297 is very thin, about 26 μm in thickness. The external index (PE/T) is about 19%. The interprismatic matrix (IPM) of PE does not incline, but is vertical to the external surface in both the longitudinal and cross sections. The PI is thicker, about 114 μm. The ratio of PE/PI is about 23%. The cross section of i2 shows that the Hunter-Schreger bands (HSB) extend slightly obliquely from the enamel dentine junction (EDJ) to medial longitudinal ridge (mr). Their inclinations gradually decrease from lingual side to the labial side. The differences of the inclinations between the lingual side and the labial side gradually become smaller or even indistinct from near the medial ridge to both the medial and lateral sides of i2 respectively.

There are two longitudinal ridges on the labial side of right i2 of V 16297. The cross section shows that the enamel increases in the two ridges in thickness (Fig. 59 A–C). The enamel of the medial ridge (mr) is similar to that of *Pa. hipparionum* in microstructure. The Hunter-Schreger bands (HSB) of the medial and lateral sides beside the medial ridge extend obliquely and labially and are convergent to the medial ridge. Within the medial ridge the directions of both Hunter-Schreger bands (HSB) and Interprismatic matrix (IPM) in the inner portion (PI) vary as following: HSB cross IPM at right angles on the lingual part near the enamel dentine junction (EDJ); then they gradually become parallel to each other near the labial part, and then gradually transit into the external portion. The enamel of the lateral ridge (lr) is similar to the other part of the i2 in the microstructures of the inner portion and does not show any distinct variation.

(5) Postcranial bones (Figs. 60–70)

Some postcranial bones are preserved in HMV 1942, including several vertebrae (5 cervical vertebrae and 4 thoracic vertebrae), two scapulae, several ribs and partial right forelimb. The partial forelimb of HMV 1942 is well-preserved and in more or less articulated position, but the other bones are more or less dislocated and damaged (see Fig. 60).

Cervical vertebrae (C's)

Atlas (At; Fig. 61; Tab. 6)   The wings of the atlas of HMV 1942 are damaged, otherwise the atlas is well preserved. As in V 16293, without the wings, the atlas is an oval bony ring, very short longitudinally, wider than high. The dorsal arch is also short and wide, and bow-shaped. The dorsal tubercle (dt) is robust, located at the middle of the dorsal side. Its anterior slope is longer and gentler, while the posterior slope is shorter and steeper. On the middle of the anterior slope there is a large concavity, which is separated into two small pits by a longitudinal ridge. The lateral vertebral foramen (lvf) is confluent with the *foramen alare* (fa), forming a large lateral vertebro-alar foramen (lvaf) on the dorsal side. The lvaf is smaller than that in V 16293, and situated more posteriorly, i.e., at the middle of the lateral sides of the dorsal arch. Medial to the right lvaf there is a tiny foramen communicating with the lateral vertebral canal. It is here considered a vestige of the lateral vertebral foramen (lvf), possibly passed by a small branch of the vertebral artery. The morphology of the ventral side of the dorsal arch and the position of the internal lateral vertebral foramen (ilvf) are also similar to those in *Ps. indigenus* gen. et sp. nov., but the groove for vertebral artery (gva) is deeper than in the latter. The ventral arch is shorter in anteroposterior direction, but thicker than the dorsal arch in vertical direction. The middle part of the ventral arch is prismatic in form. On the dorsal side of the ventral arch the odontoid fossa (odf) is long and flat anteroposteriorly, but concave transversely. The anterior side of the ventral arch is short and flat, notched between the two cranial articular foveae (craf). The lateral parts of the ventral side of the ventral arch are concave. These two lateral concavities are broader and shallower than those in V 16293, and with a pair of small circular pits near their anterolateral margins, absent in V 16293. The ventral tubercle (vt) is smaller than in V 16293. There is also a pair of small pits beside the ventral tubercle. The vertebral foramen is similar to that of V 16293. On the two lateral side of the foramen there are distinct tubercles for the attachment of the transverse ligament of atlas. The areas superior and posterior to the tubercles are concave and rough, which may serve as the attachment of the internal ligament of the atlas-axis articulation.

As in V 16293, the cranial articular foveae (craf) are also kidney-shaped. But the lateral parts of their upper borders extend more anteriorly, making the anterior border of the dorsal arch slightly more concave. The caudal articular facets (caf) are also similar to those of V 16293 in form, but their upper borders and anterosuperior corner are thinner and more prominent than those in V 16293.

Based on the preserved parts, it can be said that the fossa atlantis is large, with a narrower but deeper internal groove joining the *foramen alare* (fa) and *foramen alare inferior* (fain), the latter of which is smaller than that of V 16293. The distance between the fa and fain is about 2 mm, subequal to that of *Ps. indigenus* gen. et sp. nov. The transverse foramen (trf) is large, situated lateral to the posterolateral corner of the caudal articular facet and the lateral end of the groove for vertebral artery (gva). The broken part of the left wing shows that the *canalis transversarius* (ctr) joins the transverse foramen and the *foramen alare inferior.*

Axis (Ax; Figs. 60, 62; Tab. 7)   The axis in HMV 1942 is preserved, but the top of the spinous process is broken and damaged, and the ventral part of the vertebra is also slightly damaged. The axis is longer than the atlas. The vertebral body is an oblate cylinder, wider than high. The odontoid process (or dens, odpr) forms an oblate cone with an obtuse anterior end, where a pair of rough concavities for the attachments of the ligaments of dens is present. On the dorsal side of the dens there is a distinct transverse groove where the transverse ligament of atlas should pass through. On the ventral side of the odontoid process the articular facet for the odontoid fossa of atlas is convex transversely and slightly concave longitudinally. The anterior articular facet (aaf) of the axis is oval in outline, higher than wide, with arched and thin lateral and ventral borders and straight medial border. The articular surface is slightly convex and faces anterolaterally. The dorsal side of the body is flat, having a pair of nutrient foramina on its posterior 1/3 and a pair of rough areas near the anterior articular facets, which may be the attachment areas of the internal ligaments of atlas-axis articulation. The ventral side of the body (Fig. 60 A) is similar to that of *Ps. indigenus* gen. et sp. nov. in outline, but slightly narrower and longer in proportion. No distinct ventral spine is present. But the middle convexity of the posterior side of the ventral arch is more prominent than in the latter. On the anterior margin there are two pits separated by a weak longitudinal ridge. The vertebral fossa (vfs) is oblate in form, wider than high, with a slightly concave surface. The vertebral foramen is circular in form.

The spinous process (sp) is well developed and slightly inclines backwards. Its lower part is an isosceles triangle in cross section, with a thin anterior border and a wide posterior side excavated by a vertical groove. Its upper part forms a thin plate.

The pedicle of the vertebral arch is short. The postzygapophysis (pzy) is located at the lower part of the posterior side of the pedicle. Its articular facet (fpzy) is oval in outline, higher than wide, slightly concave, facing inferolaterally. On the posterior border of the pedicle there is a distinct caudal vertebral notch between the pzy and the transverse process.

The transverse process (trpr) forms a long triangular pyramid, extending from the lateral side of the united part of the pedicle and the body posterolaterally. It has two roots. The dorsal root forms a large plate, while the ventral one, a slender rod. The transverse foramen (trf) is large, lying between the dorsal and ventral roots.

Fourth (or fifth) cervical vertebra? (C4/C5?)   In HMV 1942 a right vertebral arch is preserved, including partial dorsal arch, pedicle, postzygapophysis and partial prezygapophysis. From the lateral view, the lateral border of the posterior articular surface is parallel to that of the anterior articular surface. The anterior and posterior ends of the posterior articular surface are at the same levels as those of the anterior articular surface. The shortest distance between the two articular surfaces is 0.9 mm. It is similar to C4 or C5 of *Ps. indigenus* gen. et sp. nov.

Sixth cervical vertebra (C6; Fig. 63 A; Tab. 8)   A C6 is preserved in HMV 1942, with its dorsal part of vertebral arch and the extremities of the transverse processes missing. The C6 is similar to that of *Ps. indigenus* gen. et sp. nov. (V 16293) in the vertebral arch and pre- and postzygapophysis. The articular surface on prezygapophysis (fprzy) is oval in outline and its anteroposterior diameter is larger than the transverse width. It is slightly concave, facing mainly mediosuperiorly and slightly anteriorly. The articular facet on postzygapophysis (fpzy) is also oval in outline, but the difference between the anteroposterior and the transverse diameters is smaller. Its articular surface is largely flat, with a slightly convex upper part, mainly facing inferolaterally and slightly posteriorly. From the lateral view, the articular surfaces on pre- and postzygapophysis are parallel to each other.

The caput (cap) and the vertebral fossa (vfs) are oval in outline. The surface of the caput is weakly convex with an oval sunken center. The surface of the vertebral fossa is weakly concave. The dorsal surface of the vertebral body is flat, possessing a middle longitudinal ridge and two lateral concavities. The ventral side of the body is convex transversely. The transverse process is well developed and two-rooted. Its ventral root is a broad plate, originating from the lateral side of the body, while its dorsal root originates from the pedicle, being shorter than the ventral root in anteroposterior direction. The lateral part of the transverse process (trpr) is also separated into two branches. The ventral branch is called *lamina ventralis vertebrae cervical* (lvvc). It is an anteroposteriorly long plate, with its lateral part bending inferiorly. The dorsal branch (*lamina dorsalis vertebrae cervical*, ldvc) is narrower transversely than the ventral one and extends mainly laterally. The transverse canal is short. The anterior and posterior transverse foramina (trf) are large and communicate with the anterior and posterior vertebral notches respectively.

Seventh cervical vertebra (C7; Figs. 60, 63 B; Tab. 8)   In the C7 of HMV 1942 the spinous process, ventral side of the body, and the lateral ends of the transverse processes are damaged. The body is a flattened cylinder in form. The caput (cap) is larger than the vertebral fossa (vfs), wider than high, with a sunken center. The vertebral fossa is concave, slightly wider than high. Near the ventral parts of the two lateral sides of the vertebral fossa there is a pair of small articular facets for the head of the first rib (fhR1). The vertebral arch is short and wide, bow-shaped in form. The pedicle, pre- and postzygapophysis, and the anterior and posterior articular facets of C7 are similar to those of C6. But its posterior articular facet is flatter. The transverse process of the C7 has only one single root, which is cylindrical in form and extends laterally. No transverse foramen is present.

Thoracic vertebrae (T's)

Among the thoracic vertebrae of HMV 1942 only four are preserved. They are fragmentary and dislocated. Based on the T1–T6 of *Ps. indigenus* gen. et sp. nov. described above, the four specimens of HMV 1942 are questionably referred to T1–T4 respectively.

First thoracic vertebra? (T1?; Fig. 64 A)   Of the T1? only the vertebral body, pedicle, right transverse process, pre- and postzygapophysis and left prezygapophysis are preserved. The body is in the form of a shortened oblate cylinder. The surface of the caput (cap) is slightly convex and that of the vertebral fossa (vfs), slightly concave. On either of the left and right lateral borders of the vertebral fossa there is an articular facet for the head of the second rib (fhR2), which is triangular in form and faces laterally. On the anterior parts of either side of the ventral surface of the body there is a concave facet, which articulates with the head of the first rib (fhR1). The pedicle is robust. The prezygapophysis extends forward from the pedicle. The articular facet on the prezygapophysis (fprzy) is oval in outline, slightly concave, facing mediosuperiorly. The articular facet on the postzygapophysis (fpzy) is also oval, but slightly convex, facing inferolaterally. The robust transverse process (trpr) extends laterally from the pedicle, situated lower than the pre- and postzygapophysis in position.

Second thoracic vertebra? (T2?; Fig. 64 B)   Of the T2? only the dorsal part of vertebral arch and spinous process are preserved. The spinous process (sp) is long and cone-shaped, compressed transversely and long anteroposteriorly, extending vertically, with ridged anterior and posterior borders. The length of the preserved part of the spinous process is 5.5 mm.

Third thoracic vertebra? (T3?; Fig. 64 C)   Of the T3? only the body and pedicle of the vertebral arch are preserved. The body is similar to that of T1? in being cylindrical in form, but slightly longer than in the latter. The dorsal side of the body is flat, while its ventral side is straight longitudinally and convex transversely. The caput (cap) and the fossa (vfs) are like those in T2?. The pedicle is robust. The prezygapophysis extends forwards from the pedicle. The articular facet on the prezygapophysis (fprzy) is long and oval in outline, flat, facing superolaterally. The postzygapophysis extends backwards from the pedicle. The transverse process is robust, extending superolaterally from the pedicle. Its dorsal side is situated slightly higher than the articular facet on the prezygapophysis.

Fourth thoracic vertebra? (T4?; Fig. 64 D)　Only right half of the vertebral arch of T4? is preserved. The spinous process (sp) is about 11 mm high, and horn-like in form. Its lower part is oval in cross section, slightly wider than anteroposteriorly long. On the lower part of the anterior side of the spinous process there is a shallow vertical groove, while the cross section of its upper part is oval, anteroposterior longer than wide. The articular facet on the prezygapophysis (fprzy) is oval, slightly convex, facing slightly laterosuperiorly. The articular facet on the postzygapophysis has a slightly concave surface, facing inferomedially. The robust transverse process extends laterosuperiorly from the pedicle and is situated much higher than the prezygapophysis and separated from the latter by a distinct sulcus. The dorsal side of the transverse process is slightly lower than that of the postzygapophysis in position.

Ribs (R's；Figs. 60, 64 D)

Three ribs are preserved in HMV 1942. Of them one is squeezed to the T4?. It looks like a R2 in form, thus labelled as "R2" here. The costal neck (cn) is straight and long, and separates the costal head (ch) from the tubercle of rib (tur) widely. On the costal neck there is a rough area near the costal head, which may be the attachment area for ligament. The costal head is larger than the tubercle of rib and has an oval articular facet for the thoracic vertebra on the top. The upper part of the body of rib is flat, with a transversely convex anterior side and a flat posterior side. The costal groove is shallow. The length, width and thickness of the proximal end of "R2" are 6.9 mm, 3.2 mm and 1.5 mm.

The other two ribs, which are squeezed together, may be the third and fourth ribs, labelled here as "R3" and "R4". The tubercle of rib of "R3" is lower than that of "R2" in height and separated from the body of rib by a distinct depression. The top of the tubercle of "R3" is oval in form, with a convex surface. The bodies of "R3" and "R4" are long and slightly curved. Their upper parts are wide and flat, becoming narrower downwards. The anterior sides of the bodies are slightly convex transversely. The posterior side of "R3" is flat and has a shallow costal groove. The costal groove of "R4" is long and distinct.

The length of the preserved part of "R3" and "R4" are 17 mm and 18.5 mm. The widths and thickness of the proximal ends of "R3" and "R4" are: 2.5 mm × 1.4 mm and 2.5 mm × 1.3 mm respectively.

Forelimbs

The preserved forelimb bones of HMV 1942 are the distal parts of right and left scapulae, head of right humerus, and the distal portion of an articulated right forelimb including distal parts of radius and ulna, and almost the whole manus.

Scapula (Sc; Figs. 60, 65 A)　Only the lower parts of left and right scapulae are preserved. The glenoid cavity (glc) is oval in outline. Its major axis is longer than the width of the neck of scapula (nsc). The basal part of the spine of scapula (spsc) originates about 4.2 mm above the lateral border of the glenoid cavity. The acromion is broken completely. The preserved lower end of the spine shows that it may be tall and rather thick. The upper part of the anterior border of the scapula bends slightly forwards. The lower parts of both the supraspinous fossa (sspf) and infraspinous fossa (ispf) are very narrow and fan-shaped. The posterior margin of the infraspinous fossa is thick. The muscular line (lm) is well developed near the posterior margin. The area between the muscular line and the spine of scapula is concave transversely, while the area posterior to the muscular line is flat or slightly concave. The subscapular fossa (sscf) is also triangular in form, with a smooth surface. There is also a well-developed muscular line near its posterior margin. Below the muscular line and on the posterior side of the neck there is a rough area for the attachment of *m. teres minor*.

Humerus (Hu; Fig. 65 B; Tab. 14)　Only the proximal part of a right humerus of HMV 1942 is preserved. The caput (head) is directed posterosuperiorly, with a hemispheric articular surface, which is larger than the glenoid cavity of scapula (caput W: 5.3 mm, APD: 6.9 mm). The neck of humerus is distinctly shown on the posterior side of the humerus. The greater tuberosity (gtu) extends anterolaterally rather than superiorly, and separated from the head of humerus by a concavity. Its top is situated lower than the head of humerus. The lesser tuberosity (ltu) is located at the anteromedial side of the head of humerus and smaller and lower than the greater tuberosity. The intertubercular sulcus (bicipital groove) is wide, without intermediate ridge on it.

Radius (Ra; Figs. 66, 67 A; Tab.15)   Only the lower part of the right radius is preserved. The shaft of radius and the distal epiphysis are still separate. The epiphysis is not fully ossified, and split into two pieces.

The lower part of the shaft is cylindrical in form, oval in cross section, wider than long (APD). The lateral side of the shaft is concave transversely. The distal end of the radius is large and about trapezoid in outline in distal view. The articular facet for the scapulo-lunar is wide and concave. The maximum width of the distal end is 4.8 mm.

Ulna (Ul; Figs. 66, 67 A)   As in the radius, only its lower part is preserved, and the shaft and epiphysis are separated.

The lower part of the ulna is also semicylindrical in form. Its distal end is wider than the shaft. From the anterior view it is slightly narrower than that of the radius. The distal articular facet for the cuneiform is a convex surface. On the lateral side of the distal end there is a distinct vertical sulcus, which may be passed by the tendon of *m. extensor carpi ulnaris*. The maximum width of distal end of ulna is 4 mm.

Manus (Ms; Figs. 66, 67)   HMV 1942 preserved a complete right manus, including all carpals, metacarpals and phalanges of five fingers. Most of these bones are preserved in original articulated status, only several are slightly displaced.

Carpal bones (Carp; Figs. 66, 67)   The carpal bones of the right manus are squeezed together. Our description has to be mainly concentrated on their dorsal sides, which are clearly shown after preparation. The proximal row consists of three carpal bones: scapho-lunar, cuneiform and pisiform, and the distal row consists of six ones: trapezium, trapezoid, centrale, magnum, unciform and falciform.

Scapho-lunar (Sc-lu)   The scapho-lunar is the largest bone in the proximal row. It is wide, but short longitudinally. The proximal articular facet for the radius is large and has a convex-concave surface: the mediovolar part is concave and the laterodorsal part strongly convex dorsovolarly. Its dorsal part extends distally to meet the distal articular facet at a right angle. Its distal facet articulates with trapezium, centrale and magnum. The articular facet with the cuneiform is concave, located on the dorsal part of the lateral side and extends dorsovolarly. Its volar part of the lateral side strongly protrudes laterovolarly. On its medial side the articular facet for the falciform is flat.

Cuneiform (Cu)   The dorsal side of the cuneiform is small and triangular in outline. Its proximal articular facet for the ulna is oval in outline, concave, facing laterosuperiorly. The articular facet for the scapho-lunar is located on the medial side, being convex. The distal articular facet for the unciform has a concave medial part and a convex lateral part, which meets the proximal articular facet at a sharp angle at the lateral border. The volar side has a flat surface, with a tiny pit on its inferolateral corner. The lateral side is small, triangle in outline, with a rough surface.

Pisiform (Ps; Fig. 66B)   From the lateral view of the wrist, the pisiform is about trapezoid in outline, with its dorsal end larger than the volar one. There are two articular facets on the dorsal side, articulating with ulna and cuneiform respectively. The lateral and volar sides are rough.

Trapezium (Trm)   Form the dorsal view of the wrist the trapezium is the smallest one among the distal row of the carpal bones. Its dorsal side is trapezoid in outline. The proximal articular facet for the scapho-lunar is relatively small, and that for the trapezoid is slightly concave and larger. The distal articular facet for metacarpal II is a slightly concave surface facing laterodistally and meets the facet for the trapezoid at an obtuse angle. On the medial side there are two articular facets: the facet for metacarpal I is larger and occupies most part of the medial side; the other facet is small and located near the proximal border of the medial side, which may be the facet for the falciform.

Trapezoid (Trd)   The dorsal side of the trapezoid is triangular in outline and slightly larger than that of the trapezium. Its distal side articulates with Metacarpal II only, with a slightly convex dorsal border. Its medioproximal side articulates with the trapezium and the lateroproximal side articulates with the centrale. The two proximal facets are relatively flat, and meet each other at a sharp angle.

Centrale (Ce)   The centrale is larger than the trapezium, trapezoid and magnum. Its dorsal side forms a transversely wide triangle in outline. Its proximal articular facet for the scapho-lunar is large and convex. Its distal side articulates with trapezoid, Metacarpal II and Metacarpal III. From dorsal view the three articular facets are situated in one and the same line. The proximal and distal articular facets meet each other at a sharp angle on the medial side. The lateral side of the centrale articulates with the magnum, with a slightly concave surface, facing laterally.

Magnum (Mg)   From the dorsal view, the magnum is smaller than the centrale and subequal to the trapezoid in size. Its dorsal side is also triangular in outline. Its medial side articulates with the centrale, with a slightly convex surface. Its lateral side is flat, articulating with the unciform. The two articular facets meet at an acute angle on the proximal side. The distal articular facet for Metacarpal III is slightly convex.

Unciform (Un)   The unciform is the largest bone in the distal row of the carpal bones. It is irregular in form, much wider than high in proximodistal direction. The proximal articular facet for cuneiform is a sinuous surface, convex medially and concave laterally. Its distal side articulates with McIII, Mc IV and Mc V, forming a V-shaped dorsal border. On its medial side the articular facet for the magnum is flat.

Falciform (Fal)   The falciform is located at the medial side of the wrist. It is irregular in form and extends mediovolarly. It is slightly smaller than the unciform, but larger than the other carpal bones of the distal row. Its lateral side articulates with scapho-lunar and trapezium.

Metacarpus (Mc)

First metacarpal bone (McI; Fig. 66 A; Tab. 16)   The McI is mostly buried under the other bones, only its dorsal part is partly exposed. So far as can be judged, McI is very short and irregular in form. The top of the proximal end inclines lateroproximally, not parallel to the distal articular facet, but meets the latter at an acute angle. On the top there are two articular facets: the medioproximal one articulates with the trapezium and the lateral one for the McII. The shaft is cylindrical in form. There is a distinct longitudinal groove on the lower part of the dorsal side. On the distal end the dorsal part of the articular facet for the first phalanx is spherical.

Second metacarpal bone (McII; Figs. 66 A, 67; Tab. 16)   The proximal end of McII is large. There are three articular facets on the top of the proximal end. The medial one for the trapezium is transversely narrow and dorsovolarly long, with a slightly convex surface, facing mediosuperiorly. The middle facet, articulating with the trapezoid, is the largest. It is transversely concave, having a distinct concavity near the dorsal margin. The lateral articular facet for the centrale is also transversely narrow and anteroposteriorly long. On the medial side of the proximal end there is a very narrow facet below the facet for the trapezium, which should be the facet for McI. On the lateral side of the proximal end there is an articular facet for McIII below the facet for the centrale, facing inferolaterally.

The shaft is straight. Its upper part is narrow and trapezoid in cross section, becoming wider in lower part. The dorsal side of the shaft is straight longitudinally and convex transversely. The tuberosity of McII (tmcII) for attachment of *m. extensor carpi radialis longus* is well developed. However, the tuberosity of McII of HMV 1942 is different from that of *Ps. indigenus* gen. et sp. nov. in form and position. It forms a depression surrounded by eminences rather than a tuberosity as in the latter and it is located higher than that of the latter in position.

The distal end is much wider than the shaft, wider than thick. The articular facet for the first phalanx is in the form of a cross-axle. Its dorsal part is hemispherical in shape, with an asymmetric arcuate upper border: the curvature of the upper border of the lateral part is more convex, while that of the medial part is gentle and extends mediodistally. The articular surface is strongly convex longitudinally, without sagittal ridge on dorsal and distal sides, but a marked sagittal ridge is developed on the volar side. A pair of wide grooves is situated bilateral to the ridge. The lateral groove is narrower and deeper than the medial one. On either of the medial and lateral sides of the distal end there is a prominent depression for attachment of the medial and lateral collateral ligament.

Third metacarpal bone (McIII; Figs. 66, 67; Tab. 16)   The McIII is slightly larger than McII in general. It is similar to McIII of *Ps. indigenus* gen. et sp. nov. in morphology.

The proximal end is enlarged and has three articular facets on the top. The articular facet for the magnum is the largest and located in the middle. Its surface is concave transversely and convex dorsovolarly. The facet for McII on the medial part has a convex surface facing mediosuperiorly. The facet for the unciform on the lateral part is narrow. On lateral part of the dorsal side the crest for ligament is also prominent, but does not extend so long as in *Ps. indigenus* gen. et sp. nov.

The shaft is straight, with its upper part trapezoid in cross section. Its lower part becomes a flattened cylinder. The tuberosity of metacarpus III (tmcIII) for attachment of *m. extensor carpi radialis brevis* is well developed. It is a large rough depression located near the medial margin in the upper part of the dorsal side.

The distal end of McIII is similar to that of McII in morphology. But the dorsal part of the distal articular facet for the first phalanx is symmetric.

Fourth metacarpal bone (McIV; Figs. 66 B, 67 B; Tab. 16)   The McIV is much shorter than McIII and subequal to McII in length. The proximal end is large. On its top there are two articular facets. The lateral one for the unciform is larger and occupies the most part of the top, with a concave dorsal border. The medial one for McIII is much narrower and faces mediosuperiorly. On the lateral side of the proximal end the articular facet for McV is concave and faces inferovolarly. On the dorsal side of the proximal end near the medial and lateral margins the two crests for ligaments are prominent, but are weaker than in *Ps. indigenus* gen. et sp. nov. and *Pa. hipparionum.*

The shaft is similar to those of McII and McIII in morphology. The distal end is slightly wider than that of McII and subequal to that of McIII in width. The dorsal part of the articular facet for PhI is symmetrical.

Fifth metacarpal bone (McV; Figs. 66 B, 67 B; Tab. 16)   The McV is very short, much shorter than McIV. The upper and lower parts of the McV are slightly twisted in different directions. McV is situated posterolateral to McIV.

McV is quite different from that of *Ps. indigenus* gen. et sp. nov. Its proximal end is robust. On the top of the proximal end there are two articular facets. The lateral articular facet for unciform occupies most part of the top. It is trapezoid in outline, and has a saddle-shaped surface, concave transversely, but convex dorsovolarly, and slightly inclines mediosuperiorly. The medial articular facet for McIV is small and has a slightly convex surface, facing mediosuperiorly. On the dorsolateral corner of the proximal end there is a well-developed tuberosity, which may be the attachment area of *m. extensor carpi ulnaris*.

The shaft is very short. Unlike McV of *Ps. indigenus* gen. et sp. nov., it is quadrihedro-prismatic in form and much slenderer than the proximal and distal ends. Its dorsal side is straight longitudinally and convex transversely and both its lateral and medial sides are concave longitudinally. The distal end is similar to those of McII–McIV in morphology, but its medial side is thicker than the lateral one in dorsovolar direction. The distal articular facet for the first phalanx is asymmetric, with the medial part longer than the lateral one in dorsovolar direction and more convex medial anterior border than the lateral one.

Phalanges of fingers (Ph)   All of the phalanges of the five fingers of right manus of HMV 1942 are well preserved, mostly in their original articulated position.

First finger (FI)

Proximal phalanx of FI (FI-Ph1; Fig. 66 A; Tab. 17)   The proximal phalanx of the first finger is slightly dislocated from its original position. Viewed from the lateral view, FI-Ph1 is trapezoid in outline, with its proximal end much larger than the distal one.

The proximal end of FI-Ph1 is larger than that of *Ps. indigenus* gen. et sp. nov. Viewed from the top, the proximal end is semicircular in outline, with its volar side wider than dorsal side. The articular facet for McI is an oval concave surface, facing dorsally. On the volar side there are two tuberosities separated by a middle groove. The medial tuberosity is more prominent and may serve as the attachment area for *m. abductor pollicis*.

The shaft is a quadrihedral pyramid in form, with its dorsal side narrower than the volar one. The extensor tuberosity, which is the attachment area of *m. extensor pollicis brevis*, is prominent, situated in the upper part of the dorsal side of the shaft.

The distal end of FI-Ph1 is thicker than the shaft in dorsovolar direction. The articular facet for FI-Ph2 is slightly concave transversely and strongly convex dorsovolarly. Its volar border is situated higher than the dorsal one and has a slightly concave upper margin. On the lateral and medial sides of the distal end there is a prominent depression respectively, where the collateral ligaments may attach.

Distal phalanx of FI (FI-Ph2; Fig. 66 A; Tab. 17)   The FI-Ph2 is subequal to that of FI-Ph1 in length, but slightly narrower than the latter. The proximal end is very thick, with the dorsovolar diameter larger than transverse width. The articular facet for FI-Ph1 is a wide oval in outline, convex transversely and concave dorsovolarly. Its dorsal margin is thicker, with a prominent extensor process on its middle, which is for the attachment of *m. extensor pollicis longus*. Its volar margin is also thick and convex.

The dorsal side of the shaft is convex transversely and longitudinal slightly concave in upper part but slightly convex in lower part. On either of the lateral and medial sides there is a prominent volar sulcus near the volar side. Towards the distal end the volar sulcus forks into two: one extends along the lateral or medial sides; the other turns to the volar side. There are two volar foramina in the volar sulcus, one located at the upper part and the other, near the forking point. On the volar side the tuberosity of ungual phalanx for attachment of *m. flexor digitorum profundus* is well developed. The tuberosity assumes a form of angular processes, which may serve for enlarging the attachment area for *m. flexor digitorum profundus*. On the volar side a transverse groove above the tuberosity joins the lateral and medial volar sulci. The distal end of FI-Ph2 is trowel-shaped, with a slightly convex dorsal side and slightly concave volar side. Unlike the FI-Ph2 of *Ps. indigenus* gen. et sp. nov., there are some longitudinal ridges and grooves near the distal border, so that the distal border forms a serrated arch. No sagittal notch is present on the distal end.

Second finger (FII)

Proximal phalanx of FII (FII-Ph1; Figs. 66 A, 67 B; Tab. 17)   FII-Ph1 is about 2/3 length of the McII, but is much longer and thicker than FI-Ph1 (see Tab. 17). The top of the proximal end is not vertical to the axis of the shaft, but inclines dorsally. The articular facet for McII is concave, semicircular in outline. On the volar side of the proximal end there are two tuberosities separated by a middle groove. The medial tuberosity is larger than the lateral one. On either of the lateral and medial sides of the proximal end there is a rough depression respectively, which is the attachment area of collateral ligaments.

The shaft is in the form of a truncated cone, becoming slenderer downwards. The extensor tuberosity, the attachment area of *m. extensor digitorum communis*, is well developed and located on the upper 1/4 of the dorsal side. The volar side is flat and has longitudinal ridge on either of the lateral and medial borders in the lower part. The distal end is much smaller than the proximal end. The articular facet for FII-Ph2 inclines superovolarly. Its surface is convex dorsovolarly and separated by a sagittal groove into two transversely convex condyles. The medial condyle is larger than the lateral one. The dorsal margin of the facet for FII-Ph2 rolls up. The dorsal side of the distal end is flat and meets the articular facet of FII-Ph2 at a sharp angle. On either of the medial and lateral sides of the distal end there is a prominent depression surrounded by eminences serving for attachment of the collateral ligaments.

Middle phalanx of FII (FII-Ph2; Figs. 66 A, 67 B; Tab. 17)   The FII-Ph2 is shorter than FII-Ph1. Its proximal end is slightly thinner in dorsovolar direction than that of FII-Ph1, but its distal end slightly thicker than in the latter.

The articular facet for FII-Ph1 is oval in outline, inclining antero-inferiorly. Its surface is dorsovolarly concave, separated by a gentle sagittal ridge. The medial concavity is slightly larger than the lateral one. On the dorsal side of the proximal end the well-developed extensor tuberosity for attachment of *m. extensor digitorum communis* is pillow-like in form. On the volar side the groove passed by the tendon of *m. flexor digitorum profundus* is wide and the bilateral tuberosities are prominent. The medial tuberosity is larger than the lateral one. On either medial or lateral side of the proximal end there is a rough tuberosity for attachment of collateral ligament, *m. flexor digitorum superficialis* and sheath. The medial tuberosity is larger than the lateral one. The shaft is in the form of a truncated cone, becoming slenderer downwards. The volar side is flat.

The distal end is more robust than the lower part of the shaft, protruding dorsally beyond the dorsal side of the shaft and is in the form of a cross-axle. The articular facet for FII-Ph3 is semicircular, convex in dorsovolar direction. Its dorsal part extends superiorly, forming a narrow and flat surface with an arcuate upper border, which is asymmetrical, more convex in medial part. The volar side of the facet for FII-Ph3 is wider and slightly concave transversely. It extends superiorly, surpassing the dorsal one in height. The part of the dorsal side of the distal end above the facet for FII-Ph3 is flat and inclines dorsosuperiorly. The depressions serving for the attachment of collateral ligaments on the medial and lateral sides of the distal end are prominently shown.

Distal phalanx of FII (FII-Ph3; Figs. 66 A, 67 B; Tab. 17)   The distal end of the FII-Ph3 is partly broken. The FII-Ph3 is claw-shaped. It is larger than FI-Ph2, but narrower and longer than FII-Ph2.

The proximal end is thicker than wide. The articular facet for FII-Ph2 is oval in outline. Its surface is strongly concave dorsovolarly and slightly convex transversely. Along the medial and lateral margins there is a pair of narrow grooves, which may serve for the attachment of collateral ligaments. The extensor process is well developed, much more developed than that of FI-Ph2. Unlike that of FI-Ph2 of HMV 1942, but like those of FIII-Ph3 and FIV-Ph3 of *Ps. indigenus* gen. et sp. nov., the extensor process of FII-Ph3 is narrow and tile-like in shape. Viewed from the dorsal and volar sides, it is about trapezoid in outline, with a narrower and more concave top.

On the dorsal side of the shaft there is a slightly convex longitudinal crest. It is distinct on the upper part, becoming weaker distally and finally disappeared. The dorsomedial and dorsolateral surfaces are in the form of a narrow triangle. They are slightly convex, smooth on their proximal parts, but roughened with distinct longitudinal ridges and groove on their distal parts. The volar sulci are well developed on the medial and lateral parts of the volar side. The medial and lateral volar sulci join with each other to form a transverse groove at their proximal ends. The volar foramen is large, and the tuberosity of ungual phalanx (tunph) is well developed. Unlike FI-Ph2, the tuberosity is separated into two parts: a convex upper part and a concave lower part with a thin lower margin. On the volar side there is a longitudinal ridge extending from the tuberosity of ungual phalanx to the distal end. The distal end of FII-Ph3 is different from that of FI-Ph2: it is thicker, forming an obtuse cone, and covered with well-developed longitudinal ridges and grooves. No central notch is present on the distal end.

The third finger (FIII)

Proximal phalanx of FIII (FIII-Ph1; Figs. 66, 68; Tab. 17)　The FIII-Ph1 is similar to FII-Ph1 in general morphology, but slightly thicker and longer than the latter. The articular facet for McIII is larger and more deeply concave, and the medial and lateral tuberosities on volar side are larger than those of FII-Ph1. The medial tuberosity is still larger than the lateral one. On the medial and lateral sides of the proximal end the attachment areas of the collateral ligaments are in the form of tuberosities rather than depressions. They are symmetric and subequal in size. The extensor tuberosity (extu) on the dorsal side is larger than that of FII-Ph1. The shaft and distal part are similar to those of the FII-Ph1 in general morphology. But unlike FII-Ph1, the articular facet for FIII-Ph2 is more convex dorsovolarly. The dorsal side of distal end above the articular facet for FIII-Ph2 is concave rather than flat. The lateral depressions on the medial and lateral sides of distal end are larger than those of FII-PH1.

Middle phalanx of FIII (FIII-Ph2; Figs. 66, 69; Tab. 17)　The FIII-Ph2 is thinner and shorter than FIII-Ph1, but slightly longer and thicker than FII-Ph2. FIII-Ph2 is close to FII-Ph2 in general morphology. Unlike FII-Ph2, the sagittal ridge on the articular facet for FIII-Ph1 is weaker, and the bilateral concavities are symmetric and equal in size. The middle groove on the volar margin of the proximal end is wider and deeper, and the bilateral tuberosities for *m. flexor digitorum superficialis* are symmetric and equal in size. The dorsal part of the articular facet for FIII-Ph3 is narrower transversely, with a rather straight anterior border.

Distal phalanx of FIII (FIII-Ph3; Figs. 66, 70; Tab. 17)　The FIII-Ph3 is much longer than FIII-Ph2, with a narrower but thicker proximal end. It is similar to FII-Ph3 in general morphology, but is thicker and longer than FII-Ph3. The extensor process is well developed. The articular facet for FIII-Ph2 (ffIII-2) is pentagonal in outline, strongly concave in dorsovolar direction. The longitudinal crest on the dorsal side is longer, slightly bending medially and extending almost to the very distal end. The dorsomedial and dorsolateral sides are narrow and triangular in form. The distal part of the dorsolateral side is slightly wider and more convex transversely, with less developed ridges and grooves than the dorsomedial one. The two well-developed volar sulci converge downwards to join with each other below the tuberosity of the ungual phalanx. The medial volar sulcus is deeper than the lateral one. The volar foramen is located in the volar sulcus. The medial volar foramen is larger than the lateral one. No lateral foramen is present. The tuberosity of ungual phalanx is well developed. No central notch is present on the distal border.

Fourth finger (FIV)

Proximal phalanx of FIV (FIV-Ph1; Figs. 66 B, 67 B; Tab. 17)　The FIV-Ph1 is similar to FIII-Ph1 in morphology and in thickness, but shorter than the latter, and shorter and thicker than FII-Ph1, about 2/3 the length of McIV. The concave articular facet for McIV is deeper than that of FIII-Ph1. On the medial and lateral sides of the proximal end the tuberosities where the collateral ligaments attach are more prominent. The lateral tuberosity is more developed than the medial one. The extensor tuberosity on the dorsal side of the shaft is also well developed. The dorsomedial and dorsolateral sides of the shaft are asymmetric: the dorsomedial side is steeper inclined than the dorsolateral one.

Middle phalanx of FIV (FIV-Ph2; Figs. 66 B, 67 B; Tab. 17)　The FIV-Ph2 is shorter and slenderer than FIV-Ph1, and also shorter than FII-Ph2 and FIII-Ph2. It is subequal to FIII-Ph2 in robustness, but wider than FII-Ph2. FIV-Ph2 is similar to FIII-Ph2 and FII-Ph2 in morphology. Unlike FIII-Ph2 and FII-Ph2, the sagittal crest in the articular facet for FIV-Ph1 on the proximal end is very weak. The tuberosities for the attachment of *m. flexor digitorum superficialis* and collateral ligaments on the medial and lateral sides of the proximal end are more developed. The shaft of FIV-Ph2 is wider and flatter in the middle part. The anterior border of the distal articular facet for FIV-Ph3 is more convex, with more curved lateral part than the medial one.

Distal phalanx of FIV (FIV-Ph3; Figs. 66 B, 67 B; Tab. 17)　The FIV-Ph3 is much longer than FIV-Ph2, about twice as long as the latter, but subequal to FIII-Ph3 in length. FIV-Ph3 is similar to FIII-Ph3 and FII-Ph3 in morphology. But the upper part of the dorsolateral side is wider and more convex transversely than that of the dorsomedial one. The distal part of FIV-Ph3 becomes flatter and more pointed. The volar side is flat, facing mediovolarly.

Fifth finger (FV)

Proximal phalanx of FV (FV-Ph1; Figs. 66 B, 67 B; Tab. 17)   The FV-Ph1 is slightly shorter than McV. It is similar to FIV-Ph1 in morphology, but slightly shorter and slenderer than the latter. Its medial side is longer than the lateral side. Viewed from the lateral side, both the top and bottom of FV-Ph1 are not perpendicular to the axis of the shaft. The proximal tuberosities on the medial and lateral sides are not as prominent as those of FIV-Ph1. However, there is a depression below the lateral tuberosity. The shaft is flat, much wider than thick. The extensor tuberosity is also well developed. The dorsomedial and dorsolateral sides are asymmetric: the dorsomedial side is narrower and inclines steeper transversely, facing mainly medially, while the dorsolateral side is wider, gently inclined, facing dorsolaterally. The volar side is flat transversely and concave longitudinally. The distal end is flat dorsovolarly, facing slightly laterovolarly. The articular facet for FV-Ph2 inclines more superovolarly, forming a sharp angle with the dorsal side. The sagittal groove of the articular facet for FV-Ph2 is distinct and the medial condyle is wider than the lateral one.

Middle phalanx of FV (FV-Ph2; Figs. 66 B, 67 B; Tab. 17)   The FV-Ph2 is shorter and slenderer than FV-Ph1. It is similar to FIV-Ph2 in morphology, but is shorter and slenderer than the latter. The pillow-like extensor process is more developed and separated by a middle notch. On the distal end the lateral depression for collateral ligament is more developed than the medial one.

Distal phalanx of FV (FV-Ph3; Figs. 66 B, 67 B; Tab. 17)   The FV-Ph3 is longer than FV-Ph2. It is similar to FIV-Ph3 in morphology, with asymmetric dorsomedial and dorsolateral sides and a slightly oblique, flat and pointed distal end. But it is slightly shorter and slenderer than the latter. Among the five ungual phalanges FV-Ph3 is smaller than FII-Ph3, FIII-Ph3 and FIV-Ph3, but larger than FI-Ph2.

2) Comparison

HMV 1942 and the other above described specimens are similar to those of *Ps. indigenus* gen. et sp. nov. rather than to those of *Pararhizomys* in skull morphology and tooth structures, as exemplified as follows: the dorsal surfaces of the nasals are straight longitudinally and hardly convex transversely; the premaxillae extend anteriorly beyond the nasals; the dorsolateral crest of premaxilla is short and joins with naso-premaxillary suture anteriorly; the dorsal part of premaxilla is narrow and triangular in outline; on the dorsal side, the premaxillo-maxillary and the maxillo-frontal sutures form a unified curved line; the infraorbital foramen is large and circular in form; the anterior border of attachment area of *m. masseter lateralis* is S-shaped; the anterior root of the zygomatic arch is more anterior in position; the jugal is long, extending to the anterior border of the orbit, and usually join with the lacrimal; the parietal is wide and hexagonal in outline; the deep temporal, masticatory and buccinators foramina are confluent, but their posterior openings are partly separate; the lower border of the horizontal ramus of mandible is convex; the mental foramen is situated higher in position; the mandibular notch is deeper; the molars are lower crowed, only M1 and m3 with mesiodistal hypsodonty; I2 is less curved and proodont, with its anterior part extending antero-inferiorly, and its posterior end lying in the middle of maxilla, just anterior to m1; no longitudinal ridge is present on labial surface, etc.

The above list of distinctive features renders it reasonable to refer the above described specimens to the new genus *Pseudorhizomys*. However, they differ from those of *Ps. indigenus* gen. et sp. nov. as mentioned in differential diagnosis (*vide supra*). It is obvious that they represent a new species of *Pseudorhizomys*, named as

Pseudorhizomys gansuensis here.

It is necessary to mention that V 16296 is slightly different from HMV 1942 and the paratypes in some features, such as: the articular facet of the condyle of mandible has a sharp anterior border, its cheek teeth are higher crowned, the lingual part of the sinusid in m3 is unforked, etc. However, based on the fact that V 16296 remain inseparable from the other specimens of *Ps. gansuensis* in overwhelming number of features, we are inclined to consider the above mentioned differences as intraspecific variation. It is also worthy of noting that V 16296 was collected from the upper part of the middle Liushu Formation, while all the other specimens of *Ps. gansuensis* were found from the lower part of the middle Liushu Formation. Thus, it is quite possible that V 16296 represents an individual of slightly higher evolutionary level within the species *Ps. gansuensis*.

Pseudorhizomys gansuensis gen. et sp. nov. is known to be represented by nine specimens. Among them four specimens (HMV 1942, V 16296, V 16298 and V 16301) differ from the other five in being larger in size and having more robust rostrum and incisors. The former four may represent males and the other five may be females.

Pseudorhizomys planus gen. et sp. nov.
(Figs. 71–72, 76; Tabs. 1, 10–12)

Holotype   V 16304, anterior part of skull, with left and right I2 and molars, known from Zhuangkeji Village, Guanghe County, but without further information about locality and horizon.

Diagnosis   Rostrum and maxillary diastema are short. Dorsal side of premaxilla is short and triangular in form. Convex lower part of S-shaped anterior border of attachment area of lateral masseter muscle protrudes anteriorly beyond the ventral slit of infraorbital foramen. Molars are higher crowned. M1 is distinctly mesially hypsodont. Sinuses in M1–3 are short transversely, not overlapping anterosinuses. M2–3 have closed anterosinuses. Mesosinus and sinus in M3 are situated opposite, or linked with each other. I2 crown is narrow, with weak longitudinal ridge on flat labial surface.

Differential diagnosis   It differs from *Ps. indigenus* gen. et sp. nov. and *Ps. gansuensis* sp. nov. in having shorter rostrum and maxillary diastema, dorsal side of premaxilla forming a short triangle, and narrower I2 with weak longitudinal ridge on flat labial surface. It further differs from *Ps. gansuensis* gen. et sp. nov. in M1–3 having shorter sinuses, M1 sinus not overlapping anterosinus, M2–3 having anterosinuses, and M3 having mesosinus and sinus being opposite to each other.

Etymology   *Planus*, flat in Latin, denoting the flat labial surface of I2.

1) Description

Skull (Fig. 71; Tab. 11)   The muzzle is short and narrow, with lateral sides slightly convergent forwards. The anterior end of the premaxilla extends anteriorly beyond that of the nasal. The dorsal surface of the nasal is straight longitudinally and slightly convex transversely. As in *Ps. indigenus* gen. et sp. nov., the dorsolateral crest of the premaxilla is short and joins the naso-premaxillary suture anteriorly. The dorsal side of the premaxilla is triangular in outline. The straight premaxillo-frontal suture extends anterolaterally. The premaxillo-maxillary and the maxillo-frontal sutures form a unified curved line. The infraorbital foramen is circular in form and has a ventral slit and a bony septum. The maxillary diastema is longer than the upper tooth row. The incisive foramen is about 1/4 the length of the diastema and located in the posterior half of the latter. The premaxillo-maxillary suture intersects the incisive foramen at the anterior 1/3 of the latter. Along the lateral borders of the incisive foramen a pair of longitudinal crests extends to the anterior border of M1. The anterior border of the attachment area of lateral masseter muscle is S-shaped, with its upper part concave and lower part convex anteriorly, the latter of which protrudes anteriorly beyond the ventral slit of the infraorbital foramen. The posterior border of the zygomatic plate is situated slightly posterior to the incisive foramen. The palatine sulcus deepens against the M1–2. The posterior palatine foramen is single and large, located medial to M1/2. The left and right tooth rows are nearly parallel to each other.

Teeth (Fig. 72; Tab. 12)　The dental formula is 1·0·0·3/. The molars are sub-hypsodont and rooted. The size decreases from M1 to M3.

The M1 crown is elliptic in outline, longer than wide, with buccal side longer than lingual one. The crown of M1 is mesially hypsodont. On the occlusal surface of M1 there are three well-developed reentrants (anterosinus, mesosinus and sinus), which are open in V 16304. The protosinus is present, but weak and shallow. The buccal two reentrants (anterosinus and mesosinus) are longer and deeper than the sinuses. The mesosinus is slightly longer than the anterosinus and bends slightly posteriorly. The short sinus extends toward paracone, but does not overlap the anterosinus. The M2 is approximately quadrate in outline and also has three reentrants on the occlusal surface. The sinus extends anterobuccally to meet the closed anterosinus. The mesosinus extends posterolingually and is longer than the sinus transversely and subequal to the latter in depth. The M3 is oval in outline, with a slightly narrower posterior side, and has three reentrants on the occlusal surface as well. The anterosinus is also closed. The short sinus and mesosinus extend transversely, and situated opposite to each other, scarcely separated from each other by a pit. The sinus is shallower than the mesosinus in depth.

The I2 bends slightly, with its anterior part extending anteroventrally rather than posteroventrally. I2 is triangular in cross section, with a flat labial side. The enamel mainly covers the labial side, only slightly expands to the two lateral sides. The labial surface is flat transversely and has a weak longitudinal ridge on it.

2) Comparison

V 16304 is similar to *Pseudorhizomys* gen. nov. but different from *Pararhizomys* in many features. These are, for example: the anterior end of the premaxilla extends more anteriorly than that of the nasal; the premaxilla has short and oblique dorsolateral crest and triangular dorsal side; the dorsal surface of the nasal is straight longitudinally; the I2 is less curved, with its anterior part extends anteroventrally, etc. It is obvious that V 16304 should be referred to *Pseudorhizomys* gen. nov. rather than to *Pararhizomys*. V 16304 is different from the above two new species (*Ps. indigenus* and *Ps. gansuensis*) in being smaller in size, having shorter and narrower rostrum, premaxilla having shorter dorsal side, and thinner and smaller I2 with weak longitudinal ridge on flat labial surface. Furthermore, V 16034 differs from *Ps. gansuensis* gen. et sp. nov. in M1–3 having shorter sinuses, M2–3 having anterosinuses, and M3 having transversely extending and opposite mesosinus and sinus, etc. V 16304 represents a new species morphologically distinct from both *Ps. indigenus* gen. et sp. nov. and *Ps. gansuensis* gen. et sp. nov. It is here called *Pseudorhizomys planus*.

Pseudorhizomys pristinus gen. et sp. nov.
(Figs. 73, 74, 76; Tabs. 1, 10–12)

Holotype　V 16305, a skull with mandible from LX 201001, 500 m south of Panyangyinwa, Guantangou Village, Hezheng County, probably lower part of the middle Liushu Formation; Late Miocene Middle Bahean ALMA[1] (see Tabs. 1, 10).

---

[1] Since the apecimen was collected from the surface of the locality, its horizon assignment is to be further verified.

Diagnosis This is a small-sized and more primitive species of *Pseudorhizomys*. Rostrum and maxillary diastema are long. Anterior border of attachment area of lateral masseter muscle is convex anteriorly, reaching to the lateral rim of ventral slit of infraorbital foramen. Masseter tubercle is located near the ventral slit. Temporal crest is longer and sharp. Mandible diastema has a shallower and longer concave posterior part. Coronoid process is short, reclines more posteriorly, with prominently bending upper part and more concave posterior border. Articular facet of condyle is kidney-shaped. Molars are lower crowned, without distinct undulating dental tract. Only m3 shows distinct distal hypsodonty and M1 has no distinct mesial hypsodonty. Anterosinuses are present on M2-3, open and less reduced on M2. Sinuses of M1-3 insert between anterosinuses and mesosinuses. Posterosinusids are present on m1-3 and protosinusid is present on m3. Labial side of I2 is convex transversely, without distinct longitudinal ridge on it.

Differential diagnosis It differs from three new species of *Pseudorhizomys* gen. nov. (*Ps. indigenus*, *Ps. gansuensis* and *Ps. planus*) in having arcuate anterior border of attachment area of lateral masseter muscle; masseter tubercle located near the ventral slit of infraorbital foramen; molars lower crowned, without undulating dental tract; M1 without distinct mesial hypsodonty.

It further differs from *Ps. indigenus* gen. et sp. nov. and *Ps. gansuensis* gen. et sp. nov. in being smaller in size, having more developed and sharper temporal crest, coronoid process being shorter longitudinally, with prominent posteriorly bending upper part and more concave posterior border; condyle having kidney-shaped articular facet, and m1-3 having posterosinusids.

It differs from *Ps. indigenus* and *Ps. planus* in M1-3 having transversely longer sinuses, inserting between anterosinuses and sinuses, M2 having less reduced and open anterosinus.

It differs from *Ps. gansuensis* in having more posteriorly inclined coronoid process, M2-3 having anterosinuses, sinusids of m1-3 being unforked, and m3 having protosinusid.

It differs from *Ps. planus* in having longer rostrum, maxillary diastema and dorsal side of premaxilla, and I2 having transversely more convex labial side.

Finally, it differs from *Ps.? hehoensis* in molars being higher crowned and m3 being distally hypsodont, and i2 having less convex labial side.

Etymology *Pristinus*, early, original, primitive in Latin.

1) Description

(1) Skull (Fig. 73 A; Tab. 11)

The skull is slightly distorted so that the left half of the skull is shifted anteriorly, the posterior part of rostrum becomes narrower and higher and the incisive foramen and the tooth rows are laterally compressed.

In dorsal view, the rostrum is long and narrow, with nearly parallel lateral sides. The nasal is mostly broken, but its basic morphology is recognizable. It is wedge-shaped, with a pointed posterior end, which is located nearly in the same line as the posterior end of the premaxilla. The serrated premaxillo-frontal suture extends straight and anterolaterally. The premaxillo-maxillary and the maxillo-frontal sutures form a laterally concave line. The dorsolateral crest of the premaxilla is long, joining with the naso-premaxillary suture. The dorsal side of the premaxilla is very narrow and triangular in outline. The zygomatic arch is slender and roundly curved. The jugal is long, extending to the anterior border of the orbit, and its anterior end reaches to the posterolateral corner of the infraorbital foramen. The lacrimal is poorly preserved. The postorbital constriction is prominent. The frontal crest is indistinct. The sagittal crest is narrow and sharp. The temporal crest is prominent, long and sharp-crested, extending to the nuchal crest.

In lateral view, the anterior end of the premaxilla extends anteriorly beyond that of the nasal. The alveolus of I2 forms a longitudinal bulge on the lateral side of the premaxilla. Between the bulge and the dorsolateral crest of the premaxilla is a narrow triangular concavity, which goes into the infraorbital foramen. The infraorbital foramen is large and circular in outline, confined to the maxilla. It has the ventral slit and small bony septum. The I2 originates at the middle of maxilla, in front of M1. Its anterior part extends anteroventrally rather than posteroventrally. The zygomatic plate is confined to the maxilla and mainly extends anterolaterally, with a concave surface. The anterior border of the attachment area of lateral masseter muscle is anteriorly convex and reaches to the lateral rim of the ventral slit of the infraorbital foramen. The masseter tubercle is distinct and located inferoposterior to the ventral end of the ventral slit. The maxillary diastema is much longer than the upper tooth row. The anterior part (before M1) of the skull is subequal to the posterior part of the cranium (including M1) in length. The glenoid fossa is long and extends to the nuchal crest, pushing the external auditory meatus downwards to below its posterior part. The postglenoid foramen is located posteromedial to the glenoid fossa and above the external auditory meatus. The mastoid process is small. The stylomastoid foramen is located between the external auditory meatus and the mastoid process.

In ventral view, the posterior border of the hard palate may be aligned with the middle of M3. The mesopterygoid fossa is subequal to the pterygoid fossa in width. The internal pterygoid process does not slant. The deep temporal foramen, masticatory foramen and buccinator foramen are confluent into one slit. The posterior foramen of the alisphenoid canal (ascp) is small and located at the posterolateral corner of the pterygoid fossa, medial to the base of the posterior end of the external pterygoid process, and medioposterior to the confluent posterior aperture of the temporal-masticatory-buccinator canals. The posterior foramen of the pterygoid canal (pcpf) is located at the medial side of the pterygoid fossa, near the base of the posterior part of the external side of the internal pterygoid process. The foramen ovale is also confluent with the middle lacerate foramen.

The basioccipital widens posteriorly and no distinct sagittal ridge can be seen. The occipital condyle has a wide ventral side. The ventral parts of the left and right occipital condyles are close to each other in position, and separated by a narrow sagittal groove. The hypoglossal foramen is large. The right auditory bulla is well preserved and roughly oval in outline, with its major axis extending anteromedially and the spine in the same direction. It is moderately inflated and no distinct concavity or ridge is present on its anteromedial part. The foramen of Eustachian canal opens anteromedial to the spine of the bulla. The jugular foramen is distinct.

The nuchal surface is semicircular in outline and slightly inclines forwards. Viewed from the lateral side, the occipital condyles are located slightly posterior to the nuchal crest. The *foramen magnum* is large and circular in outline, higher than wide. The posterior parts of the two occipital condyles are separated widely. The paramastoid process is small.

(2) Mandible (Fig. 73 B, C; Tab. 11)

The horizontal ramus of the mandible is short and thick. The upper border of the mandibular diastema is S-shaped, with a short, convex anterior part, and a longer and less concave posterior part. The masseteric fossa extends anteriorly below m1/2, with a more prominent and flaring masseteric ridge. The mental foramen is situated antero-inferior to m1, at the mid-height of the horizontal ramus, nearly at the level of the anterior end of the masseteric fossa. The coronoid process is shorter longitudinally and reclines posteriorly, with its anterior border forming an angle of about 120° with the alveolar border. Its top part bends posteriorly and its posterior border is deeply concave. The distinct temporal fovea is located on the basal part of the lingual side of the coronoid process. The mandibular notch is about 2/5 the height of the mandible. The condyloid process extends mainly posterosuperiorly, with its top bending slightly lingually. The articular facet of the condyle is kidney-shaped, spherical, with a narrow posterior part. The attachment area for *m. pterygoideus lateralis* is concave.

The posterior end of the alveolus of i2 forms a prominent bulge on the buccal side of the ascending ramus. The mandibular foramen is large and located below the posterior border of the coronoid process and at about the level of the occlusal surfaces of the lower molars. The lower border of the ascending ramus is convex ventrally, with its posterior part extending posterosuperiorly. The internal pterygoid fovea is large and deep and extends forwards to below the middle part of the condyloid process. The angular process is situated higher than the other parts of the lower border of the ascending ramus.

(3) Teeth (Fig. 74; Tab. 12)

The dental formula is 1·0·0·3/1·0·0·3. The molars are similar to those of *Ps. indigenus* gen. et sp. nov. in basic features. They are lophodont, subhypsodont and rooted. The molars decrease posteriorly in size. However, the crowns of the molars are lower than those of the species mentioned above. The distal hypsodonty is present on m3 only.

The M1 is quadrilateral in occlusal view, with the anterior side wider than the posterior and the buccal side longer than the lingual. No distinct mesial hypsodonty is present. The protosinus is very weak. There are two buccal reentrants (anterosinus and mesosinus) and one lingual (sinus). All three reentrants extend to the middle of the tooth and open at lateral sides. The anterosinus and mesosinus are subequal in transverse length, with their lingual parts bending posteriorly. The mesosinus is more curved than the anterosinus. The sinus is shorter than the two buccal reentrants in transverse direction and intersects between the latter two, overlapping them.

The M2 is similar to M1 in morphology. It is also quadrilateral in outline, but the posterior side is slightly narrower than the anterior one, and the buccal and lingual sides are subequal to each other in length. It is smaller than M1. There are also three reentrants on the occlusal surface on M2. But the anterosinus and mesosinus are less curved than those of M1.

The M3 is oval in outline, wider than long and posterior side narrower than anterior one. There are also three reentrants on the occlusal surface, all of them being closed basins. The anterosinus is the smallest and the mesosinus may be slightly larger (on left M3) or smaller (on right M3) than the sinus. The sinus inserts between the anterosinus and the mesosinus.

The I2 is similar to that of *Ps. indigenus* gen. et sp. nov. in morphology. It is proodont and its anterior part extends anteroventrally rather than bends posteriorly. It is triangular in cross section. The enamel layer covers mainly the labial side, only slightly on the two lateral sides. The labial surface is convex transversely, but the curvature is weaker than in *Ps. indigenus* gen. et sp. nov. and *Ps. gansuensis* gen. et sp. nov. No longitudinal ridge is present on the labial side.

The m1 is oval in outline, with its anterior side narrower than posterior one. On the occlusal surface of the left m1 four reentrants can be seen. The mesosinusid extends transversely and opens lingually. Its buccal part bends toward the protosinusid. The sinusid forms a closed basin and no forking can be seen at its lingual end. The protosinusid and posterosinusid become small closed basins because of heavy wear. On the occlusal surface of the right m1 only two reentrants (mesosinusid and sinusid) can be seen. These two reentrants become closed basins. The posterosinus becomes vestigial, and no vestige of protosinusid can be seen. The reason for the disappearance of the protosinusid lies in the uneven wear of the teeth, so that the right m1 is more heavily worn than the left one.

The m2 is elliptic in outline, longer than wide. On the occlusal surface there are four reentrants. Among them only the posterosinusid is closed and the others are open. The mesosinusid is the longest transversely. Its buccal part bends toward the protosinusid, but does not reach the latter. The sinusid is shorter than the mesosinusid but longer than the protosinusid. It extends transversely and no forking can be seen. The posterosinusid is the smallest.

The m3 is oval in outline, slightly longer than wide, with narrower posterior side. The right m3 is similar to m2 in having four reentrants. Among them the mesosinusid and sinusid are open. The mesosinusid is longer than the sinusid transversely and its buccal part bends anteriorly rather than towards the protosinusid. The sinusid extends posterolingually. Both the protosinusid and posterosinusid are closed because of wear. The protosinusid is larger than the posterosinusid. The left m3 is similar to the right m3 in morphology of mesosinusid and sinusid. But on its occlusal surface no protosinusid and posterosinusid can be seen. They are probably worn away.

The i2 curves less than the I2 does, with its anterior part directed anterosuperiorly. Its cross section forms a equilateral triangle. The enamel layer mainly covers the labial side, only slightly on the two lateral sides. On labial side there are two longitudinal ridges.

It is necessary to point out that the right i2 may not belong to the same individual as the right hemimandible of V 16305, because some sign of mismatching can be detected where the i2 is glued to the hemimandible. Irrespective of this, the right i2 may really belong to the same species as V 16305 does, since the right i2 does have the same features as the left i2 of V 16305. But it is slightly narrower than the left i2 (see Tab. 12). This difference may represent either intraspecific variation or a sexual one.

2) Comparison

V 16305 is similar to *Pseudorhizomys* gen. nov. in the following features: the premaxilla extending more anteriorly than the nasal; the obliquely extending dorsolateral crest of premaxilla joining with naso-premaxillary suture; the premaxillo-maxillary and maxillo-frontal sutures forming a unified curved line; infraorbital foramen circular in outline; the long jugal extending to the anterior border of the orbital; the anterior part of I2 extending not posteriorly, and the I2 originating from the middle of maxilla in front of M1; the posterior concavity of the mandibular diastema shallow; the mandibular notch deeper, etc. All these features are different from those of *Pararhizomys*. Obviously V 16305 is to be referred to *Pseudorhizomys* gen. nov.

However, V 16305 is quite different from all the three above-described new species of *Pseudorhizomys* gen. nov. (*Ps. indigenus*, *Ps. gansuensis* and *Ps. planus*) in the following three features: the anterior border of the attachment area of lateral masseteric muscle is anteriorly convex; the masseter tubercle is near the ventral slit of infraorbital foramen; and the molars are lower crowned. Furthermore, it differs from *Ps. indigenus* and *Ps. gansuensis* in being smaller in size, having more developed and sharper temporal crest, etc (*vide* differential diagnosis). It seems that V 16305 represents a new species distinct from the known species of *Pseudorhizomys*, named here as *Pseudorhizomys pristinus*.

3) Discussion

Zheng (1980) described a partial left (in fact, it is right) hemimandible (V 5183), collected from Late Miocene deposits in Bilung Basin of Xizang, as *Brachyrhizomys hehoensis* and referred it to the Rhizomyidae. Later Jacobs et al. (1985) and Flynn (2009) thought that the specimen may belong to *Pararhizomys*.

Having closely examined the hemimandible (in fact, it is a right one, not a left one as Zheng stated), we agree with the above mentioned authors' opinion that V 5183 is similar to pararhizomyines rather than rhizomyids, because the mandible in question has a well-developed masseteric ridge of masseter fossa; the lower molars are slightly bend buccally, with lingual crown higher than the buccal one, more simple occlusal structure and lacking mesolophid, etc. Within the Pararhizomyines V 5183 is more similar to *Pseudorhizomys* gen. nov., especially to *Ps. pristinus* gen. et sp. nov., because its molars are lower crowned, with four reentrants on the occlusal surface, m2–3 have longer and anteriorly bent mesosinusid, and i2 has two longitudinal ridges on labial surface. However, V 5183 is also different from that species in such features as the horizontal ramus of mandible being lower, the molars being lower crowned, lacking distal hypsodonty, and i2 having more convex labial surface, etc. All these differences tend to indicate that V 5183 may represent a valid species different from *Ps. pristinus* gen. et sp. nov. at least. Since no features other than the mandible have so far been known for V 5183, it seems better tentatively to refer V 5183 to *Pseudorhizomys*, rename it as *Pseudorhizomys* (?) *hehoensis*.

III. PHYLOGENY AND CLASSIFICATION OF PARARHIZOMYINES

1. Phylogenetic relationships of pararhizomyines at generic and specific levels

1) Interrelationships between *Pararhizomys* and *Pseudorhizomys*

Having compared *Pararhizomys* and *Pseudorhizomys*, we found that a series of features of *Pseudorhizomys*, such as the forms of the nasal and premaxilla, the position of the anterior end of premaxilla relative to that of the nasal, the longer jugal, the more anterior position of the posterior border of the hard palate, the proodont I2 lacking longitudinal ridge on labial side, the lower molar crowns with more reentrants on occlusal surface, etc. (for details see Tab. 9: 1–4, 7–9, 12, 21–24, 27), are more primitive than those in *Pararhizomys*. Among these primitive features those of I2 may play a leading role. It is possible that with increasing curvature of the I2, a series of other features change in correlation as well (see Tab. 9: 1–4, 7, 8 and 21). However, *Pseudorhizomys* may not be the direct ancestor of *Pararhizomys*, because *Pseudorhizomys* not only possesses some more derived characters, such as the short and wide parietal bone, etc. (see Tab. 9: 10, 15), but also coexists with *Pararhizomys*, at least in the Linxia Basin. In fact, the earliest species of *Pararhizomys* (*Pa. longensis*) appeared even earlier than all known species of *Pseudorhizomys*. Thus, it can be postulated that both genera might have originated from a common ancestral form living either in early Middle Miocene, or even earlier.

2) Relationships among the species of *Pararhizomys*

(1) Character argumentation within *Pararhizomys*

In searching for the phylogenetic relationship of the four species of *Pararhizomys*, the "old-fashioned (and perfectly valid) grouping rule" (Wiley et al., 1991: 15), or the "Hennig Argumentation" (idem: 45–47) is used. Essential to this practice is searching for characters commonly shared by the in-group members but represented by different states. The character polarity is determined based on comparison with closely related out-group taxa, like *Pappocricetodon*, etc. The following are the major commonly shared characters chosen for Hennig argumentation:

1. The longitudinal groove on lateral side of premaxilla: (0) shallow; (1) deep.
2. Position of zygomatic plate: (0) more anteriorly, its anteriormost point of emarginated posterior border being anterior to M1; (1) more posteriorly, opposite to middle of M1.
3. Length of mandibular diastema and lower cheek tooth row: (0) subequal in length; (1) the former longer than the latter.
4. Crown height of molars: (0) crown lower, only some molars (M1–M2 and m3) are mesiodistally hypsodont; (1) all molars are mesiodistally hypsodont, but without lingual hypsodonty; (2) all molars are mesiodistally and lingually hypsodont.
5. Mesosinus and sinus on M3: (0) obliquely positioned to each other; (1) opposite to each other.
6. Mesosinusid of m1: (0) shorter and less curved; (1) longer and strongly curved, and its buccal part extending forwards.

(2) Analysis of relationships among species of *Pararhizomys*

The genus *Pararhizomys* is known to include four species and two *Pararhizomys* spp.

Pararhizomys longensis is the earliest appeared species. Although it has a series of primitive features (Fig. 75: 3-0, 4-0, 5-0 and 6-0), it also has some low-stated apomorphies: the deeper longitudinal groove on the lateral side of the premaxilla (1-1), and posteriorly positioned zygomatic plate (2-1). *Pa. longensis* may represent the first split branch.

Pararhizomys hipparionum and *Pa. qinensis* appeared later and share one synapomorphy: the molar crown becomes higher than that of *Pa. longensis*, with mesiodistal hypsodonty and subequally deep lingual and buccal reentrants (4-1). *Pa. hipparionum* has an autapomorphic character of its own: m1 has distinctly curved mesosinusid (6-1), which is different from those of *Pa. longensis* and *Pa. huaxiaensis* (unfortunately, no lower molar is known from *Pa. qinensis* and we do not know its character state). At any rate, *Pa. hipparionum* and *Pa. qinensis* may represent a sister group.

Pararhizomys huaxiaensis appeared the latest and possesses more apomorphic character states: its molars are higher crowned, characterized by distinct mesiodistal and lingual hypsodonty (4-2), mandibular diastema is much longer than the lower cheek tooth row (3-1), and the mesosinus and sinus are situated opposite to each other in M3 (5-1). It seems that *Pa. huaxiaensis* is more progressive than the other three species of *Pararhizomys* (*Pa. longensis*, *P. qinensis* and *Pa. hipparionum*). However, *Pa. huaxiaensis* has a synplesiomorphy with *Pa. longensis*: m1 with less curved mesosinus (6-0).

As a result of the above analysis, it seems that *Pa. longensis* may represent the first split branch, being sister taxon of the other three species; then *Pa. huaxiaensis* is the sister taxon of *Pa. hipparionum* + *Pa. qinensis* (see Fig. 75). Expressed in simplified Venn diagram, this should be: (*Pa. longensis* (*Pa. huaxiaensis* (*Pa. qinensis* + *Pa. hipparionum*))).

3) Relationships among the species of *Pseudorhizomys*

(1) Character argumentation within *Pseudorhizomys*

Similarly to *Pararhizomys*, the following 12 major characters are chosen for *Pseudorhizomys*:

1. Anterior border of attachment area of *m. masseter lateralis*: (0) anteriorly convex; (1) S-shaped, upper part anteriorly concave and lower part anteriorly convex.
2. Temporal crest: (0) longer and distinct, with a narrow and sharp posterior part; (1) shorter, with a low and gentle posterior part.
3. Posterior concavity of mandibular diastema: (0) long and shallow; (1) short and deep.
4. Labial side of I2: (0) transversely wide, convex and smooth; (1) transversely narrow and flat, with longitudinal ridge.
5. Crown height of molars: (0) molars without mesiodistal hypsodonty; (1) only m3 distally hypsodont; (2) M1 and m3 mesiodistally hypsodont.
6. Sinus on M1–M2: (0) long, inserted between anterosinus and mesosinus; (1) short, inserted between anterosinus and mesosinus only on M1; (2) very short, no insertion at all.
7. Anterosinus on M2–M3: (0) developed on M2–M3, but reduced to a closed basin on M3; (1) present on M2–M3, but reduced to a closed basin on both M2 and M3; (2) reduced and disappeared on M2–M3.
8. Sinus and mesosinus on M3: (0) sinus long, extending obliquely, overlapping mesosinus partly; (1) sinus short, situated opposite to mesosinus.
9. Posterosinusids of m1–m3: (0) present; (1) absent.

10. Sinusids of m1–m3: (0) not forking; (1) forking.

11. Protosinusid of m3: (0) present; (1) reduced or disappeared.

12. Position of sinusid and mesosinusid on m3: (0) sinusid situated posterior to mesosinusid; (1) sinusid situated opposite to lingual part of mesosinusid.

(2) Analysis of relationships among species of *Pseudorhizomys*

The genus *Pseudorhizomys* is known to include five species (*Ps. indigenus*, *Ps. gansuensis*, *Ps. planus*, *Ps. pristinus*, and *Ps.? hehoensis*). Among them, *Ps. gansuensis*, *Ps. pristinus* and *Ps.? hehoensis* are the species appeared earlier (Late Miocene, middle Bahean ALMA). Using the increase of the molar crown height and simplification of the molar crown structure as general evolutionary tendencies in pararhizomyine rodents, *Ps.? hehoensis*, if it really belongs to this genus, may represent the most primitive species of *Pseudorhizomys*, because its molars are brachyodont without mesiodistal hypsodonty (*vide supra*: 5-0), and have more reentrants (9-0, 11-0), etc. *Ps. pristinus* may represent the next more primitive species just above *Ps.? hehoensis*, because its molars are higher crowned, m3 is distally hypsodont (5-1), but at the same time it has some features more primitive than the other three species (*Ps. indigenus*, *Ps. gansuensis* and *Ps. planus*) (1-0, 2-0, 3-0, 6-0, 7-0, etc.), while *Ps. indigenus*, *Ps. gansuensis* and *Ps. planus* share such synapomorphies, such as, the attachment area of lateral masseter muscle having an S-shaped anterior border (1-1), the temporal crest becoming gentle (2-1), the posterior part of mandibular diastema being shorter and more deeply concave (3-1), the molars being higher crowned and M1 becoming mesially hypsodont (5-2), and the posterosinusid of m1–3 reduced or disappeared (9-1), etc. Thus the latter three species seem to be more advanced than *Ps.? hehoensis* and *Ps. pristinus*. Among the three more advanced species, *Ps. indigenus* and *Ps. planus* are to be considered as a sister pair based on the following synapomorphies: the anterosinus of M2–M3 reduced into a closed basin (7-1), the sinus becoming short transversely on M1–M3, not reaching anterosinus on M1–2 (6-2), and opposite to mesosinus on M3 (8-1). *Ps. gansuensis* appeared earlier than *Ps. indigenus*. This is in accord with the primitive features observed in *Ps. gansuensis*, for example, more complex molar crown structure (6-1, 8-0), etc. However, *Ps. gansuensis* also has some derived features of its own: the anterosinus of M2–M3 and protosinusid of m3 disappeared (7-2, 11-1), and m1–m3 having forked sinusid (10-1). Thus, *Ps. gansuensis* may represent a separate branch split earlier.

To sum up, among the known species of *Pseudorhizomys*, with exception of *Ps.? hehoensis*, *Ps. pristinus* may represent the most primitive branch and have split the earliest, and the other three species may have split into two branches: *Ps. indigenus* and *Ps. planus* forming a sister-group and *Ps. gansuensis* representing another branch of its own. Since *Ps. gansuensis* appeared earlier than *Ps. indigenus* (middle Bahean ALMA), the two latter branches may have split at least in early Bahean or even earlier. The tentative interrelationships among the species of *Pseudorhizomys* are shown in Fig. 76.

The cladistic relationships of the species of *Pseudorhizomys*, expressed in simplified Venn diagram, would look like as follows: (*Ps.? hehoensis* (*Ps. pristinus* ((*Ps. indigenus* + *Ps. planus*) *Ps. gansuensis*))).

2. Phylogenetic relationships between pararhizomyines and some closely related muroid families

1) Current opinions on classification of *Pararhizomys*

The systematic position of *Pararhizomys* has long been a puzzle. While establishing the genus *Pararhizomys*, Teilhard de Chardin and Young (1931: 12) regarded it as closely related to the *Rhizomys* group of the Spalacinae, but they refrained from referring it to any known family of the superfamily Muroidea. Later, *Pararhizomys* had either been referred to the family Rhizomyidae (Simpson, 1945; Young et Liu, 1950; Kowalski, 1968), or to the subfamily Rhizomyinae of the Muridae (McKenna et Bell, 1977). Flynn (1982, 1990) removed *Pararhizomys* from the Rhizomyidae to the Spalacidae. While dwelling on the systematic status of *Pararhizomys*, Zhang et al. (2005) admitted two possibilities: either it was a member of the Rhizomyidae, or of the Muridae, without making any final decision. This point of view was supported by Li (2010). Recently, Flynn (2009) abandoned his original opinion, admitting that *Pararhizomys* might be a member of the Rhizomyidae. The problem now concerns not only the single genus *Pararhizomys*, but a group of rodents consisting of at least 9 species in 2 genera.

2) Cladistic analysis of pararhizomyines and closely related muroid families

(1) Choice of taxa in analysis

Based primarily on the myomorphous zygo-masseteric structure of the skull, a widely recognized criterion to separate the muroid rodents from all the other rodents, the *Pararhizomys*-group (*sensu lato*, containing 9 species of 2 genera) is unquestionably to be referred to the Muroidea. However, the opinions on the subdivision of the Muroidea are quite different (*vide* Ellerman, 1940; Simpson, 1945; Wood, 1955; Schaub, 1958; McKenna et Bell, 1997; Michaux et al., 2001; Jansa et Weksler, 2004; Norris et al. 2004; Steppan et al., 2004; Flynn, 2009; Jansa et al., 2009; Blanga-Kanfi et al., 2009; Lin et al., 2014; de Bruijn et al., 2015 etc.). While compiling the *Palaeovertebrata Sinica*, the editors responsible for the fascicles of Rodentia gradually reached an agreement to subdivide the Muroidea into nine major families: Cricetidae, Muridae, Arvicolidae, Gerbillidae, Myospalacidae, Spalacidae, Tachyoryctoididae, Rhizomyidae, and Platacanthomyidae (personal communication with Profs. Li, Qiu, et al.). Of the nine families only Spalacidae, Tachyoryctoididae and Rhizomyidae are close to *Pararhizomys*-group in having a series of commonly shared characters. In addition to the myomorphous zygo-masseteric structure mentioned above, the list of commonly shared characters include: the long glenoid fossa extending posteriorly toward the nuchal crest, the sciurognathous mandible, loss of P4/p4, i2 possessing uniserial microstructure, the posterior end of the alveolus of i2 forming a prominent bulge, cheek teeth being lophodont, M1(m1) lacking a separated anterocone (id), the protoconid and metaconid connected with anterolophid in m1, etc.

Since the key problem of the present study is to search the phylogenetic position of *Pararhizomys*-group in Muroidea, we use genera as operational units in our cladistic analysis. As a result of the above consideration, we choose, for the in-group taxa, only the most representative genera of the three families: Spalacidae, Tachyoryctoididae and Rhizomyidae. Tachyoryctoididae are currently known by only three genera, *Tachyoryctoides*, *Argyromys* and *Eumysodon*. Of these only *Tachyoryctoides* is represented by several incomplete skulls with mandibles, and it is thus chosen in this study. In the Spalacidae the late Oligocene genus *Vetuspalax* is known to be the earliest representative. However, it is only known from isolated cheek teeth showing more derived features (Bruijn et al., 2013, 2015). Although *Debruijnia* also appeared earlier than *Heramys* and have more primitive molars than the latter, it is too poorly known to be informative in our comparative purpose. Of the Spalacidae only *Heramys* and the living genus *Spalax* are chosen. Of the Rhizomyidae a fossil genus, *Miorhizomys*, and two living genera, *Rhizomys* and *Tachyoryctes*, are selected.

As for the out-group, *Pappocricetodon*, the earliest and most primitive representative of the Muroidea, should have been an ideal candidate, if its fossil material were better preserved. However, the material of *Pappocricetodon* is known to include only some teeth, partial mandible and a rostrum, affording little information on polarity (Tong, 1992, 1997; Wang et Dawson, 1994; Dawson et Tong, 1998). We finally made a decision to use *Paramys* as an out-group as well. The reasons for this choice are twofold. The first is the antiquity of the genus, and the second is its relative completeness of information. The basal position of *Paramys* and the multitude of plesiomorphic characters available for comparison are widely accepted by authoritative rodent specialists for the rodents in general, and for the muroids in particular. The rich representation in fossils of *Paramys* and the thorough description by Wood (1962) are highly valued as well. *Cocomys* and *Exmus* are the earliest ancestral forms in Asia, and well preserved skulls were described (Li et al., 1989; Wible et al., 2005). However, their skulls and teeth are so widely different from our *Pararhizomys*-group that the coding based on character-to-character comparison seems very difficult to proceed.

As a result, the following 10 genera have been chosen: *Paramys*, *Pappocricetodon*, *Heramys*, *Spalax*, *Miorhizomys*, *Rhizomys*, *Tachyoryctes*, *Tachyoryctoides*, *Pararhizomys* and *Pseudorhizomys*.

(2) Choice of characters and state coding

In order to search and find more reliable most parsimonious trees (MPT), we choose as many characters of skull, mandible and teeth as possible. Altogether 54 characters are chosen: 32 of skull and mandible (0–31) and 22 of teeth (32–53). Among the 54 characters 22 are binary, and 32 are multistate characters, of which only 4 (45, 48, 50 and 53) are inversely evolving characters, i.e., from weak (state 0) →well developed (state 1) →reduced or vestigial (state 2). For these 4 particular multistate characters we refrain from coding them as binary (present or absent), but treat them as 3-state characters as demonstrated above, since we believe such evolutionary process can be authenticated by closer observation on fossils.

The sources of the characters and character states are either taken from literature or from our own observations on specimens. *Paramys* is from the monograph of Wood (1962) and the papers of Wahlert (1974, 1985b); *Pappocricetodon* is from the papers of Tong (1992, 1997), Wang and Dawson (1994), Dawson and Tong (1998); The Spalacidae and Rhizomyidae are mainly from Ellerman (1940, 1941), Flynn (1982, 1990, 2009), Carleton and Musser (1985), Hofmeijer and de Bruijn (1985), Zheng (1993), Ünay (1999), Sarica and Sen (2003) and Wesselman et al. (2009). The Tachyoryctoididae are from Bohlin (1937, 1946), Kowalski (1974), Li and Qiu (1980), Bendukidze et al. (2009), Wang and Qiu (2012), and Daxner-Höck et al. (2015). Some skull characters of *Tachyoryctoides* are from our own observations of the specimens collected recently from Nei Mongol.

The characters and their states used in cladistic analysis are listed as follows (numbering starting from 0 as required by WinClada):

0. Length of rostrum: (0) subequally long and wide; (1) much longer than wide.

1. Height of middle-posterior part of skull: (0) lower, skull height at M2 (exclusive of M2 crown height) less than maxillary diastema; (1) medium high, skull height at M2 subequal to maxillary diastema; (2) higher, skull height at M2 longer than maxillary diastema.

2. Morphology of nasal: (0) lateral sides (exclusive of posteriormost part) nearly parallel to each other; (1) lateral sides convergent posteriorly without forming pointed end; (2) lateral sides strongly convergent, forming a pointed end.

3. Position of posterior end of nasal: (0) much posterior to the posterior end of premaxilla; (1) nearly in the same transverse line as that of the premaxilla; (2) much anterior to that of the premaxilla.

4. Dorsolateral crest of premaxilla: (0) absent; (1) short, present in posterior part of premaxilla; (2) long, present throughout the premaxilla.

5. Incisive foramen: (0) long, length approximately equal to 1/2 length of maxillary diastema, situated closer to I2 than to the first cheek tooth; (1) short, about 1/4–1/3 length of maxillary diastema and located near the middle of maxillary diastema; (2) tiny, shorter than 1/5 length of maxillary diastema,

situated near the middle of maxillary diastema.

6. Infraorbital foramen: (0) small and round, without ventral slit, located at inferoposterior corner of rostrum and antero-inferior to orbit; (1) large and round, without ventral slit, located above zygomatic plate and in front of orbit; (2) round or transversely oval, located in upperposterior part of lateral side of the rostrum and anteromedial to zygomatic plate, with ventral slit; (3) triangular, without ventral slit, located in upper end of the posterior part of the lateral side of the rostrum and above zygomatic plate.

7. Zygomatic plate: (0) absent; (1) small, wider than long, facing downwards; (2) larger, wider than long, facing antero-inferiorly; (3) very large, subequally long and wide, facing antero-infero-laterally.

8. Attachment area of anterior end of lateral masseter muscle: (0) confined to maxilla; (1) extending from maxilla to premaxilla.

9. Posterior extension of zygomatic arch: (0) not reaching nuchal crest, separated from the latter by a notch; (1) connecting with nuchal crest.

10. Jugal bone: (0) long, anterior end reaching anterior border of orbit, closer to or meeting lacrimal; (1) short, anterior end far away from the anterior border of orbit and lacrimal.

11. Posterior border of hard palate: (0) anterior to M3; (1) roughly in line with M3 or its posterior border; (2) posterior to M3.

12. Palatine sulcus: (0) not deepening around the posterior palatal foramina; (1) very deep and narrow.

13. Parapterygoid fossa: (0) absent; (1) present.

14. Anterolateral wall of cranium: (0) smooth and convex; (1) having vertical ridge near the fronto-squamosal suture.

15. Pterygoid fossa: (0) much smaller than mesopterygoid fossa, its anterior border by far posterior to the posterior border of the hard palate; (1) subequal to or slightly smaller than mesopterygoid fossa in size, its anterior border situated slightly posterior to the posterior border of the hard palate; (2) large and deep; (3) large and deep, directly opening into the cranium.

16. Masticatory, buccinator and deep temporal foramina: (0) masticatory foramen confluent with deep temporal one, but separated from the buccinator foramen, their canals long; (1) masticatory foramen confluent with deep temporal foramen, but separated from buccinator foramen, their canals short; (2) all three foramina confluent into one pit, their canals short.

17. *Foramen ovale*: (0) a separate foramen, located on lateral side of alisphenoid bone; (1) confluent with middle lacerate foramen.

18. External auditory meatus: (0) represented by an aperture; (1) embryonic tube in shape, margin of aperture irregular; (2) forming a short tube; (3) forming a long tube.

19. Position of external auditory meatus: (0) situated inferoposterior to glenoid fossa; (1) below posterior end of elongated glenoid fossa; (2) posterior or posterosuperior to glenoid fossa, abutting on nuchal crest.

20. Glenoid fossa: (0) equally long and wide, or slightly longer than wide, its posterior end is far away from nuchal surface; (1) longer than wide, separated from nuchal crest by a notch; (2) much longer than wide, its posterior end close to or reaches the nuchal crest.

21. Nuchal crest and nuchal surface: (0) nuchal surface flat or slightly convex, nuchal crest and nuchal surface sub-vertical; (1) nuchal surface flat, nuchal crest and nuchal surface extending in anterosuperior direction; (2) nuchal crest extending in antero-inferior direction, nuchal surface prominently convex, extending in anterosuperior direction.

22. Distance between occipital condyles on ventral side: (0) widely separated from each other; (1) close to each other, separated by a very narrow longitudinal groove.

23. Paramastoid process: (0) very small; (1) wide and large.

24. Condyloid process of mandible: (0) extending posterosuperiorly; (1) extending mainly superiorly.

25. Position of posterior end of angular process: (0) under or posterior to posterior border of condyle of mandible; (1) anterior to posterior border of condyle of mandible.

26. Form of angular process: (0) pointed and angular, extending posteriorly; (1) lobate, often with a small pointed tip at its posterosuperior angle.

27. Posterior end of i2: (0) not forming distinct bulge on mandible; (1) forming a bulge on buccal side of ascending ramus, situated lower than condyle, without forming a groove between bulge and condyle; (2) bulge robust, with its top lower than that of the condyle and separated from the latter by a groove; (3) bulge subequal to or higher than condyle in position, separated from the latter by a deep groove.

28. Masseteric ridge of mandible: (0) weakly crested or linear; (1) well developed, flaring laterally.

29. Attachment area of temporal muscle on mandible (= temporal fovea): (0) situated on the upper part of the coronoid process, extending to the lingual side of the condyloid process; (1) enlarged, extending downwards to the base of the coronoid process; (2) further enlarged, extending antero-inferiorly to form a scar at the base of the anterolingual margin of the coronoid process.

30. Medial pterygoid muscle fovea: (0) large and deep; (1) small and shallow.

31. Length of the mandibular diastema: (0) shorter than lower molar row; (1) subequal to, or longer than, lower molar row.

32. Labial surface of I2: (0) without distinct longitudinal ridge; (1) with longitudinal ridge.

33. Expansion of labial enamel layer on lateral side of i2 in labiolingual direction: (0) forming a wide band, much wider than that on medial side; (1) enamel layer bands on both sides equally narrow.

34. Upper premolars: (0) with two upper premolars (P3 and P4); (1) with one upper premolar (P4); (2) without premolar.

35. Crown height of molars: (0) brachyodont; (1) moderately high-crowned, mesiodistally hypsodont; (2) moderately high-crowned, unilaterally hypsodont [lingual side is higher (lower) than that of buccal one on upper (lower) molar]; (3) hypsodont.

36. Crown pattern and mesoloph of M1–M3: (0) bunodont, without mesoloph; (1) bunolophodont, with short mesoloph; (2) lophodont, with well-developed mesoloph; (3) lophodont, mesoloph reduced or lost.

37. Anterocone of M1: (0) absent; (1) small, conical, with anterior cingulum; (2) anterocone joining with anteroloph to form a transverse loph, lacking anterior cingulum.

38. Anterosinus of M2: (0) present; (1) absent.

39. Anterosinus of M3: (0) present; (1) absent.

40. Metalophs of M1 and M2: (0) joining with posterior arm of protocone; (1) joining with hypocone or anterior or posterior arms of hypocone; (2) confluent with posteroloph so that the posterosinus is reduced or absent.

41. Metaloph of M3: (0) joining with protocone; (1) joining with anterior arm of hypocone; (2) confluent with posteroloph so that the posterosinus is reduced or lost.

42. Mesosinuses of M1 and M2: (0) anteroposterior width $\geqslant$ transverse length; (1) anteroposterior width $<$ transverse length; (2) anteroposterior width $\leqslant$ 1/2 transverse length.

43. Mesosinus of M3: (0) anteroposterior width $\geqslant$ transverse length; (1) anteroposterior width$<$ transverse length; (2) anteroposterior width $\leqslant$ 1/2 transverse length.

44. Entoloph of M3: (0) present; (1) absent.

45. Sinus of M3: (0) small and shallow; (1) wide and deep; (2) reduced or absent.

46. Anteroconid of the first lower cheek tooth (p4 or m1): (0) absent; (1) conical; (2) transformed into transverse lophid (= anterolophid).

47. Metaconid and protoconid of the first lower cheek tooth (p4 or m1): (0) no connection between the two cuspids; (1) directly connected (= metalophid); (2) anteriorly connected (= anterolophid); (3) doubly connected (= anterolophid and metalophid).

48. Protosinusid of m2: (0) absent; (1) distinctly present; (2) becoming vestigial or disappeared.

49. Sinusids of lower molars: (0) anteroposteriorly wider and transversely shorter, extending transversely; (1) anteroposteriorly wider and transversely shorter, extending posterolingually; (2) anteroposteriorly narrowed, extending transversely.

50. Anterosinusids of lower molars: (0) absent; (1) well developed; (2) becoming vestigial or disappeared.

51. Mesosinusids of lower molars: (0) anteroposteriorly wide; (1) anteroposteriorly narrow, forming reentrants extending transversely or anterobuccally.

52. Posterosinusids of lower molars: (0) well developed on m1–m3; (1) developed on m1–m2, reduced or absent on m3; (2) reduced or absent on m1–m3.

53. Mesolophids of m1 and m2: (0) absent; (1) distinctly present; (2) reduced or vestigial.

For data matrix see Tab. 20.

(3) Constructing phylogenetic trees

Since the data matrix to be analyzed is small, containing only 10 taxa and 54 characters, simple parsimony analysis is performed in our search for shortest trees. WinClada (ver. 1.00.08; Nixon, K. C., 1999-2002) is used, mainly in preference to its fully menu-driven interactive programs and its various ways of presentation of the search results. The Multiple TBR+TBR searching strategy under "Heuristics" in WinClada is performed in our search for the MPT's (most parsimonious trees).

Using WinClada's "Heuristics" under the default additive (= ordered) option of the character state coding, one single MPT is obtained (Fig. 77A), with the following tree statistics: Tree Length=119, CI=78 and RI=74. Another run of the same "Heuristics" searching under the nonadditive (= unordered) option, also produced one single MPT of the same topology, but with slightly different statistics: Length=114, CI=82, RI=73. Under alternative conditions of doubled weighting of the skull (+ mandible) characters and the tooth characters, different topologies have been obtained. However, the best among the MPT's shows the same topology as that obtained under the additive option without weighting, with the only differences being that the tree length is much larger (L 188, CI=78, and RI=71). In similar ways we also applied "Implicit enumeration" of TNT (Goloboff, P. A., et al., 2008) to test the feasibility of WinClada's "Heuristics." Two MPT's with slightly different topologies have been obtained for each run. One topology is identical to that obtained by WinClada, whereas the other topology shows that *Miorhizomys* is closely allied to the spalacid clade (*Spalax* (*Tachyoryctoides* (*Pseudorhizomys*+*Pararhizomys*))) rather than with the rhizomyid clade (*Tachyoryctes*+*Rhizomys*) as obtained by WinClada. The sister relationship of *Miorhizomys* with the spalacids seems rather unorthodox in view of traditional morphology.

Although the resultant topology under the two coding options is the same, the statistics (L, CI, RI) and the distribution of the characters and the character states are quite different. Taken at its face value, the statistics under the nonadditive coding option looks better than those under the additive coding (*vide supra*). However, we consider the nonadditive option produced inferior results. The number of character states used in analysis is smaller, and the number of non-sequential character transformations (for example, 0→2, or 1→3, instead of

0→1, or 1→2, etc.) is inherently greater under the nonadditive option than those in the additive one. Despite of the increased evolutionary assumptions in the additive option, we think this is justified because our fossil records tend to show the sequence of character acquisition through time, which offers reasonable evidences for character evolution in ways that coded in our matrix. Therefore, the single MPT under the additive coding option by WinClada (Fig. 77) is adopted as the working hypothesis for further cladistic analysis.

WinClada affords three types of optimization: unambiguous (Fig. 77 A), ACCTRAN (Fig. 77 C), and DELTRAN (Fig. 77 B). The first type may provide the most unambiguous optimization in terms of algorithm, but the number of characters is too small to be used for character optimization. The ACCTRAN inherently "favors reversals over repeated origins when the choice is equally parsimonious" (Wiley et al., 1991: 61). This has the effect of pushing the character transformations root-ward, as exampled in our case (Fig. 77 C). This pattern of distribution of characters is unfavorable due to the fact that the emphasis of our analysis is in the middle part of the tree rather than the root region of the tree. We prefer DELTRAN (Fig. 77 B) because it is primarily designed in favour of explaining homoplasies in terms of parallel evolution rather than reversals, which is widely accepted as occurring frequently in rodent evolution.

(4) Character analysis based on DELTRAN MPT (Fig. 77 B)

The statistics of the characters obtained using WinClada command "character diagnoser" for DELTRAN MPT are given in the Chinese text, where the character numbers are given in boldface (e.g. **6**); if with *(e.g. ***35**), it means the character is listed also in the tree under unambiguous optimization; if in italic font (e.g. *47*), parallelism; if underlined (e.g. **40**), reversal. The character number is then followed by state transformation (e.g. 0→1), then CI and RI, finally uninf (uninformative) and/or autap (autapomorphy). The readers are referred to this part of the Chinese text (*vide* pp. 137–139).

Based on the "character diagnoser" given in the Chinese text, at least the following information seems important and useful for formal classification. It can be summarized in the following.

Information provided by characters on the nodes:

1) At the node 0→1, there are 13 characters, 10 of which are synapomorphic, 3 are homoplastic. Two synapomorphies are skull characters: enlarged infraorbital foramen (6-1) and presence of small zygomatic plate (7-1). Three homoplasies (40-1, 41-1, 50-1), shared with *Tachyoryctoides*, are the results of reversals. The comparative multitude of commonly shared characters of the node 0→1 shows that *Pappocricetodon* may really have closer relationships with the other 8 in-group taxa.

2) Node 1→2 has 10 characters, all synapomorphic. Of them 5 synapomorphies have their CI and RI as high as 100. *Heramys* as the sister taxon of the clade comprising the other 7 in-group taxa is thus strongly supported. However, *Heramys* has so far been universally held as an ancestral form of *Spalax* and its 2 homoplastic characters (29-2 and 45-2) are parallel to those of *Spalax*. This may partially be explained by the paucity of fossil material of *Heramys*. As better material becomes available, especially skull of *Heramys*, systematic position of *Heramys* may possibly be changed, and directly placed as a sister taxon of the spalacid clade.

3) Node 2→3, where the synapomorphic characters supporting monophyly of the rest 7 in-group taxa, i.e., (*Miorhizomys* (*Rhizomys*+*Tachyoryctes*)) and (*Spalax* (*Tachyoryctoides* (*Pseudorhizomys*+*Pararhizomys*))) clades, are supported by 20 character transformations, the most numerous in the DELTRAN MPT. Thus, it is the strongest supported node in the cladogram.

4) Node 3→4, the node supporting the clade (*Miorhizomys* (*Rhizomys*+*Tachyoryctes*)) has only 5 shared characters, of which 3 are synapomorphic and 2 are homoplastic. One homoplasy, the very broad zygomatic

plate (7-3), is parallel with the clade (*Tachyoryctoides* (*Pseudorhizomys+Pararhizomys*)), the other homoplasy, the highly reduced protosinusid of m2 (48-2), is parallel with *Spalax*. Taken as a whole, the support of this node is rather weak.

5) Node 3→6, where characters supporting the clade (*Spalax* (*Tachyoryctoides* (*Pseudorhizomys+ Pararhizomys*))) total 12, the next most numerous in the cladogram. Of the 6 skull (+mandible) characters, 2 are homoplasies: 0-1 (rostrum much longer than wide) is parallel with *Tachyoryctes*, and 21-1 (nuchal surface flat, nuchal crest and nuchal surface extending anterosuperiorly) is parallel with *Rhizomys*. Of the 6 dental characters, there is only one homoplasy: 33-1 (enamel layer bands on both sides equally narrow in i2) is parallel with the clade *Rhizomys+Tachyoryctes*. All the rest 9 characters are synapomorphic, of them 7 with highest value of CI and RI (100). Compared with node 3→4, the support of the clade (*Spalax* (*Tachyoryctoides* (*Pseudorhizomys+ Pararhizomys*))) is certainly much stronger.

6) Node 6→7, supporting the clade (*Tachyoryctoides* (*Pseudorhizomys+ Pararhizomys*)), has only 3 shared character, 2 synapomorphic (13-1, possession of parapterygoid fossa, and 28-1, masseteric ridge of mandible well developed), 1 homoplastic (7-3, very large zygomatic plate), which is parallel with the node 3→4 (*vide supra*). This is the least supported node in the cladogram.

7) Node 7→8, supporting the sister group relationship of *Pseudorhizomys* and *Pararhizomys*, has 4 synapomorphies (2-2, posterior end of nasal pointed; 4-1, presence of laterodorsal crest of premaxilla; 15-1, pterygoid and mesopterygoid fossae about equal in size; 18-1, embryonic external meatus) and 2 homoplasies (1-0, middle part of skull low, a reversal from 1-1 of the node 2→3; 52-2, posterosinusids strongly reduced in lower molars, parallel with *Spalax*). The support of this node is medium in degree.

Some information from the terminal taxa:

The support for combining the three fossil taxa (*Heramys*, *Miorhizomys* and *Tachyoryctoides*) with their respective sister clades is generally low in degree. *Heramys* has only 2 homoplastic characters of its own: 29-2 (attachment area of temporal muscle on mandible enlarged) and 45-2 (sinus of M3 reduced) are parallel with *Spalax*. This is consistent with the notion that *Heramys* may have closer relationship with *Spalax* than with the more inclusive clade comprising all the rhizomyines (including *Miorhizomys*) and spalacines (*sensu lato*) as suggested when dealing with the node 1→2 (*vide supra*).

Miorhizomys does not possess supporting characters of its own.

Tachyoryctoides is a very special case. It possesses 7 characters of its own, all homoplastic. Three of them (40-1, 41-1, 50-1) are results of reversals from the basal node 0→1 (*vide supra*). This led us to think that the possibility to shift *Tachyoryctoides* to a more basal place in the cladogram cannot be excluded, if more fossil material of this genus could be discovered. So far as is known, the geological range of the *Tachyoryctoides*-group fossils is from Late Oligocene to Early Miocene, while that of the *Pararhizomys*-group fossils is from early Late Miocene to Early Pliocene. There is a considerably long interval (Middle Miocene) between the two groups. This seems also to support the notion that *Tachyoryctoides* may have a more basal position.

3) Classification of pararhizomyines and closely related muroid families at and above generic level

Currently, the ways of the cladistic systematists to construct classification based on resultant MPTs are highly variable. However, as a whole, two major groups can be acknowledged. One is composed of those who strictly adhere to the principles and rules of the orthodox cladistics, even at the cost of abandoning Linnaeus system of classification, such as the "PhyloCode" school. The other is more flexible. While trying to abide by the central tenet of the cladistics, systematists of this group apply certain conventional measures to save the Linnaeus system. Wiley et al. (1991) made a useful overview of the viewpoints of the second group. Some of the rules and conventions enumerated by Wiley et al. (1991: 102–108) are fully in accord with our views.

Rule 1: Only monophyletic groups will be formally classified.

Rule 2: Classification must be capable of expressing the sister group relationships among the taxa classified.

Convention 1: The Linnaean system of ranks will be used.

Convention 2: Minimum taxonomic decisions will be made to modify existing classifications.

Convention 3: Taxa forming an asymmetrical part of a phylogenetic tree may be placed at the same rank and sequenced in their order of branching.

In addition, as a special case of the above convention 3, McKenna and Bell (1997: 28–30) proposed that the apparent ancestor (not necessarily limited to species) can be listed first in the higher-ranked taxa to which the ancestor belong.

Based on the above rules and conventions, the following tentative classification is recommended (excluding *Pappocricetodon*):

Superfamily Muroidea

    Stem genus *Heramys*

    Family Rhizomyidae Winge, 1887

        Miorhizomys

        Rhizomys

        Tachyoryctes

    Family Spalacidae Gray, 1821

        Subfamily Spalacinae Gray, 1821

        Spalax

        Subfamily Tachyoryctoidinae Schaub, 1958

        Tribe Tachyoryctoidini Schaub, 1958

        Tachyoryctoides

        Tribe Pararhizomyini tribe nov.

        Pararhizomys

        Pseudorhizomys

The adopted option is slightly different from those proposed by other systematists chiefly in ranking, but not in essence. For example, Bugge (1974: 75) raised the rank of the two major subclades of the cladogram (Fig. 77) to family, thus, Spalacidae and Rhizomyidae of the superfamily Rhizomyoidea. On the other hand, McKenna and Bell (1997: 173–174) acknowledged only Spalacinae and Rhizomyinae as two separate subfamilies of the widely embracing family Muridae, which contained more than 20 such subfamilies.

Unfortunately, Bugge's important finding in cephalic arterial systems and the patterns of head blood supply in living rodents have not been incorporated in the present classification. This was simply because of the unavailability in fossil materials to carry out comparative study of these features. Nevertheless, some skull foramina, through which the cephalic arteries pass, well preserved in fossil specimens, may provide us with important phylogenetic information as well.

From the above description of the basicranial portions of some specimens of *Pararhizomys* and *Pseudorhizomys* it seems more or less certain that the two above-mentioned fossil genera are more similar to those of the living *Rhizomys* and *Spalax* in presence or absence, position and size of these foramina.

1) No distinct foramen transmitting the stapedial artery into the bulla has been found at the medioposterior part of the bulla wall in all the specimens of Pararhizomyini. Therefore, the stapedial artery (Fig. 3, st) may be lacking in this group of rodents, just as in living *Spalax* and *Rhizomys*, as amply demonstrated by Bugge in a series of his papers in the 1970's–1980's.

2) The alisphenoid canal and the *foramen ovale* are both strongly reduced in Pararhizomyini rodents. The first is tiny and hardly considered as present in some cases, whereas the latter is supposed to be confluent with the middle lacerate foramen. According to Bugge, the blood supply of the distal continuation of the stapedial artery (infraorbital and supraorbital arteries) changed drastically in *Spalax* and *Rhizomys*. Their blood is supplied either by the a5' anastomotic branch or the a4 one from the internal carotid artery (*vide* Fig. 3). Therefore, there are no true alisphenoid canal and no separate *foramen* ovale for the exit of the stapedial artery. Although not identical, these features observed in Pararhizomyini tend to show the same evolutionary tendency occurring in the above discussed fossil and living groups of rodents.

In short, the available evidence in cephalic arterial system seems to support rather than oppose to combining the Spalacinae and Tachyoryctoidinae, including Pararhizomyini, into one family Spalacidae, and Spalacidae and Rhizomyidae are the two closest families within the superfamily Muroidea.

IV. SOME BIOLOGICAL ASPECTS OF PARARHIZOMYINES

1. Restoration of some muscles of pararhizomyines

1) Skull muscles

Because Pararhizomyini belong to the Muroidea and have a myomorphous type of skull, for the muscle restoration of the Pararhizomyini skull we mainly refer to anatomical dissections by Green (1955), Young et al. (1983), and Howell (1926), supplemented by observation of morphology on skulls of some recent muroids. The restoration is confined to the muscles directly involved in chewing, gnawing and digging.

M. masseter   All the pararhizomyines have very large infraorbital foramina and oblique zygomatic plates, distinct masseteric tubercles, and large masseter fossa with well-developed masseteric ridge. These features show that the pararhizomyines may have well developed masseters (including medial, lateral and superficial layers). The masseteric ridge in pararhizomyines is flaring and more developed than those in *Rhizomys*, *Myospalax* and *Spalax*. Probably, the pararhizomyines have more developed medial masseter muscle than in the above three living forms. On the other hand, the angular process in pararhizomyines is also well developed and extends posteriorly, which are similar to that in *Rattus*, but different from those in *Rhizomys*, *Myospalax* and *Spalax*. It shows that their superficial and lateral masseters are longer and extend in a direction different from those in the three living forms.

M. temporalis   In pararhizomyines the temporal fossa is large so that a distinct sagittal crest is formed by the upper borders of the left and right temporal fossae. The temporal fovea on the medial side of coronoid process of mandible is accordingly large and concave. Therefore, *m. temporalis* in pararhizomyines must be well developed.

M. pterygoideus   The pararhizomyines have both *m. pterygoideus medialis* and *m. pterygoideus lateralis*. Unlike in *Rhizomys*, the internal pterygoid process in pararhizomyines is larger and higher than the external pterygoid process and the internal pterygoid fovea on the mandible is large and deep, but the external pterygoid fovea on the medial side of the condynoid process is small and shallow. Therefore, *m. pterygoideus medialis* may be much more developed than *m. pterygoideus lateralis* in pararhizomyines.

M. digastricus   In pararhizomyines the paramastoid process (the origin of *m. digastricus*) is much smaller than that *in Rhizomys*, and the digastric fovea on the mandibular symphysis is also small. It shows that *m. digastricus* in pararhizomyines may be much smaller than that of *Rhizomys*.

M. buccinator   All the pararhizomyines have a special feature: the premaxilla has a prominent dorsolateral crest. It may be the attachment area for *m. buccinator*. If this is tenable, the *m. buccinator* must be particularly well developed in pararhizomyines.

2) Forelimb muscles

The forelimb of *Pseudorhizomys* is well preserved, thus, serving as the main sources of our knowledge about the forelimb muscles of the Pararhizomyini. In addition to the literatures of Greene (1955), Yang et al. (1985) and Howell (1926), we also refer to Vinogradov and Grambarian (1952) for tsaganomyids, supplemented by observation of the forelimb bones of some recent rodents (*Rattus*, *Rhizomys*, *Myospalax*, *Aplodontia*, etc.).

M. supraspinatus   The supraspinous fossa of the scapula and the greater tuberosity of the humerus in *Pseudorhizomys* are similar to those of *Rattus* and *Rhizomys* in morphology. The *m. supraspintus* of *Pseudorhizomys* may be well developed as that of *Rhizomys*.

M. infraspinatus   The infraspinous fossa of *Pseudorhizomys* is similar to those of *Rhizomys* and *Aplodontia* in having developed *linea muscularis* on the posterior part of the infraspinous fossa, and the greater tuberosity of humerus is well developed, possessing a prominent concavity for this muscle on the posterior part of the lateral side. The *m. infraspinatus* of *Pseudorhizomys* is also well developed.

M. subscapularis   The subscapular fossa on scapula and the lesser tuberosity of humerus in *Pseudorhizomys* are similar to those in *Rhizomys* and *Aplodontia* in morphology. The *m. subscapularis* of *Pseudorhizomys* is thought to be similarly developed as in the two latter forms.

M. deltoideus   Although the spine of scapula and the acromion of *Pseudorhizomys* are not well preserved, its clavicle is well preserved and the well-developed deltoid tuberosity of the humerus extends prominently laterally and is located in the middle part of the shaft of the humerus. Therefore, the *m. deltoideus* must be well developed in *Pseudorhizomys*.

M. teres major   Although the upper part of the posterior border of the scapula is broken, there is a distinct rough concavity (= *tuberositas teres major*) for insertion of *m. teres major* behind the crest of lesser tuberosity of the humerus. It seems that *Pseudorhizomys* may have a well-developed *m. teres major*.

M. teres minor   A *linea muscularis* near the posterior border of the scapula and a rough convexity on the posterior side of the neck of scapula are present. On the other hand, on the posterior part of the lateral side of the greater tuberosity of the humerus there is a prominent convexity, apparently serving for the attachment of *m. teres minor*. It seems that *m. teres minor* may also be well developed in *Pseudorhizomys*.

M. pectoralis superficialis   *Pseudorhizomys* has a well-developed clavicle, and the crest of the greater tuberosity of humerus is well developed as well. Probably, *Pseudorhizomys* may have a developed *m. pectoralis superficialis*.

M. pectoralis profundus   Both the coracoid process and the scapular tuber of the scapula, and the greater tuberosity of the humerus and its crest are developed, indicating that *Pseudorhizomys* may have a more developed *m. pectoralis profundus*.

M. coracobrachialis   The coracoid process of the scapula is present, and a rough area for attachment of *m. coracobrachialis* on medial side of the middle part of the shaft of the humerus is obvious. Accordingly, *Pseudorhizomys* may have a more developed *m. coracobrachialis*.

M. triceps brachii   The preserved part the ulna's olecranon in *Pseudorhizomys* shows that *Pseudorhizomys* may have a developed olecranon. Meanwhile there is a rough tubercle on the neck of scapula and a roughened tubercle on the inferolateral part of the greater tuberosity of the humerus. *Pseudorhizomys* seems to have a well-developed *m. triceps brachii*.

M. biceps brachii   The scapular tuber, the coracoid process of the scapula and the tuberosity of radius in *Pseudorhizomys* are all well developed, and the intertubercular sulcus (bicipital groove) is also wide. All of these tend to show that *m. biceps brachii* may be well developed here.

M. brachialis   The insertion of *m. brachialis* is not distinct on the humerus of *Pseudorhizomys*. However, there are a wide musculo-spiral groove on the humerus, a distinct smooth groove on proximal part of the medial side of the radius for passing of *m. brachialis,* and a rough concavity (tuberocity of ulna, tuul) for insertion of the *m. brachialis* below the coronoid process of the ulna. It is possible that *Pseudorhizomys* has a fleshy *m. brachialis*.

M. pronator teres   In *Pseudorhizomys* there are a concavity on the medial epicondyle of the humerus and a rough concavity at the middle of the anteromedial border of the shaft of the radius for insertion of the *m. pronator teres*. It means that *m. pronator teres* is present in *Pseudorhizomys*.

M. supinator   On the lateral epicondyle of the humerus in *Pseudorhizomys* the attachment area for *m. supinator* is small. There is also a roughened area for the attachment of the *m. supinator* on the medial margin of the middle part of the anterior side of the shaft of the radius. The *m. supinator* is also present in *Pseudorhizomys*.

M. extensor carpi radialis   The *extensor carpi radialis* of *Rattus* include two muscles: a long one and a short one. According to Green (1955) and Yang et al. (1983), both muscles take their origins on the lateral epicondyles of the humeri, and insert on distal parts of McII and McIII respectively. However, Howell (1926) mentioned that the two muscles of the wood rat originate upon the lateral epicondyloid ridge of the humerus and insert upon dorsomedial parts of McII and McIII respectively, whereas according to Vinogradov and Gambarian [1952: Fig. 16 (11, 12)], these two muscles take their origins on upper parts of the lateral epicondylar crests of the humeri in *Myospalax*, *Spalax* and *Pseudotsaganomys* (= *Tsaganomys*). On the upper part of the lateral epicondylar crest of the humerus of *Pseudorhizomys* there are two concavities. The upper one may be the attachment area for *m. extensor carpi radialis*. If this proves true, the origin of *m. extensor carpi radialis* may be more similar to those of wood rat, *Myospalax* and *Spalax* than to that of *Rattus* in position. Both McII and McIII of *Pseudorhizomys* have developed metacarpal tuberosities. The metacarpal tuberosity of McIII forms a well-developed depression and located near the dorsomedial border of the proximal part of McIII, similar to those in *Rhizomys*, *Myospalax* and *Aplodontia* in position. The metacarpal tuberosity of McII of *Ps. indigenus* is a roughened tubercle and located on dorsomedial margin of the middle part of the shaft as in *Rhizomys*, *Myospalax* and *Aplodontia*, while that of *Ps. gansuensis* is a rough depression on dorsomedial margin of the upper part of the dorsal side of the shaft. If it proves true that *m. extensor carpi radialis brevis* inserts on the metacarpal tuberosity of McIII and *m. extensor carpi radialis longus* inserts on the metacarpal tuberosity of McII, then the insertions of *m. extensor carpi radialis longus* is closer to those of *Rhizomys*, *Myospalax* and *Aplodontia*, but higher than that of *Rattus* in position. In any case, *Pseudorhizomys* may have a more developed *m. extensor carpi radialis*, which is more similar to *Rhizomys*, *Myospalax* and *Aplodontia* in moment of force. Furthermore, *m. extensor of carpi radialis longus* of *Ps. indigenus* may be slightly different from that of *Ps. gansuensis* in moment of force, based on their differences in position and shape of the insertion of the muscle.

M. extensor carpi ulnaris   Although the olecranon of the ulna of *Pseudorhizomys* is broken, there are an attachment area on the lateral epicondylar crest of the humerus and well-developed tuberosity and depression on the lateral side of the proximal end of McV. Probably, *Pseudorhizomys* has a well-developed *m. extensor carpi ulnaris*.

M. flexor carpi radialis   In *Pseudorhizomys* there are a distinct attachment area for *m. flexor carpi radialis* on the medial epicondyle of the humerus and a roughened area on the volar sides of the proximal ends of McII and McIII. Probably *Pseudorhizomys* has a well developed *m. flexor carpi radialis*.

M. flexor carpi ulnaris   The attachment area for *m. flexor carpi ulnaris* is distinct on the medial epicondyle of the humerus, and the pisiform is developed in *Pseudorhizomys*. *Pseudorhizomys* may have a more developed *m. flexor carpi ulnaris*.

M. extensor digitorum communis   There are two distinct concavities on the dorsal side of the lateral epicondylar crest of humerus in *Pseudorhizomys*. According to Vinogradov and Gambarian [1952: Fig. 16 (13)], the lower concavity may be for the origin of *m. extensor digitorum communis*. In *Pseudorhizomys* the extensor processes of the third phalanges of FII–FV and the extensor tuberosities of the first and second phalanges of FII–FV are all well developed. It seems that *Pseudorhizomys* may have a well-developed *m. extensor digitorum communis*.

M. extensor digitorum lateralis   There are three distinct concavities on the lateral epicondyle of the humerus of *Pseudorhizomys*. The upper one may serve as the insertion of *m. extensor digitorum lateralis*. The third phalanges of FIV and FV of *Pseudorhizomys* have developed extensor processes, which may also be the attachment areas for *m. extensor digitorum lateralis*. Thus, *Pseudorhizomys* may have a more developed *m. extensor digitorum lateralis* as well.

M. extensor pollicis longus The distal phalanx of FI (FI-Ph2) of *Pseudorhizomys* has a distinct extensor process. It shows that *m. extensor policis longus* is present in *Pseudorhizomys*.

M. extensor pollicis brevis In *Pseudorhizomys* there is a prominent roughened convexity on the lower 3/4 of the dorsal side of the radius, and a prominent extensor tuberosity on the first phalanx of FI (FI-Ph1). It shows that *m. extensor pollicis brevis* is prominent in *Pseudorhizomys*.

M. abductor pollicis According to Howell (1926) and Green (1955), *m. abductor pollicis* of wood rat and *Rattus* arises on falciform and ends on the medial side of the base of FI-Ph1. *Pseudorhizomys* has a developed falciform and the medial side of the base of FI-Ph1 has a prominent tubercle. Thus, *Pseudorhizomys* has a developed *m. abductor pollicis*.

M. flexor digitorum superficialis In *Pseudorhizomys* the medial epicondyle of humerus is well developed, and there are roughened areas on the two sides of proximal parts of the volar sides of the second fingers of FII–FV (FII-Ph2–FV-Ph2). Therefore, *Pseudorhizomys* has a developed *m. flexor digitorum superficialis*.

M. flexor digitorum profundus In *Pseudorhizomys* the medial epicondyle of humerus is well developed and on the posterior sides of the radius and ulna there are wide attachment areas for *m. flexor digitorum profundus*. Like *Rhizomys* and *Ailuropoda melanoleuca*, all the distal phalanges of five fingers of *Pseudorhizomys* (including FI-Ph1) have well-developed tuberosities of ungual phalanges, which are more developed than those of *Rhizomys*. Probably, *Pseudorhizomys* has a well-developed *m. flexor digitorum profundus* with five ends as in wood rat, *Rhizomys* and *Ailuropoda melanoleuca*.

M. flexor pollicis brevis There is a distinct attachment tubercle for muscle on the lateral side of the proximal end of the volar side of FI-Ph1. It seems that *m. flexor pollicis brevis* is also present in *Pseudorhizomys*.

M. adductor pollicis In *Pseudorhizomys* there are distinct ridges serving as the insertion of *m. adductor pollicis* on Mc II and Mc III, and a distinct tubercle for this muscle on the lateral side of the proximal end of FI-Ph1. It shows that *Pseudorhizomys* has a *m. adductor pollicis*.

2. Functional analysis of skeleton muscles and behavior of pararhizomyines

1) Functional analysis of *Pseudorhizomys*

Taken as a whole, *Pseudorhizomys* shows the following morphologic characters: in skull dorsal profile straight, orbit small, maxillary and mandibular diastemata long, incisive foramen small; in mandible masseteric ridge of the mandible well developed and flaring outwards, posterior end of i2 forming a prominent bulge on buccal side of ascending ramus; in skeleton cervical vertebrae thick and short, limbs (at least forelimbs) robust and short with well-developed tuberosities serving as the attachment of muscles, etc.

According to Hildebrand (1982), Stein (2000) and Hildebrand et Goslow (2001), all these features are reminiscent of subterranean way of live. These authors recognized three kinds of breaking up soil commonly used by subterranean rodents: (a) scratch digging, mainly used by Geomyidae, *Ctenomys*, *Bathyergus* and *Myospalax*; (b) chisel-tooth digging, mainly used by Rhizomyidae, *Spalacopus*, *Cryptomys* and *Heterocephalus*; and (c) head-lift digging, used by *Spalax* and *Ellobius*.

(1) Functional analysis of skull

The skull of *Pseudorhizomys* is triangular in general shape, having larger temporal fossae, developed sagittal and nuchal crests, and laterally expanded zygomatic arches; the nuchal surface slightly inclining forwards toward its top, the premaxilla extending anteriorly beyond the nasal, I2 of proodont type (stretching infero-anteriorly), arising posteriorly in front of M1; the maxillary and mandibular diastemata long and the i2 long, etc. All these features are characteristic of the chisel-tooth digging rodents, as exemplified in *Rhizomys*, *Cryptomys* and *Heterocephalus*. In contrast, in *Myospalax* and *Geomys* of the scratch digging type the skulls do not have distinct sagittal crests, the zygomatic arches less strongly expand laterally, I2 is orthodont (lower tip extending slightly posteriorly) and more curved, i2 is shorter. On the other hand, the skulls of *Spalax* and *Ellobius* of the head-lift digging type are even more triangular in shape, having strongly inclined nuchal surfaces, and more backwardly curved I2, etc. Based on the above comparison, it is doubtless that *Pseudorhizomys* belongs to the rodents of chisel-tooth digging type.

Flynn et al. (1987) pointed out that the thick outer enamel layer and low band-inclination are functionally related to burrowing, and dependent on degree of induration of soil. Although the PE thickness in *Pseudorhizomys* is less than that of *Rhizomys*, the inclination of inner portion is variable. Perhaps, the enamel type of the incisor of *Pseudorhizomys* is more advantageous for preventing the incisor from breaking down during digging. The alternative explanation may be that the soils encountered by *Pseudorhizomys* could be relatively soft. Hua et al. (2015) stated that the microstructure of the incisor was related to the fracture-resistance of the food processing. *Rhizomys* mainly gnaw on fibrous bamboo. The microstructure of *Pseudorhizomys* indicates that this animal may live on the less hardened plant, rather than bamboo.

Rhizomys differs from *Pseudorhizomys* also in some other features. The horizontal ramus of mandible is higher, probably with stronger masseteric muscles in *Rhizomys*. This may indicate that *Rhizomys* possesses a stronger biting strength for breaking up soil. The internal and external pterygoid processes in *Rhizomys* are almost equally developed, meaning both medial and lateral pterygoid muscles may be strong. In addition, the molars of *Rhizomys* are unilaterally hypsodont. These features may suggest a stronger transverse movement of the mandible relative to skull, strengthening the transverse chewing in *Rhizomys*.

On the other hand, the molars in *Pseudorhizomys* are mesiodistally hypsodont, with almost horizontal occlusal surface. This could occur only when mastication proceeds in anteroposterior direction. This point of view also finds support from other features of *Pseudorhizomys*. The external pterygoid process, and hence the external pterygoid muscle, are much less developed than the medial ones. The external pterygoid muscle is known to exert mainly to move the mandible laterally. As a result, the transverse movement of the mandible and hence the transverse mastication in *Pseudorhizomys* must be weaker than in *Rhizomys*. Furthermore, in *Pseudorhizomys* the horizontal ramus of the mandible is lower, and the angular process extends more posteriorly and the glenoid fossa of the skull is longer than in *Rhizomys*. This will cause changes in the angle of the exerting forces of the superficial and lateral masseter muscles toward a more horizontal direction. All of these show that the masseters in *Pseudorhizomys* may have more strength to pull the mandible anteriorly than those in *Rhizomys* for chewing and breaking up the soil during digging.

(2) Functional analysis of forelimbs

The forelimbs of *Pseudorhizomys* are short. The humerus is longer than the radius (32.6 mm vs 28 mm), and the radius is longer than the manus (~25 mm in average, *vide* Tabs. 14–17). The scapula has an elliptical glenoid cavity and a well-developed coracoid process. The attachment areas for *m. infraspinatus* are well developed on scapula and humerus. The robust humerus has an oval head, a well-developed deltoid tuberosity on the middle part of the shaft, a well-developed medial epicondyle and a well-developed lateral epicondylar crest. The radius has an oval articular facet for humerus, and the ulna has a deeply concave semilunar notch. All phalanges have well-developed extension processes and tuberosities, and the ungual phalanges are robust and long, etc. All the above listed features show clearly that *Pseudorhizomys* are rodents adapted to use their forelimbs to dig and remove soil.

Lehmann (1963: 64) made a summary of the characters of the forelimbs of the rodent adapted for digging. Those for the scratch digging mode are: humerus having a larger deltoid process and wider epicondyles; ulna having a longer olecranon process, large coronoid process and deeper *fossa lateralis ulnae*. However, *Pseudorhizomys* doesn't have the above mentioned characters as those in the living rodents of scratch digging type (*Myospalax* and *Geomys*). The humerus of *Pseudorhizomys* has a narrower distal end, with a deltoid process located higher in position, and a less developed medial epicondyle (see Tab. 21); of the ulna the coronoid process is smaller and the *fossa lateralis ulnae* is shallower, bordered by a weak lateral ridge; the five ungual phalanges of the manus are normally developed, not particularly enlarged and lengthened as in *Myospalax* and *Geomys*, neither prominently widened as in *Spalax*. Obviously, *Pseudorhizomys* doesn't entirely rely on their claws to dig soil.

It is interesting to note that all of above-mentioned features of *Pseudorhizomys* are more or less similar to those of *Rhizomys*. According to Stein (2000: Tab.1.1), *Rhizomys* have two digging modes: the primary one is chisel-tooth digging and the secondary one is claw digging. Probably, *Pseudorhizomys* is similar to *Rhizomys* in this regard: primary mode using chisel tooth and secondary mode using claws.

However, we also notice some differences between *Pseudorhizomys* and *Rhizomys* in forelimb structures. In *Pseudorhizomys* the medial epicondyle of humerus is larger than in *Rhizomys*, and the FII-Ph3–FV-Ph3 are falciform, with well-developed longitudinal grooves and ridges on dorsal surface, rather than flatter shovel-like as in *Rhizomys*. It is evident that the claws of *Pseudorhizomys* are more suitable to soil digging than those of *Rhizomys*. In addition, in *Pseudorhizomys* all the extensors and flexors of the forelimbs are more developed. Therefore, the forelimbs of *Pseudorhizomys* can not only remove soil, but also dig soil.

In summary, based on the morphology of the skull and forelimbs, *Pseudorhizomys* should be subterranean rodents. They use two modes of breaking up soil: primarily using chisel tooth and secondarily using claws.

2) Functional analysis of *Pararhizomys*

The skull of *Pararhizomys* is similar to that of *Pseudorhizomys* in general structure, such as, the skull being about triangular in outline, with a narrow and long rostrum and expanded zygomatic arches, and a flat dorsal side; the orbit being small; the long glenoid fossa extending to the nuchal crest; the sagittal crest and nuchal crest being well developed and the nuchal surface being wide and inclining anteriorly; the external pterygoid process being much less developed than the medial one; the diastemata being long; the masseteric ridge of mandible being strongly flaring laterally, and the angular process being prominent, extending posteriorly; the i2 being procumbent and its posterior end extending posteriorly below the condyle, forming a prominent bulge, etc.

In the postcranial bones of *Pararhizomys*, the cervical vertebrae are short and robust, axis has a high and large spinal process; the scapula has an elliptic glenoid cavity and a large coracoid process; the humerus has an oval head; and the metacarpals are short, etc.

All of above-mentioned features indicate that *Pararhizomys* is also a group of subterranean rodents, and its primary digging mode is using their chisel teeth as in *Pseudorhizomys*. *Pararhizomys* are different from *Pseudorhizomys* in a number of skull features (see Tab. 9). The most noticeable features are: the strongly curved I2 with its anterior end bending inferoposteriorly, and its posterior end reaching near the incisive foramen; a distinct dorsolateral crest on the labial side of premaxilla; the anterior ends of both nasal and premaxilla are nearly in the same vertical line; the zygomatic plate is located more posteriorly, etc. The presence of a strongly curved I2 prevents *Pararhizomys* from using their upper incisors for digging soil. Thus, *Pararhizomys* may primarily use their lower incisors in digging, whereas their upper incisors are used for cutting through roots, tubers and soft parts of plant, etc.

V. STRATIGRAPHIC AND PALEOGEOGRAPHIC DISTRIBUTIONS OF PARARHIZOMYINES

In Linxia Basin all the fossils of *Pararhizomys* and *Pseudorhizomys* were collected from the lower and middle parts of the Liushu Formation of Bahean ALMA (Late Miocene). However, they were collected from slightly different stratigraphic levels (see Tabs. 2, 10). *Pa. longensis* was collected from the basal part of the Liushu Formation and belongs to the Guonigou Fauna (about 11.1 Ma, see Deng, 2005; Qiu et al., 2013; Deng et al., 2013). *Pa. hipparionum* was collected from the lower level of the middle part of the Liushu Formation and belongs to the Dashengou Fauna (about 9.5 Ma, op. cit). *Pa. huaxiaensis* was collected from the upper level of the middle Liushu Formation and belongs to the Yangjiashan Fauna (about 8.3 Ma, *op. cit)*.

Pseudorhizomys is known to include four species from the Linxia Basin: *Ps. indigenus*, *Ps. gansuensis*, *Ps. planus* and *Ps. pristinus*. As explained above, specimens of *Ps. planus* and *Ps. pristinus* were collected during surface prospecting, their exact stratigraphic positions remain uncertain, although their provenance from the Liushu Formation is unquestionable. *Ps. indigenus* was collected from the upper level of the middle Liushu Formation and belongs to the Yangjiashan Fauna. *Ps. gansuensis* was collected from the upper and lower levels of the middle Liushu Formation and belongs to the Yangjjiashan and Dashengou faunas. Thus, it can be concluded that the pararhizomyines in Linxia basin are the members of the Guonigou through Yangjiashan faunas, stratigraphically spanning from the basal to the top of the middle Liushu Formation, representing the major time span of the Bahean ALMA, roughly 11–8 Ma in age.

In the areas other than the Linxia Basin, the pararhizomyines are known to include four *Pararhizomys* species (*Pa. hipparionum, Pa. qinensis, Pa. huaxiaensis* and *Ps.? hehoensis*) and two *Pararhizomys* spp. *Pa. hipparionum* has been widely known from Fugu and Qin'an counties of Shaanxi Province, Baogeda Ula of Nei Mongol of China and Altan Teli of Mongolia. Biostratigraphically, it ranges from late Bahean to Baodean ALMA, about 8–6 Ma in age (Kowalski, 1968; Zhang et al., 2005; Li, 2010). *Pa. qinensis* was collected from the lower part of the Bahe Formation of Lantian, Shaanxi Province. Its age had first been thought as early Bahean ALMA, about 10 Ma (Zhang et al., 2005), but recently it was corrected as medial Bahean, 9.37 Ma in age (Zhang et al., 2013). If it is tenable that the specimen from Qingyang, Gansu, described by Teilhard de Chardin (1942) as "*Prosiphneus licenti*" is to be relocated to *Pa. huaxiaensis* (*vide supra*, p. 200), the occurrence of *Pa. huaxiaensis* may extend to the medial Baodean ALMA, around 7–6 Ma in age. *Ps.? hehoensis* was collected from Bilung Basin in Biru County of Xizang. Its age had previously been considered as Early Pliocene (Zheng, 1980), but has now been corrected as medial Bahean ALMA (Qiu et Qiu, 1990 and 1995; Deng, 2006; Qiu et al., 2013; Deng et Ding, 2015). *Pararhizomys* sp. I was collected from Shengou of Qinghai Province. Its age was thought as early Bahean ALMA (Qiu et Li, 2008; Li, 2010), but has now been corrected as medial Bahean ALMA (Qiu et al., 2013). *Pararhizomys* sp. II was collected from Pliocene Gaotege section of Nei Mongol (Teilhard de Chardin et Young, 1931; Li, 2010). The latest paleomagnetic dating of the lower Gaotege Fauna is 4.18–4.63 Ma (Qiu et al., 2013). This is the last occurrence of the pararhizomyine rodents.

To sum up, the tribe Pararhizomyini is known to include nine named species of two genera and two indeterminate *Pararhizomys* spp. Most of them were collected from China, especially from the Linxia Basin of Gansu Province, where most localities and the most abundant fossils were found. Among the known pararhizomyines *Pa. longensis* is the earliest representative in age (early Bahean ALMA). *Pa. qinensis, Pa. hipparionum, Ps. gansuensis, Ps.? hehoensis* and *Pararhizomys* sp. I may represent the major stage of evolutionary radiation during the medial Bahean ALMA. In addition, *Ps. pristinus* may also appear in the same time period, or slightly earlier. *Pa. hipparionum* is evidently the widest spread and longest living species of the tribe (from medial Bahean through Baodean). The occurrence of *Ps. gansuensis* is so far limited to the Linxia Basin. *Pa. qinensis* and *Ps.? hehoensis* may have appeared slightly earlier than *Pa. huaxiaensis* and *Ps. indigenus*, which were found only at the top of the medial Bahean ALMA. *Ps. indigenus* is known to exist in Linxia Basin in late Bahean ALMA only. *Pa. huaxiaensis* may have spread wider, including Linxia Basin and Qingyang district of Gansu Province, and ranges longer in age from late Bahean to medial Baodean. *Pararhizomys* sp. II from Gaotege is the latest appearance of this group of rodents.

The paleogeographic and geological ranges of the pararhizomyines are limited within a short time period from the early Bahean to Baodean ALMA in the central areas of Eastern Asia, from 31°–48° N and 92°–115° E (Fig. 78). The genus *Pararhizomys* is known to spread wider in area (from eastern Gaotege of Nei Mongol in China to western Altan Teli of Mongolia and Shengou of Qinghai Province, China) and its age ranges longer, from early Bahean ALMA to Early Pliocene. The geological and geographic ranges of the genus *Pseudorhizomys* are more limited. Except for *Ps.? hehoensis*, all the other species of *Pseudorhizomys* are found only in the Linxia Basin and its age is known to range from medial to late Bahean ALMA only.

VI. PALAEOECOLOGIC ENVIRONMENT OF LINXIA BASIN WHEN PARARHIZOMYINES EXISTED

As far as is known, most of the subterranean rodents usually inhabit open area with moderately moist to dry soil suitable for plant growth. They tend to live in porous soils, or at least in well-drained soils of poor water-holding capacity (Lacey et al., 2000: 2; Bush et al., 2000: 185). In the Linxia Basin the fossil pararhizomyines are collected from the Liushu Formation yielding abundant fossils of *Hipparion* fauna. In this fauna a large number of animals were adapted for living in open steppes, showing that the Linxia Basin was a temperate and arid steppe environment in Late Miocene (Deng, 2011). Based on data of carbon isotopes obtained from tooth enamel of ungulates of the Chinese *Hipparion* fauna, Hou et al. (2006) suggested that the western China, Linxia Basin included, was steppes dominated by C3 grasses during the Late Miocene and Early Pliocene. Ma (1998) pointed out that based on the pollen assemblages of the Liushu Formation during 8.5– 6 Ma the Linxia Basin was a vast arid or semi-arid steppes affording the inhabitants not only enough spaces, but also abundant food. The pararhizomyines, having subhypsodont molars (on average), adapted themselves for gnawing stems, leaves and roots of plants. Obviously, the environment of the Linxia Basin in Late Miocene is suitable for the pararhizomyines to live.

As far as the individual specimens are concerned, the occurrence of the pararhizomyine fossils in the Linxia Basin is sparse. With the exception of *Pa. longensis*, whose fossils representing 3 individuals from the same site (Guonigou), all the other 14 individuals were collected from separate sites as a single representative. This is quite different from the cases for the other major groups of animals of the *Hipparion* fauna in general, and in the Linxia Basin in particular, where hundreds of skulls of various ungulates, like *Chilotherium*, *Hezhengia*, etc, have been found. Two possibilities may explain this phenomenon: one is the selective preference for collecting large animals by the local fossil bone diggers; another one may be related to habits of the pararhizomyines, which might be solitary rather than living in groups. In fact, this also holds true for the 8 localities from the areas other than the Linxia Basin. Each locality also yielded only one single individual of pararhizomyines (Teilhard et Young, 1931; Kowalski, 1968; Zheng, 1980; Zhang et al., 2005; Qiu et Li, 2008; Li, 2010). As for the exceptional *Pa. longensis,* the two paratypes (V 16292.2 and V 16292.3) are only represented by rostra. Probably they were transported from other places and buried together with the type skull. Thus, we postulate that the pararhizomyines led a solitary life, as does living *Rhizomys*.

To sum up, 7 species belonging to 2 genera lived during a time span of about 3 m.y. in Late Miocene in Linxia Basin, whereas 9 species of these two genera plus two *Pararhizomys* spp. lived in a time span of about 7 m.y. in the central areas of the East Asia. Flynn (1985; 1990: 174) stated: "The burrowing habitus of rhizomyids contributes to low adult mobility and permits genetic divergence, and the fossil record is consistent in showing a higher rate of splitting in fossorial lineages than in those less modified for burrowing." The rapid cladogenesis may also play a predominant role in evolutionary history of the pararhizomyines, which may be directly related to their burrowing habitus, resulting in high species diversity.

VII. CAUSE OF ORIGIN AND EVOLUTION OF PARARHIZOMYINES IN ASIA

The pararhizomyines are known to spread in Late Miocene–Early Pliocene in central region of East Asia, primarily north of Tibetan Plateau. Their origin and evolution are closely correlated to paleoclimatic changes of Asian Continent and the uplifting of the Tibetan Plateau in Cenozoic Era.

From late Early Eocene to Early Oligocene the global climate indicated gradual and stepwise cooling and increased aridity (Wolfe, 1978; Berggren et Prothero, 1992; Zachos et al., 2001). According to Ma et al. (2003: 342–344), during the Eocene–Oligocene Epochs the Indian Plate began contacting Eurasian Plate, driving the seawater out of the whole Asian Continent, leading to increased aridity in Asian Continent. During the Late Oligocene the uplift of the southern Tibetan Plateau began to block the transport of moisture from southern oceans and induced intensified desertification in the interior of Asia (Qiang et al., 2011). Thus, the woodland-grassland adapted giant rhinoceros fauna developed (Qiu et Wang, 2007). At that period, however, the Tibetan Plateau was not lifted so high to hinder the mammals from migrating between south and north sides of Tibetan Plateau (Qiu et al., 2001; Qiu et Wang, 2007; Deng et al., 2015). This fauna includes many large and small mammals favoring arid climates (such as, to list a few, ctenodactylids, tsaganomyids and *Tachyoryctoides*).

From late Late Oligocene to Middle Miocene the global climate returned to warming again. During late Early Miocene–early Middle Miocene, the climate of East Asia became warmer and moister. In Middle Miocene Tunggurian ALMA (15–11 Ma) the climate became rather moist in the Asian Continent (Liu et al., 2009). During this period, with the uplift of Tibetan Plateau, the climate of Asia was transformed from a zonal pattern into a monsoonal one (Zhang et al., 2007). Since the humidity of the southwest and southeast monsoon was induced, the climate in the monsoonal area including Linxia Basin became warmer and more humid (Shi et al., 1999; Li et al., 2001; Guo et al., 2008). The forest-type mammals appeared and greatly diversified. Among them *Platybelodon*, a shovel-tusked elephant dwelling near water and feeding on aquatic plants, and semiaquatic castorids flourished. In the meantime the arid-adapted animals might have become extinct (such as tsaganomyids and *Tachyoryctoides*) or might have migrated into other areas (such as ctenodactylids).

Since the late Middle Miocene the global climate had gradually become cool and arid again, and the Antarctic ice-sheets reestablished (Zachos et al., 2001). As a result, the desertification of the interior of Asia and the Asian monsoon became further intensified (Kutzbach et al., 1993). Meanwhile, the uplift of the southern Tibetan Plateau and the shrinking of the Paratethys not only further blocked the transport of moisture from southern oceans and induced the intensified desertification in the interior of Asia, but also further strengthened the Asian monsoon, resulting in the reduction of the arid zone to the northwest China and Mongolia (Wang, 1990, 2005; Sun et Wang, 2005).

During the Late Miocene (11–6 Ma) the Tibetan Plateau uplifted again and further strengthened the Asian monsoon (Prell et al., 1992; Wang, 2005; Sun et Wang, 2005; An et al., 2006). The habitats in the interior of Asia, the Linxia Basin included, became steppes dominated by C3 grasses (Hou et al., 2006). In this background of environmental changes the mammals then living in the interior of Asia had changed greatly. Some of the mammals favoring water might be extinct completely (such as *Anchitheriomys*); some others may have survived in very early Late Miocene, but became extinct soon after (such as *Platybelodon*,

Hystricops and *Monosaulax*). Only a few forms might have survived longer, but lived in restricted areas (such as *Chalicotherium*). Instead, some arid or semi-arid steppes-adapted members of *Hipparion* fauna appeared (Deng, 2004a, 2011). In this environmental background the pararhizomyines and myospalacids originated and diversified in the central area of East Asia, occupying the ecological niche left open by the extinct *Tachyoryctoides* and others.

Based on the symbiotic fossil pollen-spores assemblages and primitive "Schizothoracinae" in the same levels and localities as those of the pararhizomyines, the altitude of the north part of Tibetan Plateau (including Bilung Basin and Qaidam Basin) has been postulated as rather low (~2000 m a.s.l.) in Late Miocene (Huang et al., 1980; Wu et al., 1980; Chang et al., 2010).

As a matter of fact, the habitats of the pararhizomyines are restricted to the eastern portion of the vast expanse of the Asian Late Miocene steppe (*sensu lato*). No fossil pararhizomyines have ever been reported from the areas west to Altay Mountains. They have not been found in Southeast Asia and in the areas north to 48°N, and east and south to Da Hinggan Mountains. Probably, during that period the climate of the area north to 48°N was too cold and those of the east and south areas are too damp and warm for the pararhizomyines.

In Pliocene the climate of the interior of Asia had become colder and more arid. The environments became too harsh for the pararhizomyines as well. So did for the periphery of the central area of East Asia. They could not migrate to the peripheral area either. Thus, since late Pliocene the pararhizomyines became extinct completely. The main reason for their extinction may lie in their inability to adapt to the newly changed environments.

When environmental conditions change radically, there are at least three possible options for the inhabitants to take: 1) the animals will remain where they are, if they are genetically strong and flexible enough to adapt to living under the changed conditions. 2) The species, or at least a part of them, may migrate to other places and survive there. 3) The animals finally become extinct. The myospalacids may belong to the first case. They survived and became flourished in Pliocene and early Quaternary. The ctenodactylids may belong to the second case. They did not survive the Middle Miocene in northern Asia, but succeeded in migrating from Asia to Africa during Early Miocene and remain there even today. The pararhizomyines apparently belong to the third case. Unable to adapt to living under warm and humid conditions, and as subterranean living rodents, apparently incapable for long-distance journey, they could not escape from the final fate of extinction.

VIII. CONCLUSION

1. The present study shows that during Late Miocene seven species of two genera of the pararhizomyines, including one new genus and six new species, existed in the Linxia Basin. They were collected from three different stratigraphic levels. *Pa. longensis* appeared the earliest, and was collected from the basal part of the Liushu Formation of early Bahean ALMA. The lower part of the middle Liushu Formation of medial Bahean ALMA yielded *Pa. hipparionum*, *Ps. gansuensis* and probably *Ps. pristinus*. The upper part of the middle Liushu Formation of late Bahean ALMA yielded *Pa. huaxiaensis*, *Ps. gansuensis* and *Ps. indigenus*. Unfortunately, the locality and horizon of *Ps. planus* are uncertain. In areas other than the Linxia Basin, there are four forms of pararhizomyines: *Pa. qinensis*, *Ps.? hehoensis*, plus two *Pararhizomys* spp. In total, the tribe Pararhizomyini is known to include 9 species plus two *Pararhizomys* spp. within two genera.

2. Our cladistic analysis shows that the phylogenetic interrelationships between the *Pararhizomys*-group and the closely related Muroid taxa may be summarized as (in simplified Venn diagram): *Pappocricetodon* ((*Heramys* (*Miorhizomys* (*Rhizomys+Tachyoryctes*))) + (*Spalax* (*Tachyoryctoides* (*Pseudorhizomys+ Pararhizomys*)))). In our tentatively proposed classification, two families, Rhizomyidae and Spalacidae, are recognized. The Spalacidae contain two subfamilies: Spalacinae, and Tachyoryctoidinae, the latter of which contains two tribes, Tachyoryctoidini and Pararhizomyini, which comprises two genera, *Pararhizomys* and *Pseudorhizomys*.

3. The morph-functional analysis of the skull and forelimbs shows that the pararhizomyines may be subterranean rodents. They may use two modes of breaking up soil: primary digging mode using chisel-teeth and the supplementary mode using their claws. However, *Pararhizomys* mainly used their lower incisors to dig, but *Pseudorhizomys* might use both the upper and lower incisors to dig.

4. The geographic distribution of the pararhizomyines is confined in the eastern part of the Eurasian Steppe. The paleoclimate, and symbiotic fauna and flora show that the pararhizomyines are rodents adapted to living in drier and warm open steppe. The origin, diversification, and extinction of the pararhizomyines are closely correlated with the changes of the paleogeography, paleoclimate and paleoenvironment, caused primarily by the uplift of Tibetan Plateau in Cenozoic Era.